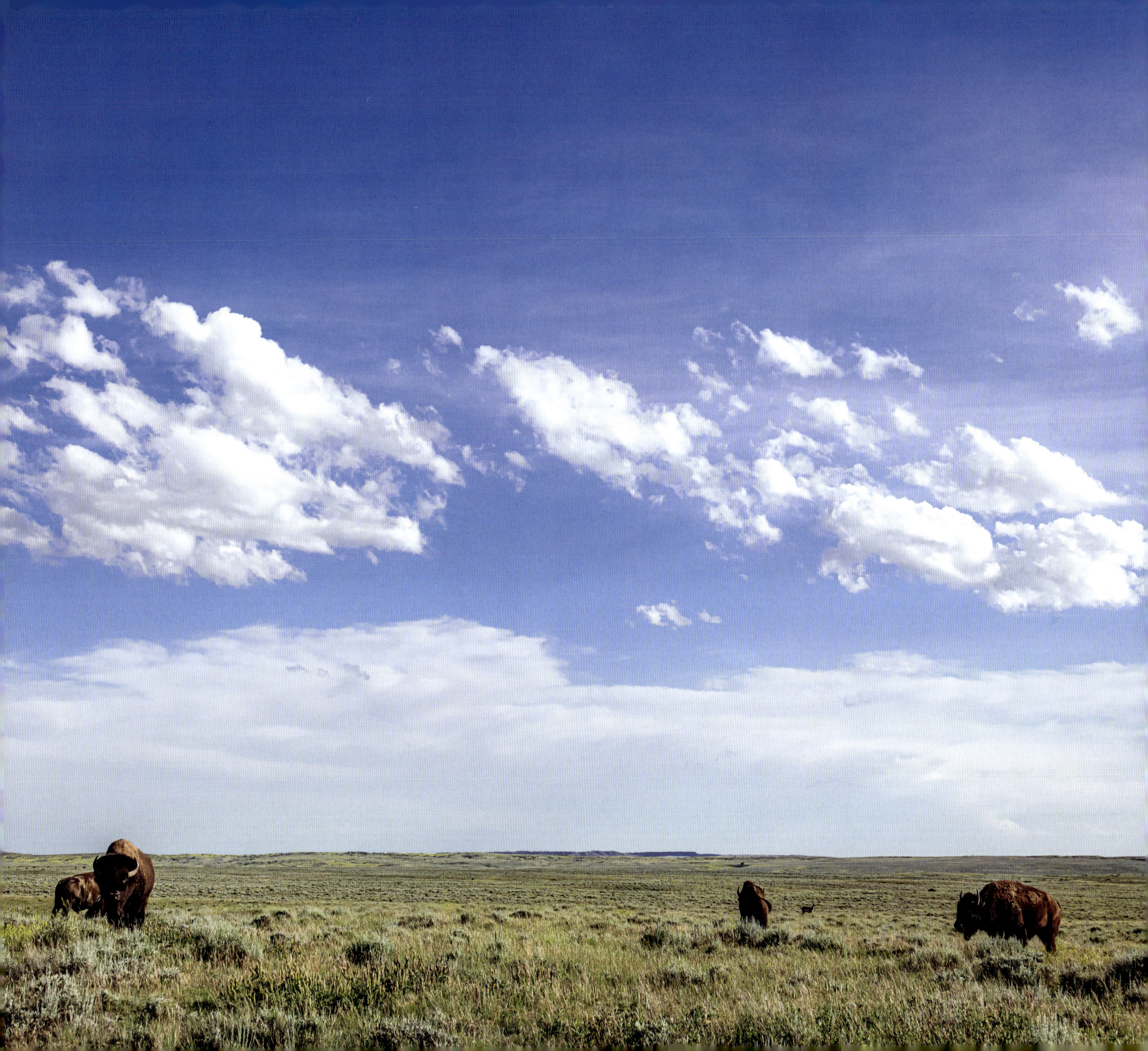

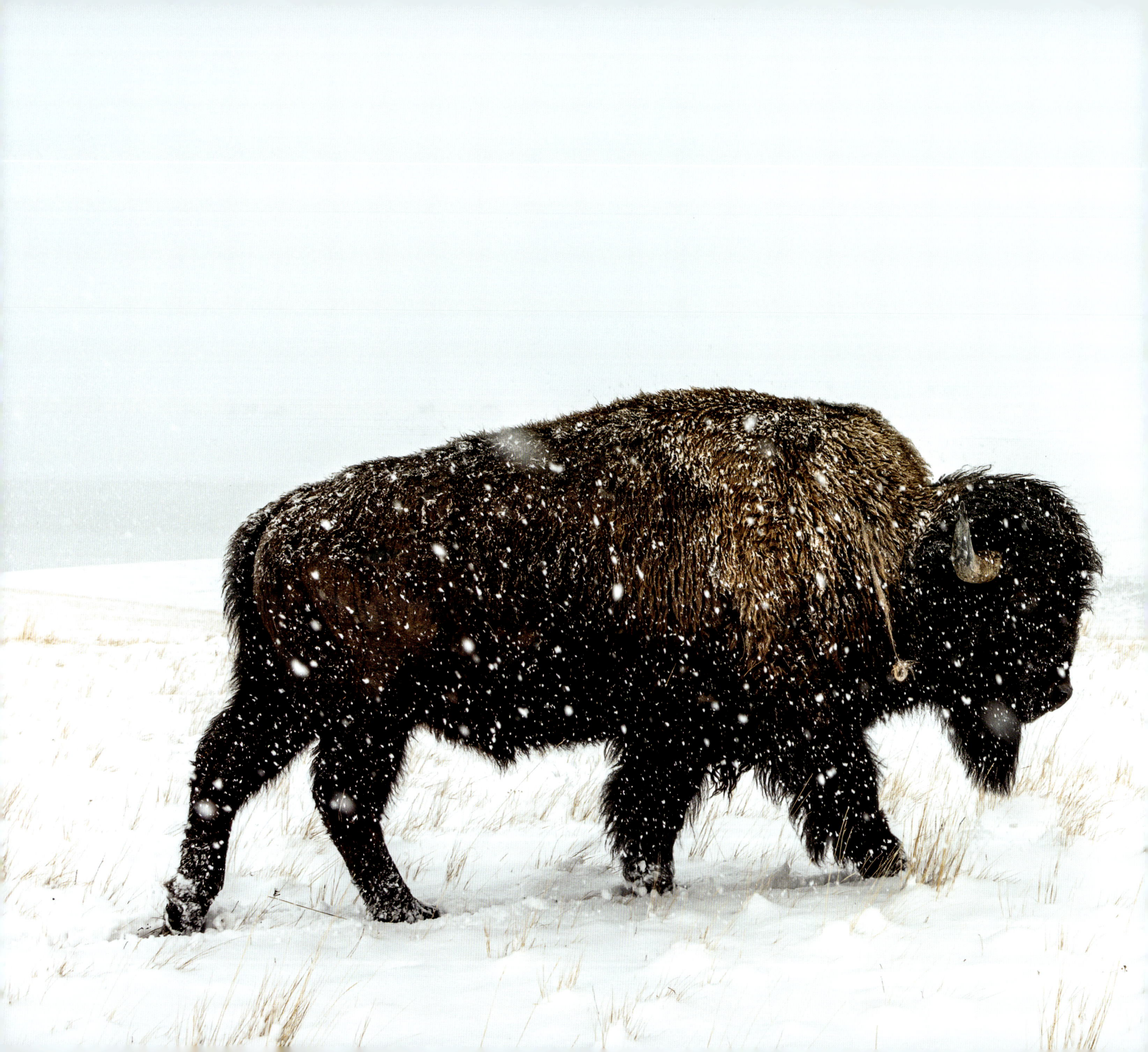

BISON

PORTRAIT OF AN ICON

AUDREY HALL | CHASE REYNOLDS EWALD

JOHN HEMINWAY | HENRY REAL BIRD

Gibbs Smith

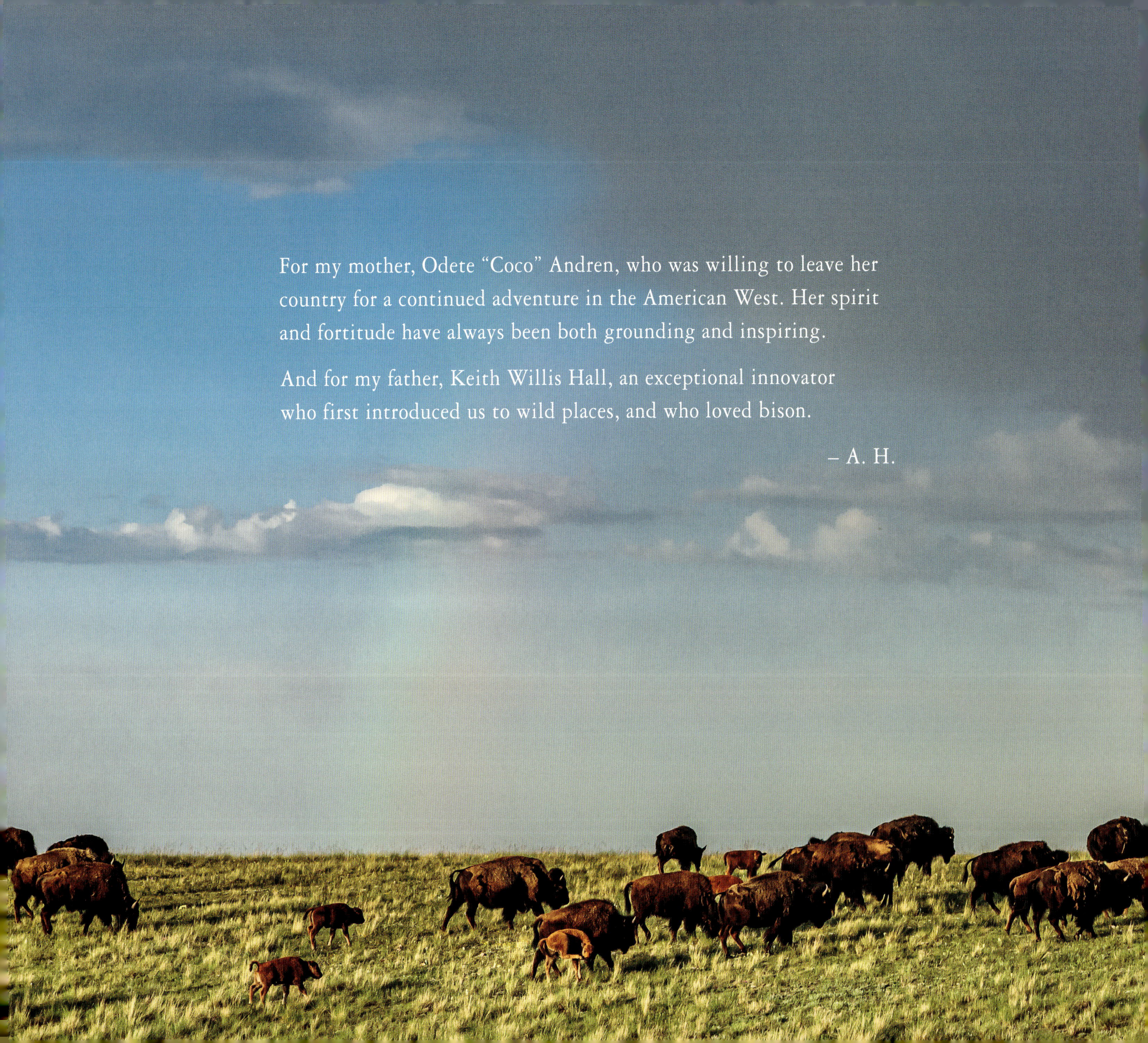

For my mother, Odete "Coco" Andren, who was willing to leave her country for a continued adventure in the American West. Her spirit and fortitude have always been both grounding and inspiring.

And for my father, Keith Willis Hall, an exceptional innovator who first introduced us to wild places, and who loved bison.

– A. H.

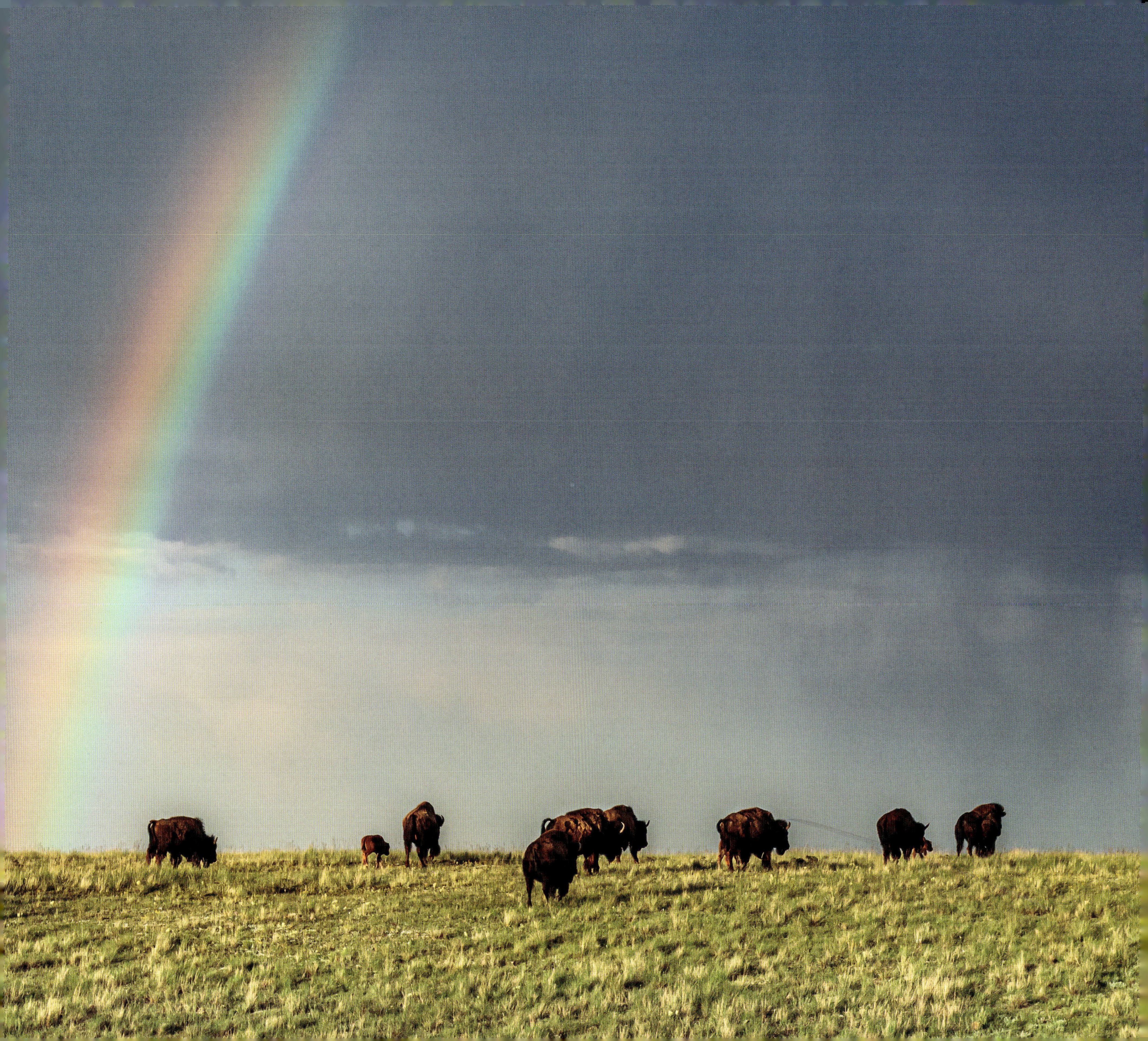

CONTENTS

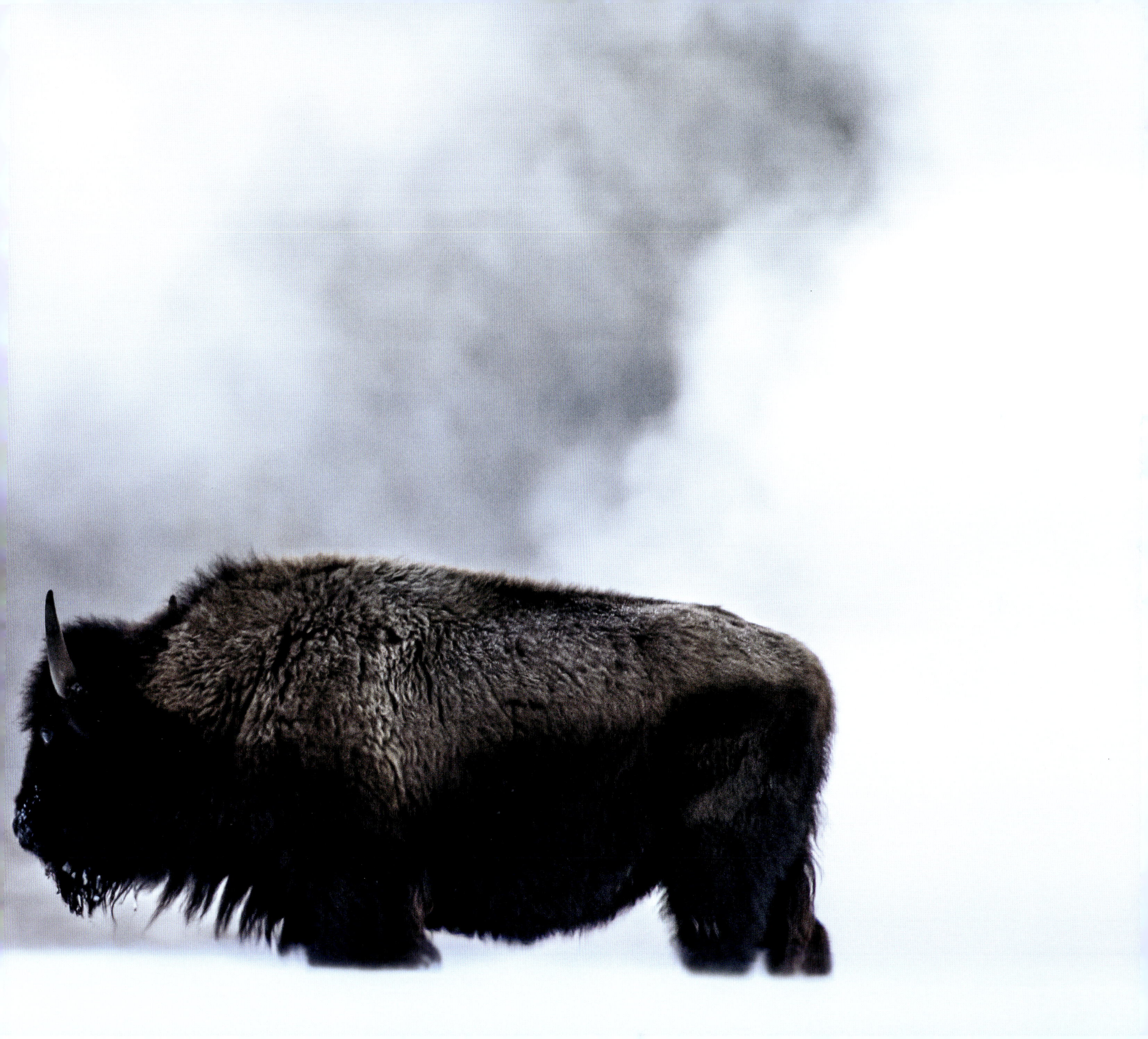

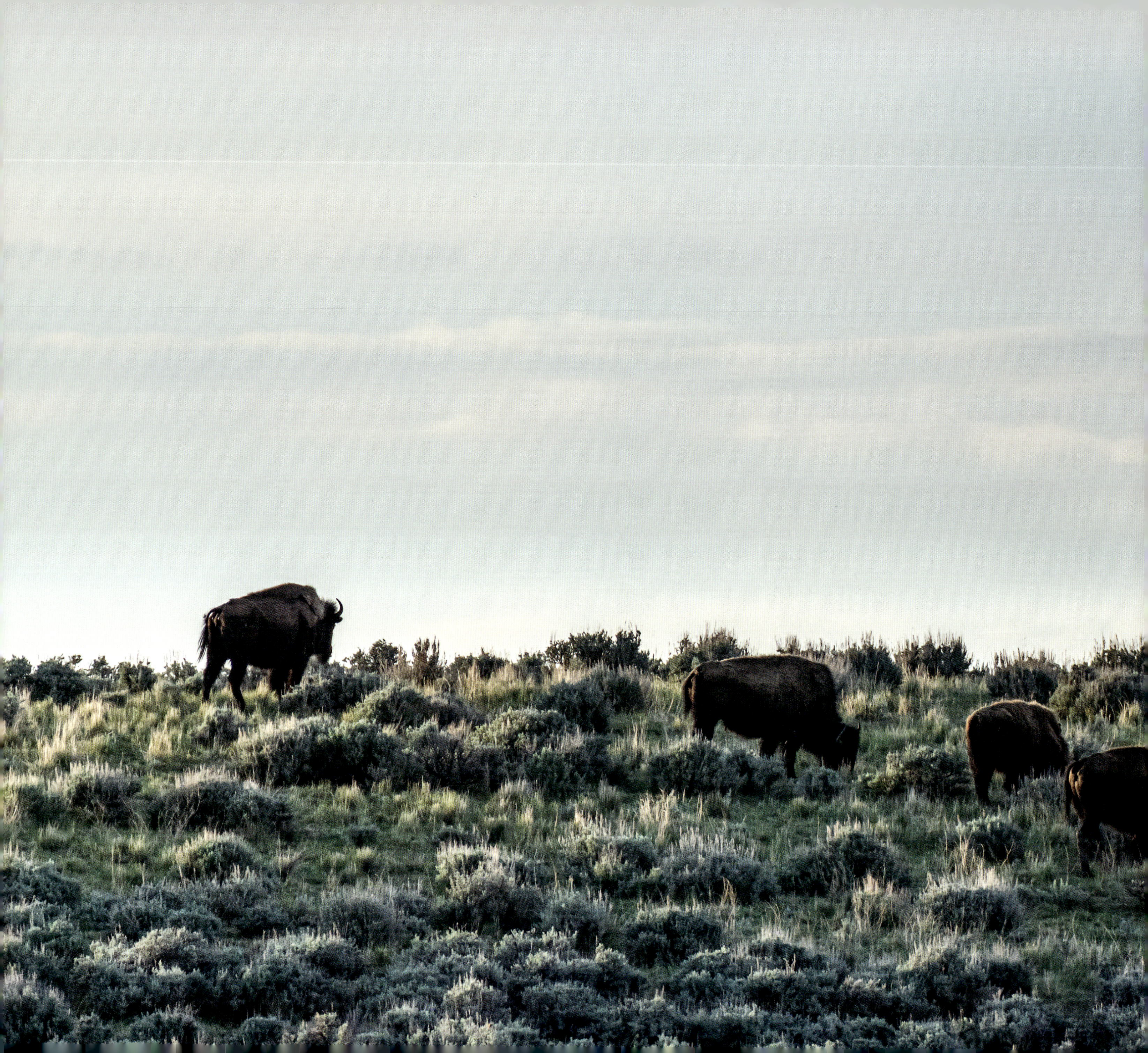

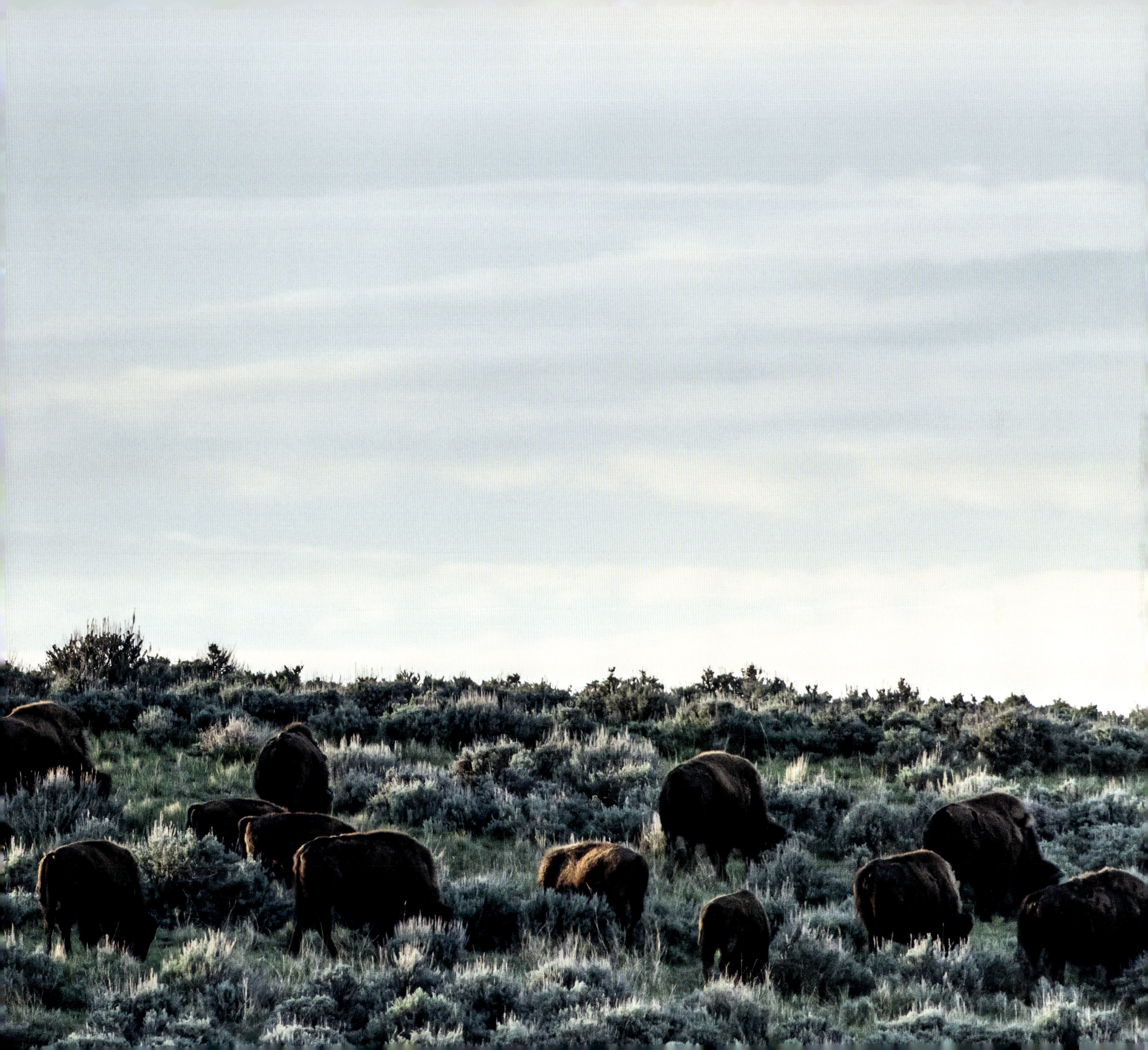

The American bison is thundering back into our consciousness a century after our thoughtless treatment nearly sent it into oblivion. Now a twenty-first-century renaissance of public awareness for this great animal—aided mightily by inspiring, artful books like this—is taking hold. Such celebrations make us feel better and unite us around a beloved emblem of the West. It reminds us how the undeniable rightness of restoring bison to public and private land can move us all toward a more meaningful relationship with Mother Earth.

—TODD WILKINSON, ENVIRONMENTAL JOURNALIST AND AUTHOR OF
LAST STAND: TED TURNER'S QUEST TO SAVE A TROUBLED PLANET

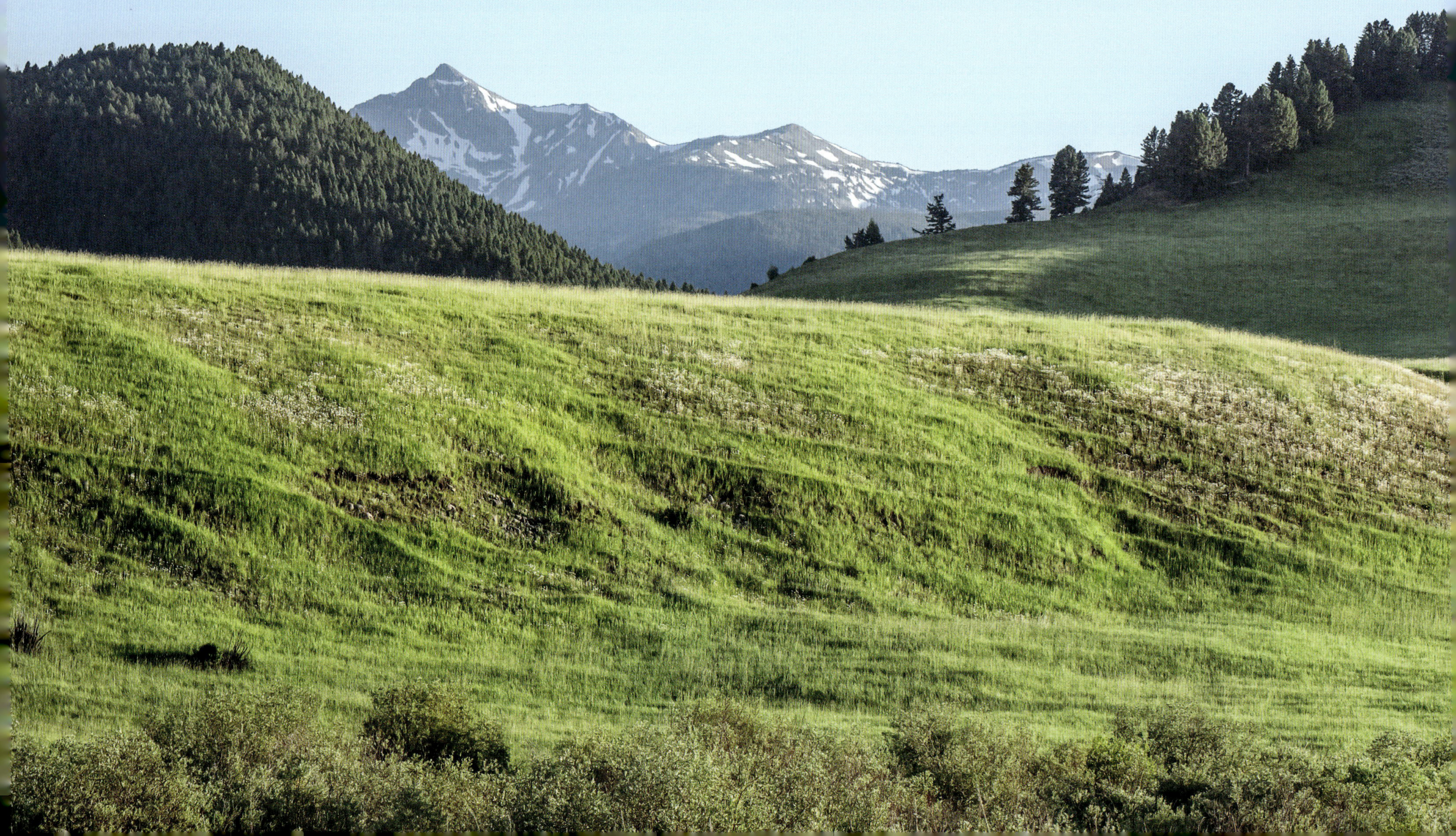

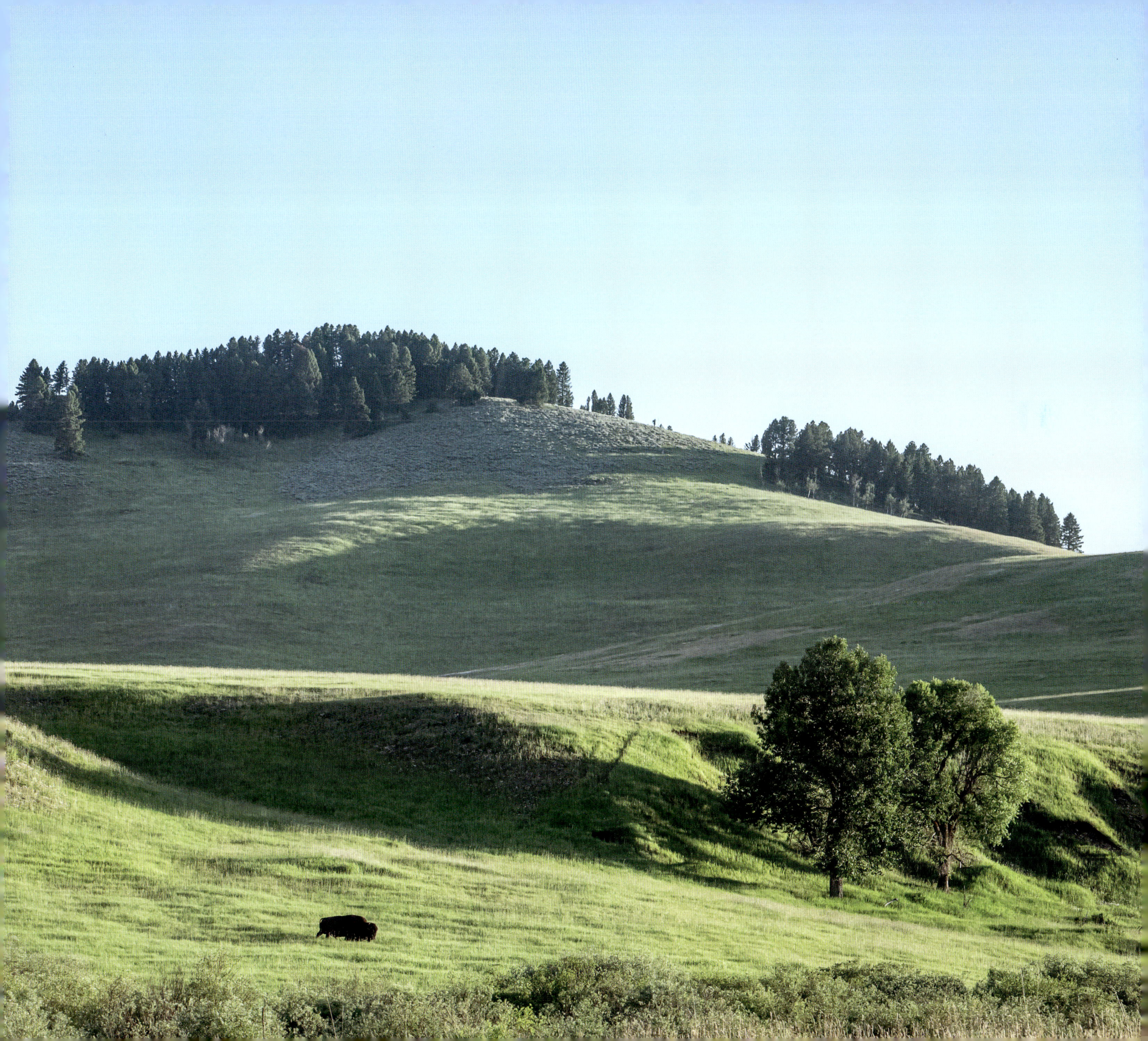

Foreword

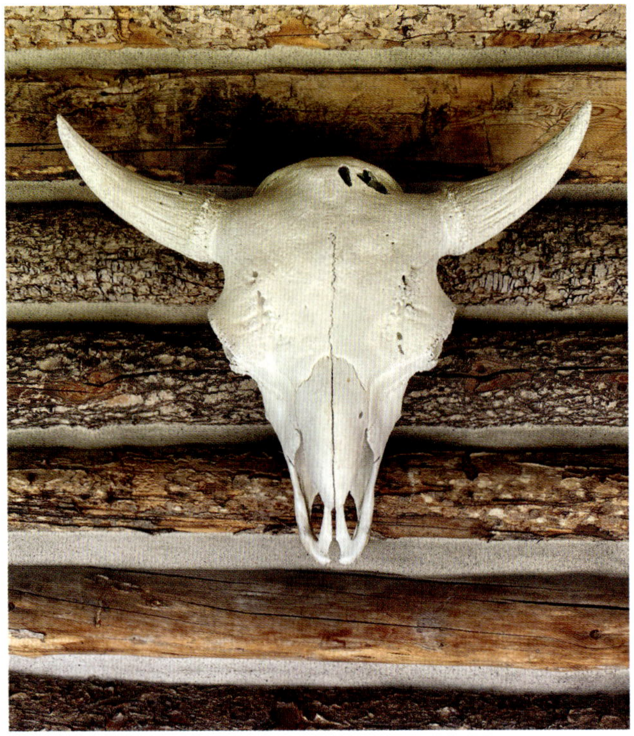

There are two ways to admire a bison. The first is to marvel, maybe chuckle, at its size and lumbering gait. The other is to consider its tempestuous history—one that echoes the worst and best of humanity and leaves us spellbound when we learn how one enchanted creature overcame insuperable odds to survive.

While I am a member of both camps, these days I lean towards the second. Yes, I marvel at the girth and heft of the bison, but when seen in multiples, I measure bison in the infinite. If you have ever enjoyed the company of elephants and hummingbirds, you will look at bison not only for their singularity (reminiscent perhaps—after much squinting—of a wildebeest), but as icons of a continent, totems of the nineteenth century, and solemn admonishments to all of us inmates of the twenty-first century to do better.

As evidence of my many decades awestruck by the bison, I need only cast my eyes around my domestic surroundings. Some objects are subtle, even unobtrusive. Others make me stand back and relive the pleasure of their company. One object that surely will never fail to inspire me will be this book, as it takes its place alongside other objects that speak to the bison's dazzling history.

Consider an oft-repeated legend that would seem pure hyperbole if it weren't true: bison in North America once numbered in the many millions—some say sixty million. There were so many, in fact, that for one winter week in three Kansas counties their breaths formed a three-county storm cloud.

Oh, to have witnessed such a sight! Somewhere out in the far reaches of the Internet there must be a photo of those millions and the storm cloud aloft, but, so far, I have not seen it. The best I can offer is an oil painting I acquired during a book tour in Virginia in the '80s. Painted at the turn of the twentieth century by Canadian artist John Innis, it portrays bison forging ahead in winter. Even now it transports me as a study in animal grit.

Two rooms away, I note the hat holder my mother gave me when I was sixteen. Mounted on a rough pine board are three

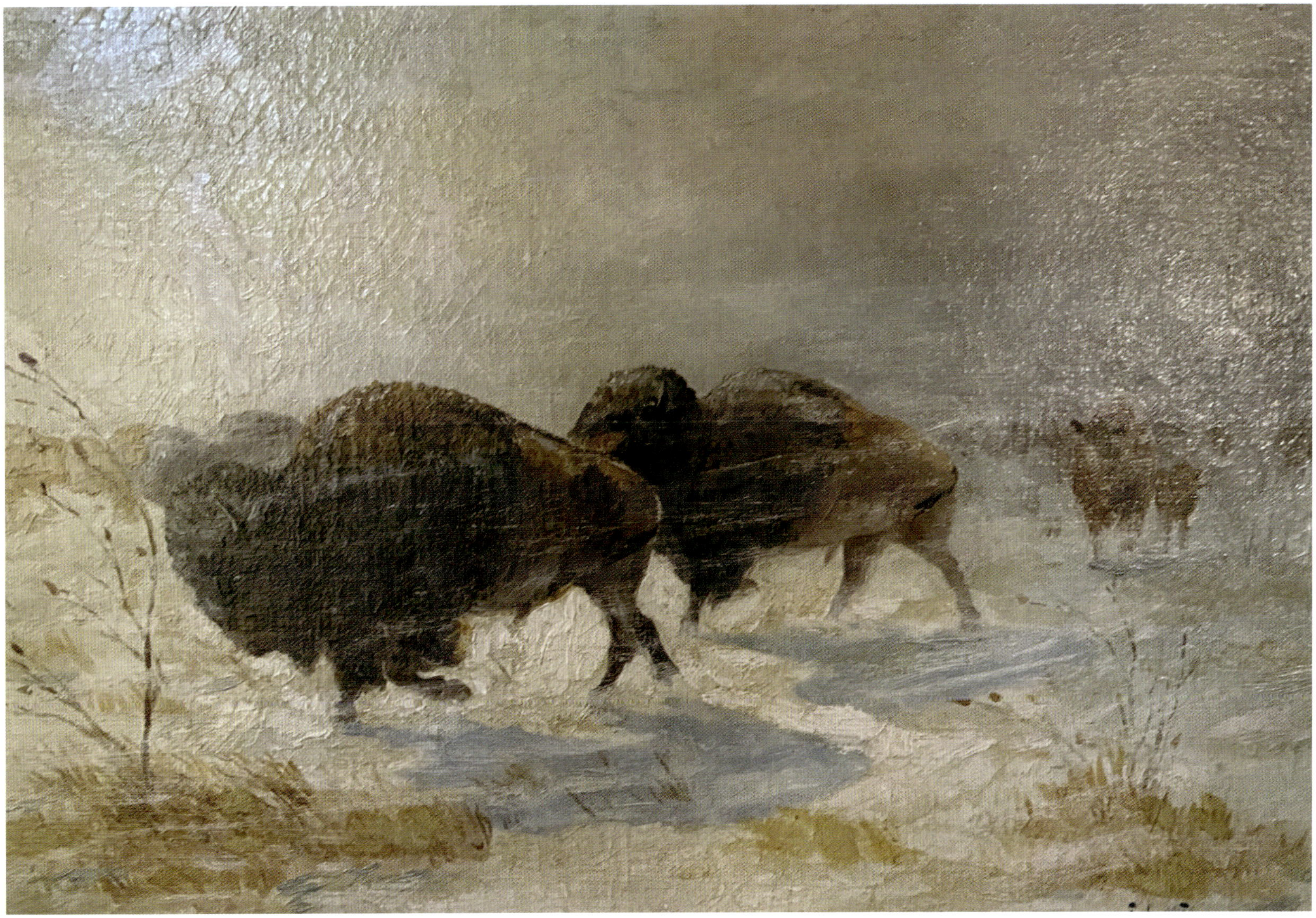

bison horns, in descending size. One can imagine they belonged to the same family, father, mother, and child, but that may be a stretch. On the back side of the board this inscription is carved: "Kiowa, Kansas, December 1889." I can only imagine these bison were taken by a party of hunters, firing in quick succession, with the intention of feeding a banquet of men at Christmas. By all accounts, bison hunting was no great feat, given the quarry's size and its predisposition to milling about in confusion as companions fell dead to either side. Even train passengers could pop them off through open windows from the comfort of their velveteen seats. No wonder that by the turn of the nineteenth century, buffalo were near extinction.

Hanging in the mudroom today, that tribute to profligacy makes me squirm, even when the rack is festooned with hats.

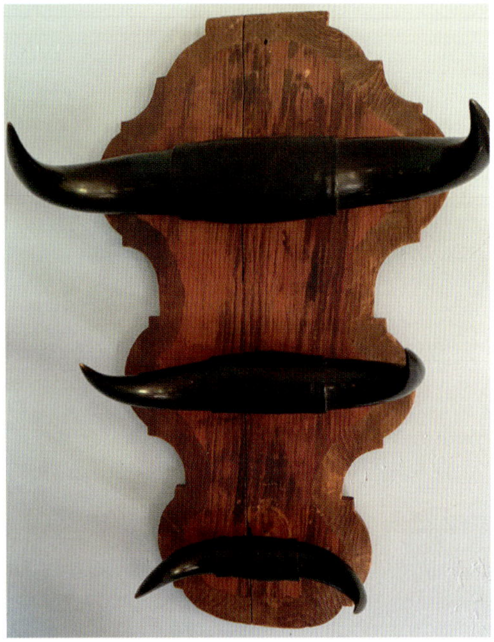

As you will learn in Chase's essay, the saving of the American bison is credited to several well-heeled New York patricians, Theodore Roosevelt and William Hornaday among others. Together, in 1905, they founded the American Bison Society, dedicated to preserving a nearly defunct bloodline. Today the purebred bison, mostly confined to Yellowstone National Park, the American Prairie Reserve, and Indian reservations in Montana and the Dakotas, is, in large part, their legacy.

But as with most stories of hope this one is not without risk. I am thinking now of the annual slaughter of Yellowstone's purebred bison on a quarter-mile-square patch of land north of the park. Each winter when the snow is shoulder deep in Yellowstone, many of its bison trundle beyond its boundaries to forage for grass. On many occasions they settle upon a minuscule piece of real estate, Beattie Gulch, where, with the approval of the National Park Service and the U.S. Forest Service, they are gunned down, sometimes in the thousands. The exercise is deemed "park management," otherwise—so the argument goes—Yellowstone would be overrun by bison.

This exceptional book, made possible through the creativity of Audrey Hall and Chase Reynolds Ewald, is, to me, a plea for all of us to be equally creative in planning for the bison's future. Genetically pure, one-hundred-percent perfect bison, saved from extirpation over a century ago, are not to be squandered. They are a message sent to us from the great wilderness that once was real America. That epic gift provides untold benefits—not least of all the prospect that a hundred years from now, a boy or girl will stand back, like me, shake his or her head, and gasp.

I still can gasp, especially when I look up from my work to admire a photograph. It is Audrey's totemic image of a bison, bound up in snow. To me, it suggests not just heft, but permanence. Nothing in my office moves me more.

So, I commend *Bison: Portrait of An Icon* to all of you, in the expectation that you will agree it is a book for the ages, as well as, and most importantly as, a call to action.

—JOHN HEMINWAY

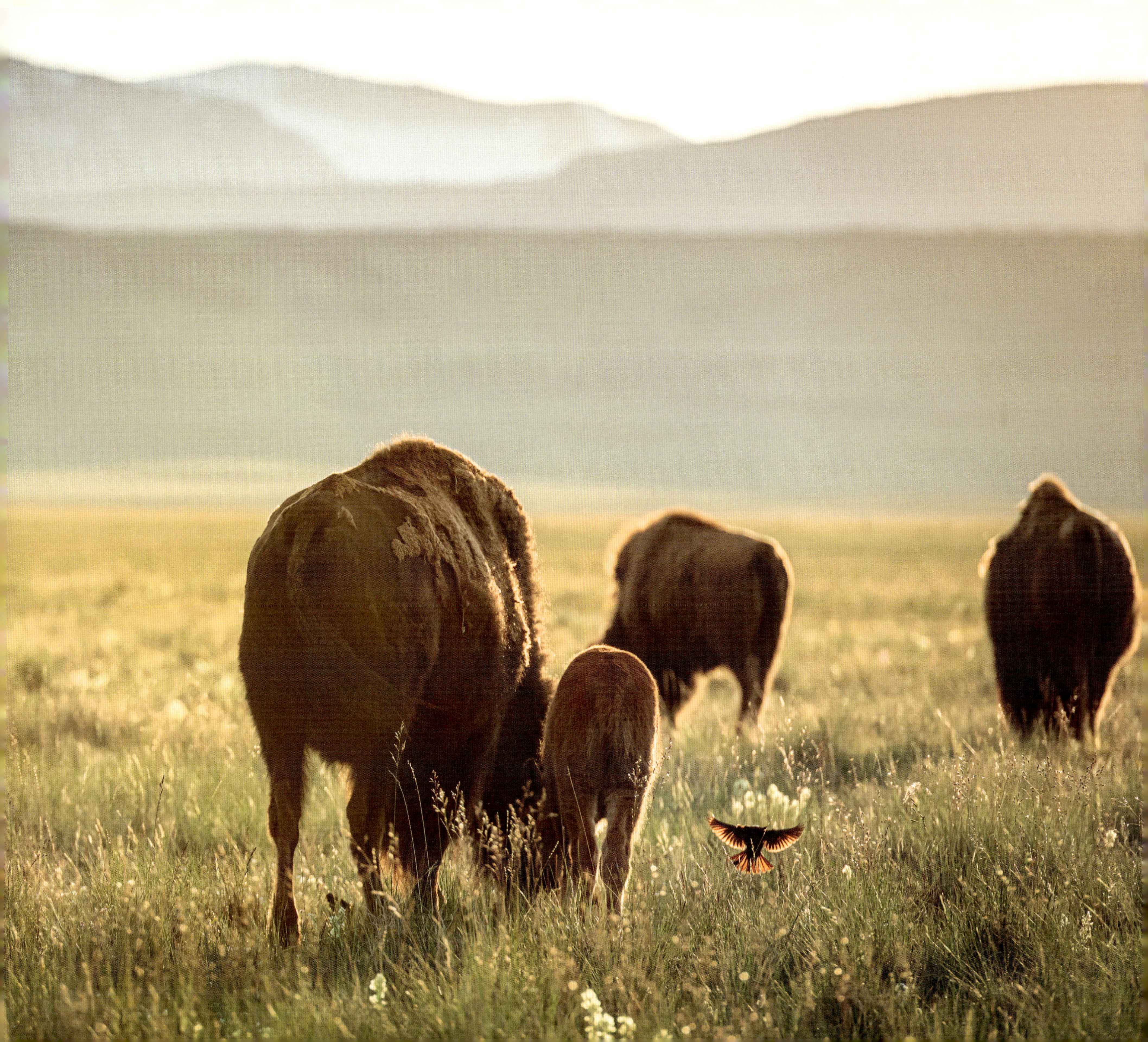

Introduction

When I was little, my family spent summers in the remote mining town of Cooke City, Montana. Nestled on the southern side of the Absoraka-Beartooth Mountains, it is a few minutes from the Northeast entrance of Yellowstone National Park. Almost every evening before sunset, we'd venture into the park for a picnic dinner. The landmarks on the gently winding two-lane highway from Cooke City into the Lamar Valley are forever etched into my mind.

First, the imposing mass of Barronette Peak, a wide and vertical sheet of jagged rock, its face streaked with delicate waterfalls. It towered over us as we drove by, and the mountain still makes me feel small.

Soon after that we'd pass Ice Box Canyon, where frozen pinnacles drop deep into the narrow slot of blue and out of sight. There's no guard rail along this stretch of road. To this day, I always hug the double yellow line tightly as I drive past.

Trout Lake Trailhead came next, a place where I learned to fish with my father. The hike is short and steep. We'd haul a boat up to the lake strapped on to a contraption he invented, a hybrid bicycle welded to an old-fashioned wheelbarrow design made of lightweight aluminum. Curious fisherman would stare in disbelief as my dad would slowly push his small dinghy up the hill. This was back when you could do this sort of thing in Yellowstone, like feeding the bears.

I could always smell the Soda Butte before I saw it; a soft stench of sulfur lingers in the air around the geothermal feature. We would make a stop here so I could scramble around its base, the queen of my own little odiferous hill. I'm pretty sure this isn't allowed now either.

Finally our destination: a picnic area in the heart of the lush and vibrant Lamar Valley, known to many as the Serengeti of the lower forty-eight states. A set of picnic tables were arrayed along the river, often surrounded by a herd of bison. The access no longer exists, undoubtedly because someone realized that picnicking among a herd of wild bison is probably not something to encourage. Over time, prairie grasses and sagebrush reclaimed the road and their territory, but if you look carefully, you can still see the ghost of the picnic area.

We came to this spot on the river specifically for an *al fresco* dinner with the Lamar Valley bison. The herd speckled the green valley floor like they still do today.

And so this wild and enduring icon of the American West became firmly intertwined with my earliest childhood memories, growing more and more interconnected with every visit to Yellowstone, saturating my earliest recollections. Like a whiff

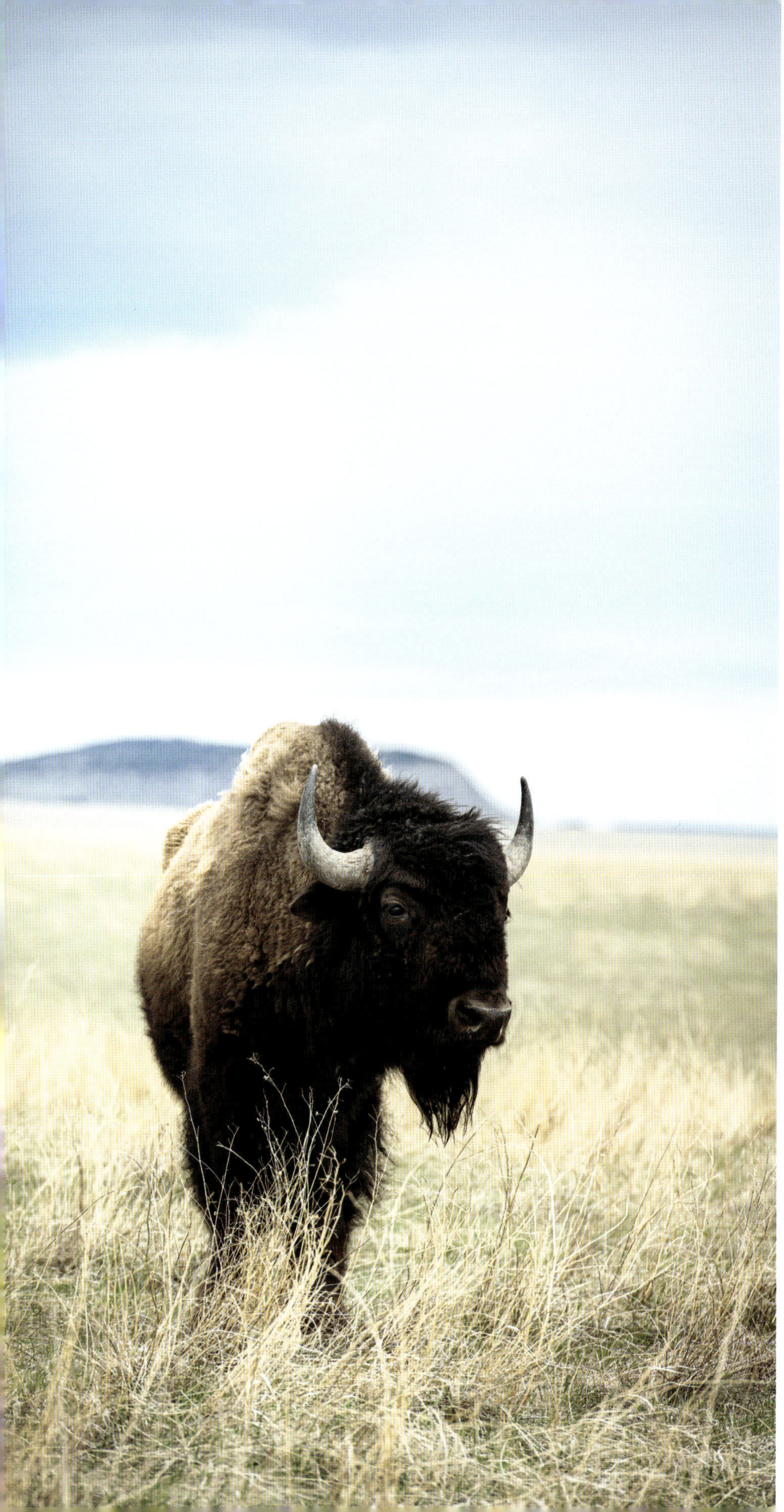

of fresh pine, each majestic bull on the skyline and every bison traffic jam immediately transports me back to my early years in the mountains.

As an artist, I've found bison crossing a horizon into my work, showing up again and again in the images I make.

There are now new landmarks as I travel and explore the world through my art. Life's road signs directed me to this book through a series of assignments, commissions, coincidences, and timing leading to the support of a few key players who encouraged the project and the process of digging deeper into the bison's world.

Putting the finishing touches on this book in the midst of the coronavirus pandemic of 2020, I was socially distant but never alone, watched over by these majestic animals.

Spending time solo in remote landscapes, watching bison both up close and from afar, sometimes waiting for hours in pouring rain or brutal sun, and often with ravenous hordes of mosquitos, I developed a sense of place beyond that of the external observer, and an overwhelming sense of what could be lost.

Moving through the herds with biologists, park rangers, and ranch managers instills a hope for the future. This diverse cast taught me much about the incredible nature and impact that bison play in the fantastical landscapes of the American West, their contributions to restoration and maintenance of these ecosystems, and the important role they play in Native Americans' lives and rituals. I thank them for their willingness to share their time and knowledge with me.

For this book, bison are my muse and the landscape my studio. I will be forever grateful for their mysterious and profound presence, as well as to those working to ensure their survival.

—AUDREY HALL

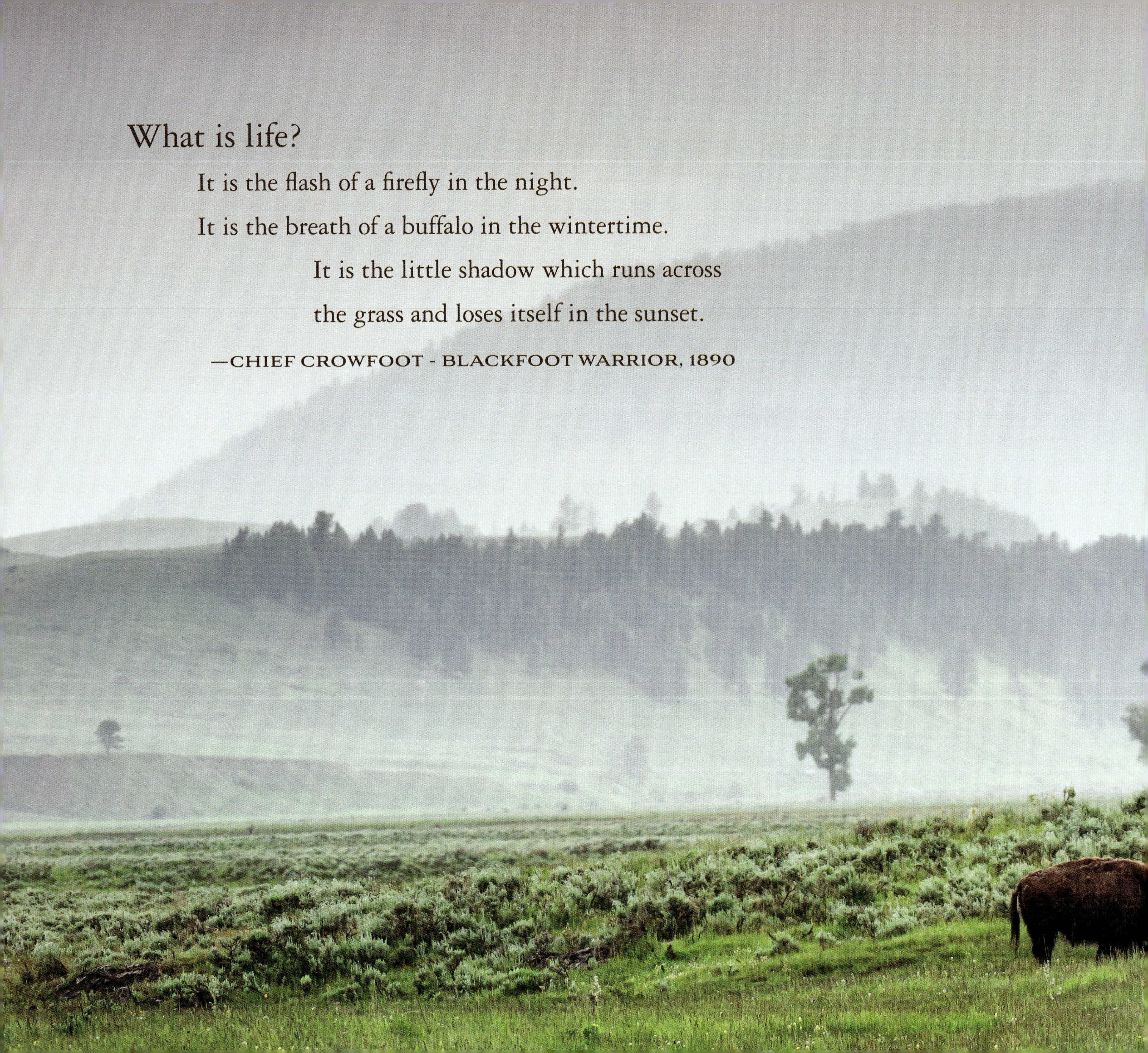
What is life?
It is the flash of a firefly in the night.
It is the breath of a buffalo in the wintertime.
It is the little shadow which runs across
the grass and loses itself in the sunset.
—CHIEF CROWFOOT - BLACKFOOT WARRIOR, 1890

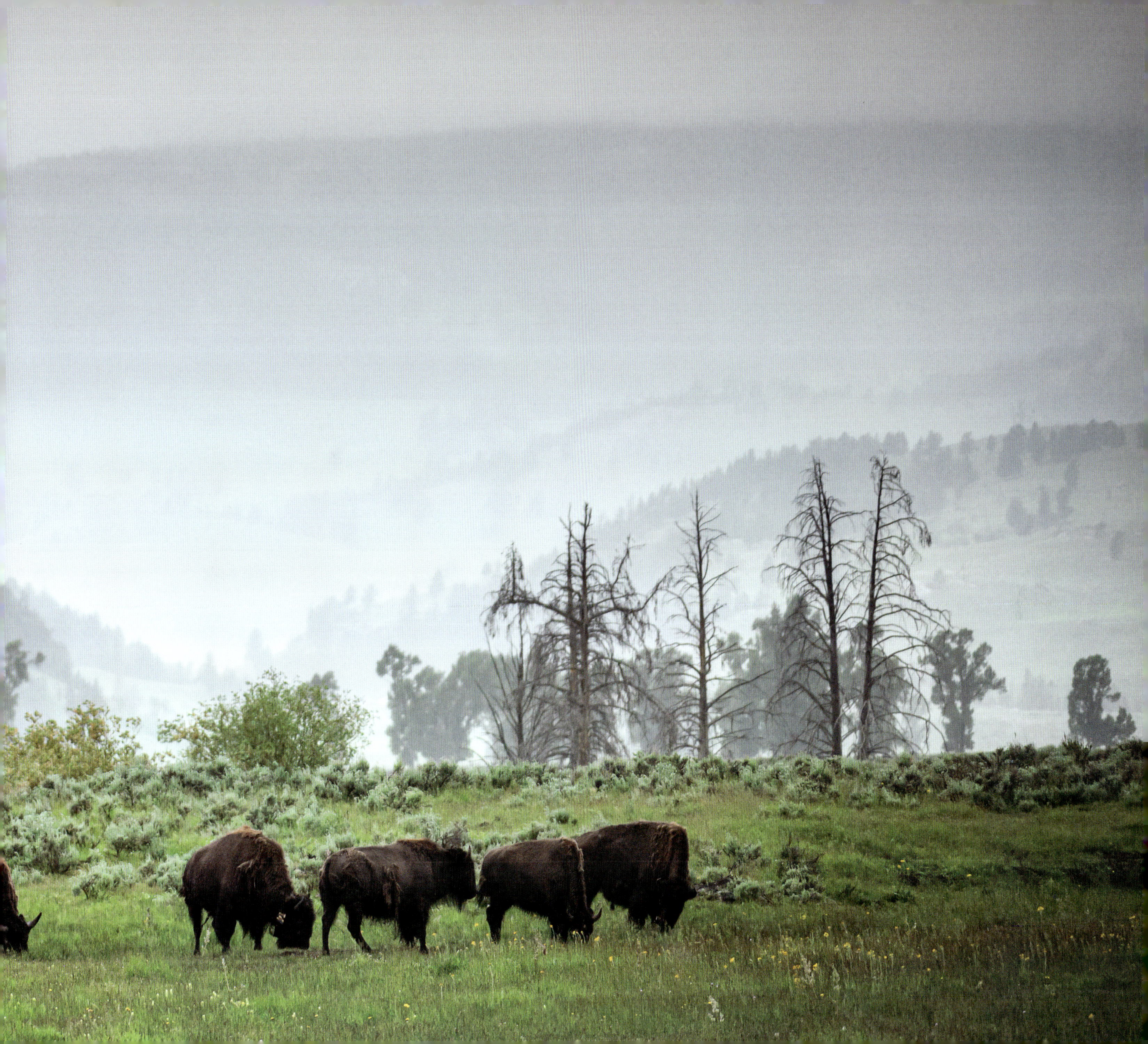

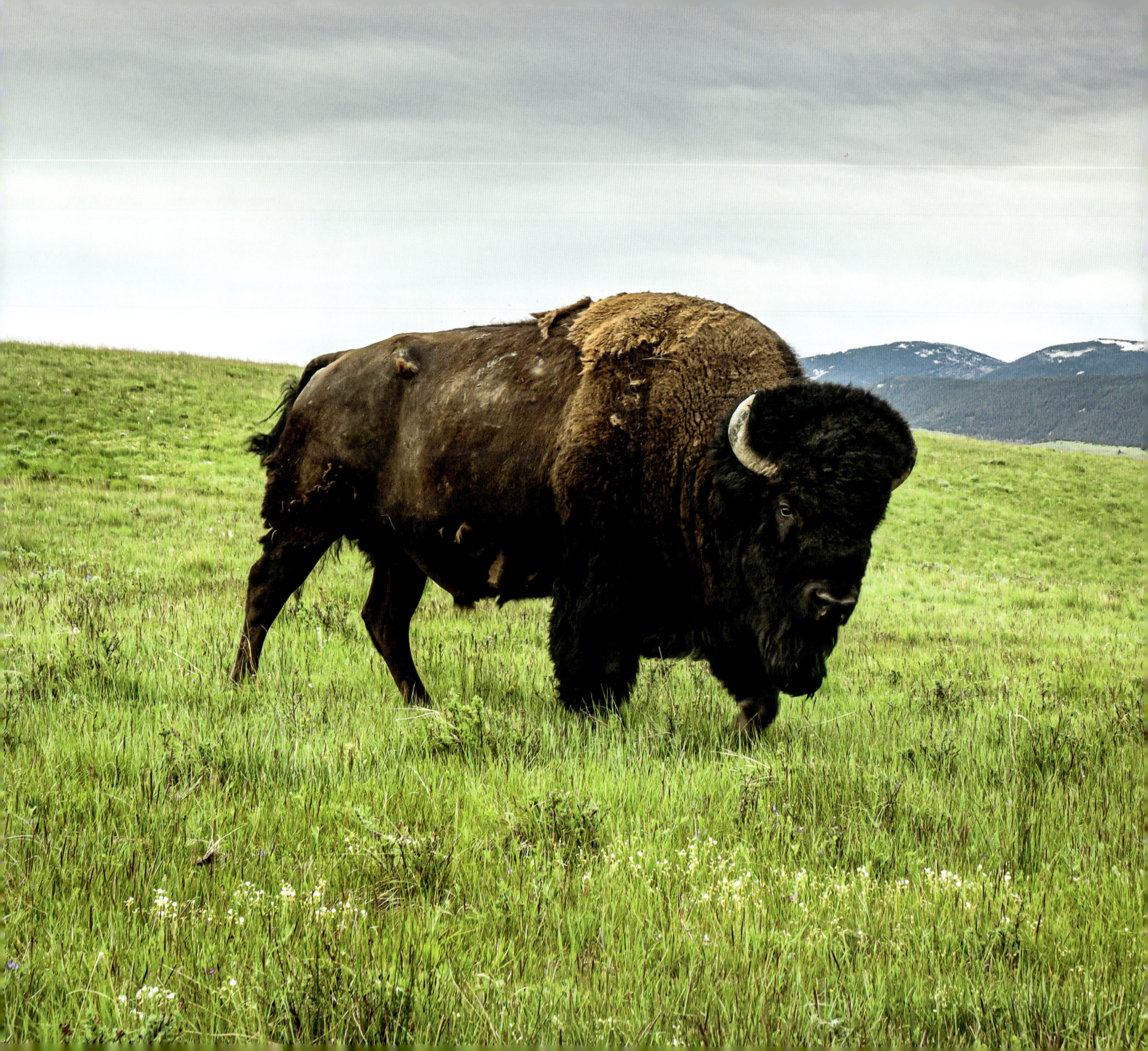

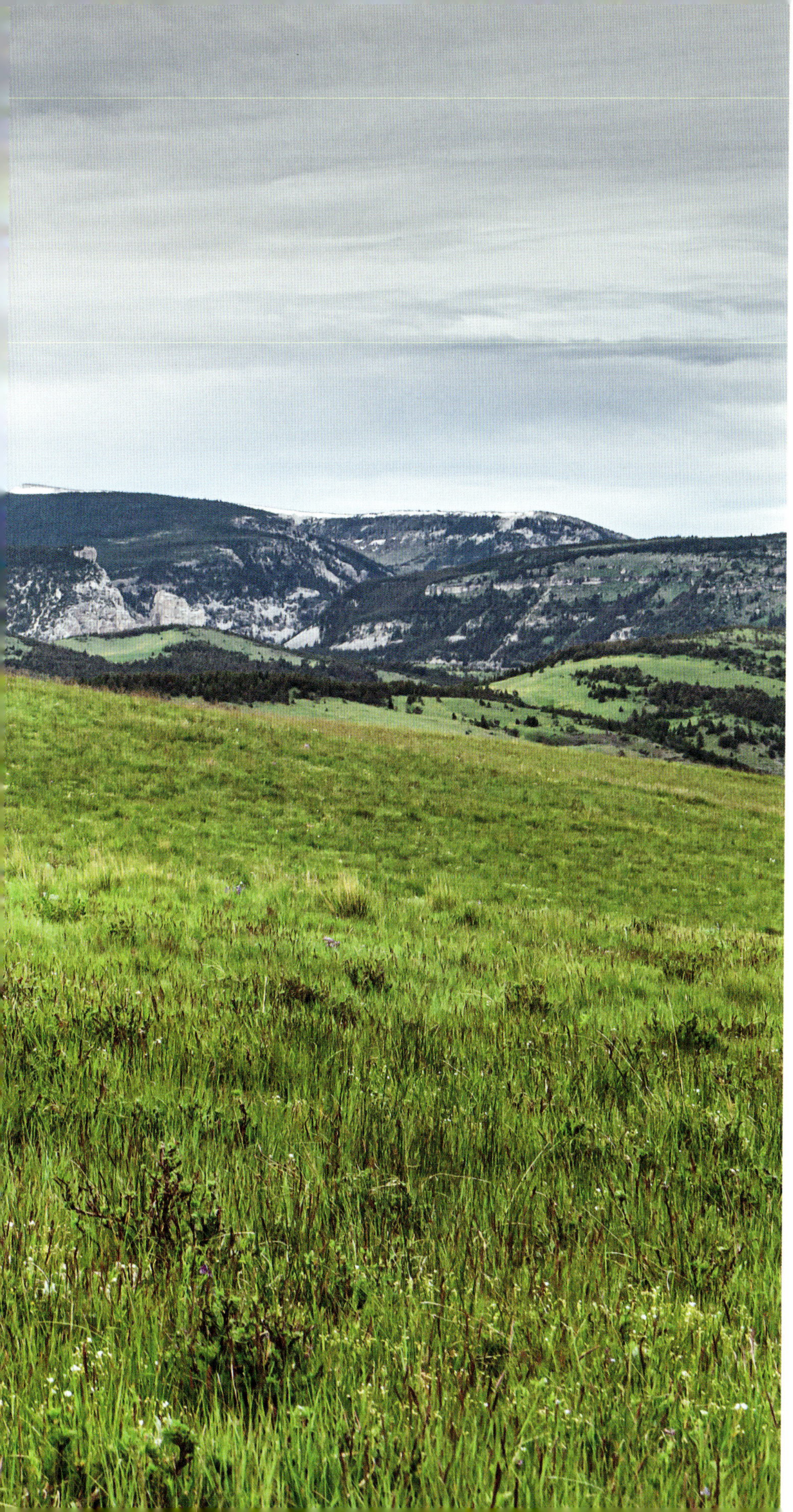

Buffalo Talk

On the second night of a Sun Dance,
in a dream I looked over my shoulder
to finally turn for a glance
at a buffalo bull big as a boulder.

I looked him in the eye to pause
and turned to run for my life
out on the flat plains (no coulees or draws
at that moment in time, no help from a knife).

The bull swooped me up from behind my knee
to toss me high up in the air.
It was like I knew nothing to be,
lying on the ground without a care.

Then the bull approached from my left
to brush softly against my shoulder
and licked my cheek in his warm breath.

"I am here to watch you
and help you in your life," he said.

—HENRY REAL BIRD

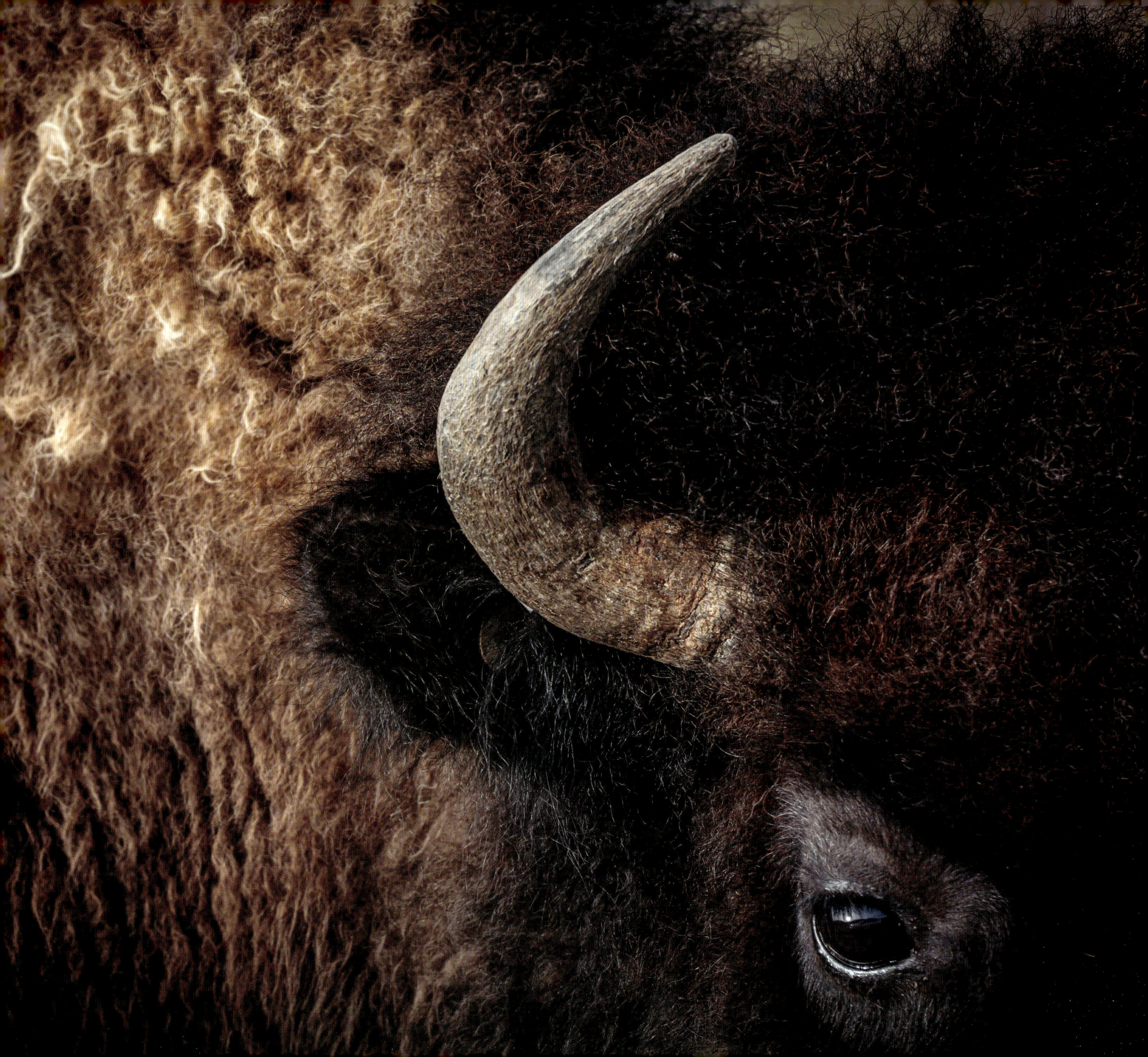

BISON: PORTRAIT OF AN ICON

IMAGINE THE SCENE. You're picnicking at a table next to the Yellowstone River on a sunny summer day. Suddenly aware of a disturbance, you turn, only to see a mass of heaving, snorting, splashing creatures swimming forcefully, inexorably, and directly at you. Retreating to cover is not an option; you're already becoming engulfed in a powerful mass of animals whose soft grunting and collective rumbling and scrabbling hooves momentarily drown out the sounds of birdsong and rushing river. They pass by so closely you could reach out and touch their prehistoric-looking leathery sides. Then they move on, ignoring you, singularly focused on staying with the herd, following their leader to some unknown destination.

Bison are creatures of contradiction. They may look like a cartoonist's vision of a prehistoric animal, with their dense, humped, compact bodies on overly short legs; their curved horns and jaunty beards; their robes' worth of fur (impossibly luxurious in places, but in others hanging off their bodies in ragged strips); and those ageless, knowing eyes. But bison are nimble, fast, and strong. They're capable of sprinting thirty-five miles an hour, clearing fence-high obstacles, fording glacier-fed rivers with their calves during the height of spring run-off, and—as anyone can witness by watching the warning video on Yellowstone National Park's website—tossing a tourist with merely a flick of their massive heads so that he lands, overhead and upside down, in the branches of a lodgepole pine.

The animals are tough, durable, and resilient. They're smart (despite having not received the memo on the dangers of men with guns as soon as they might have). They have matriarchal leaders who employ the strength of the group to protect its young. And they evolved in perfect adaptation to life on the Great Plains.

These evolutionary strategies include a metabolism that slows down in the winter so they need less forage, and a habit of walking into and through storms instead of drifting with them which prolongs exposure. The immense size of the herds and their patterns of movement and intermittent grazing behavior actually helped shape the grassland ecosystem itself.

Yet in a short couple of decades late in the nineteenth century, these iconic creatures barely survived an assault that took their numbers from an estimated twenty-five to sixty million to fewer than one thousand. Bison came within a horn's width of disappearing forever, and this makes their comeback all the more remarkable.

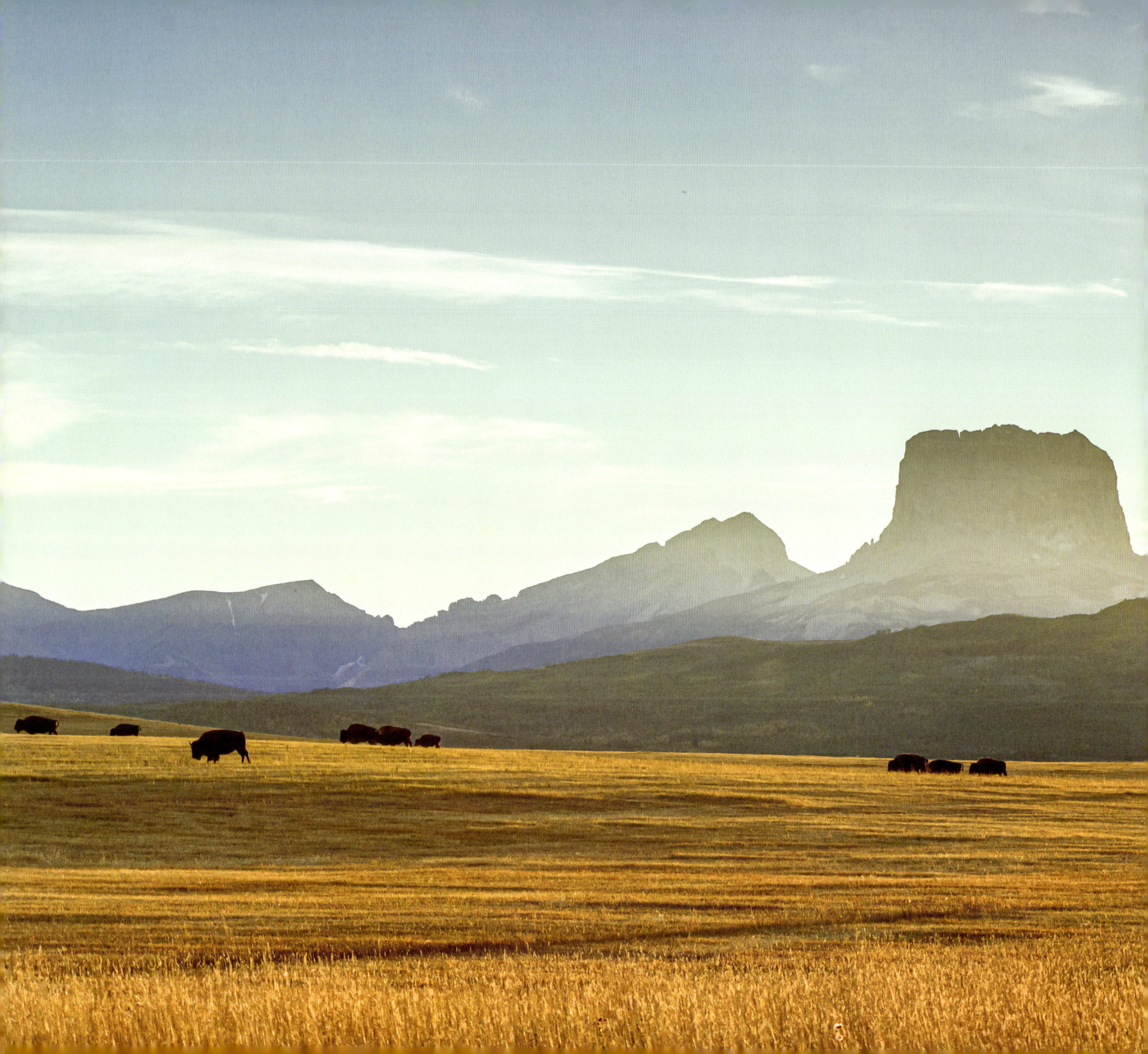

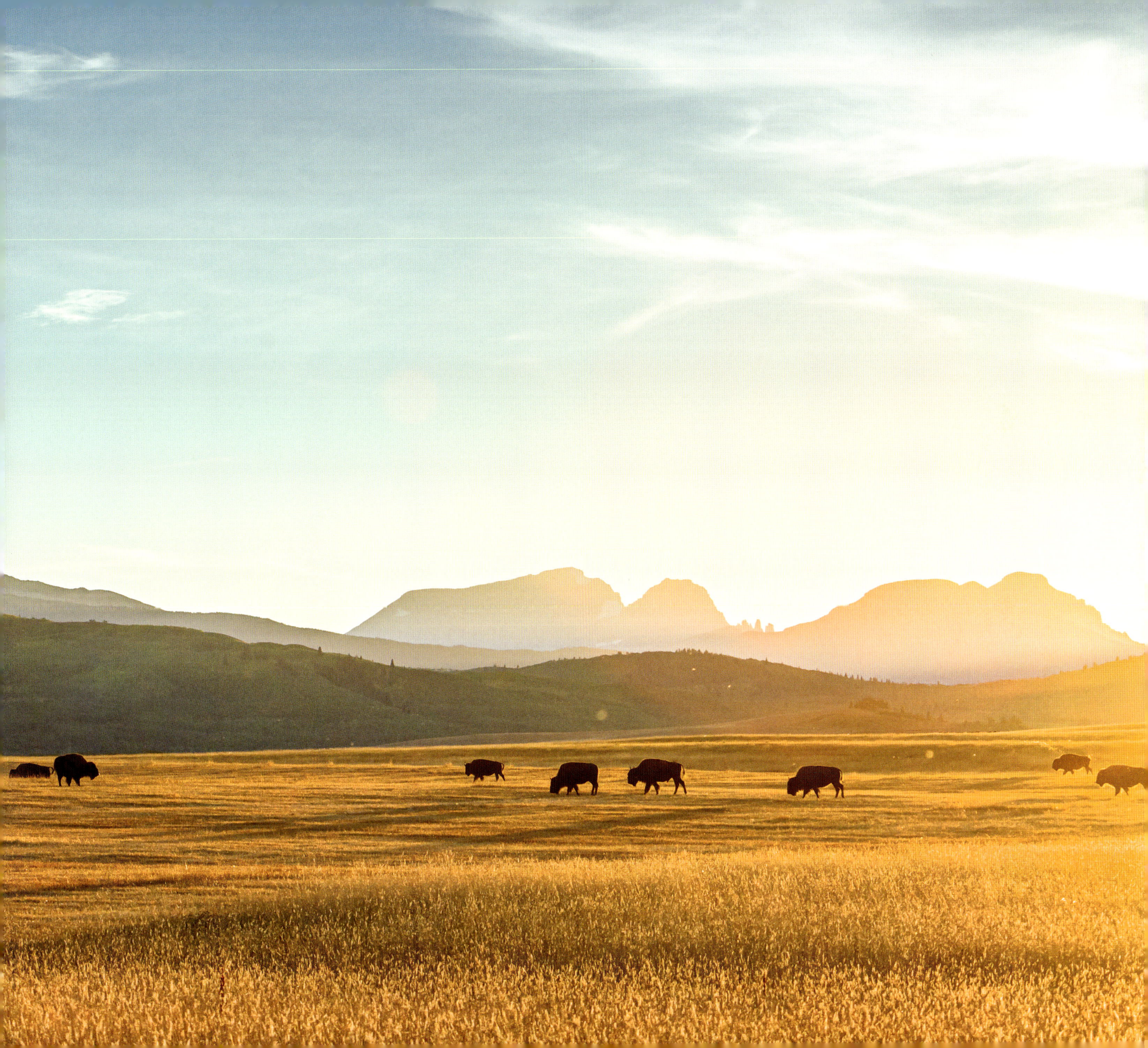

Bison are capable of sprinting
thirty-five miles an hour,
clearing fence-high obstacles,
and fording glacier-fed rivers
with their calves during the
height of spring run-off.

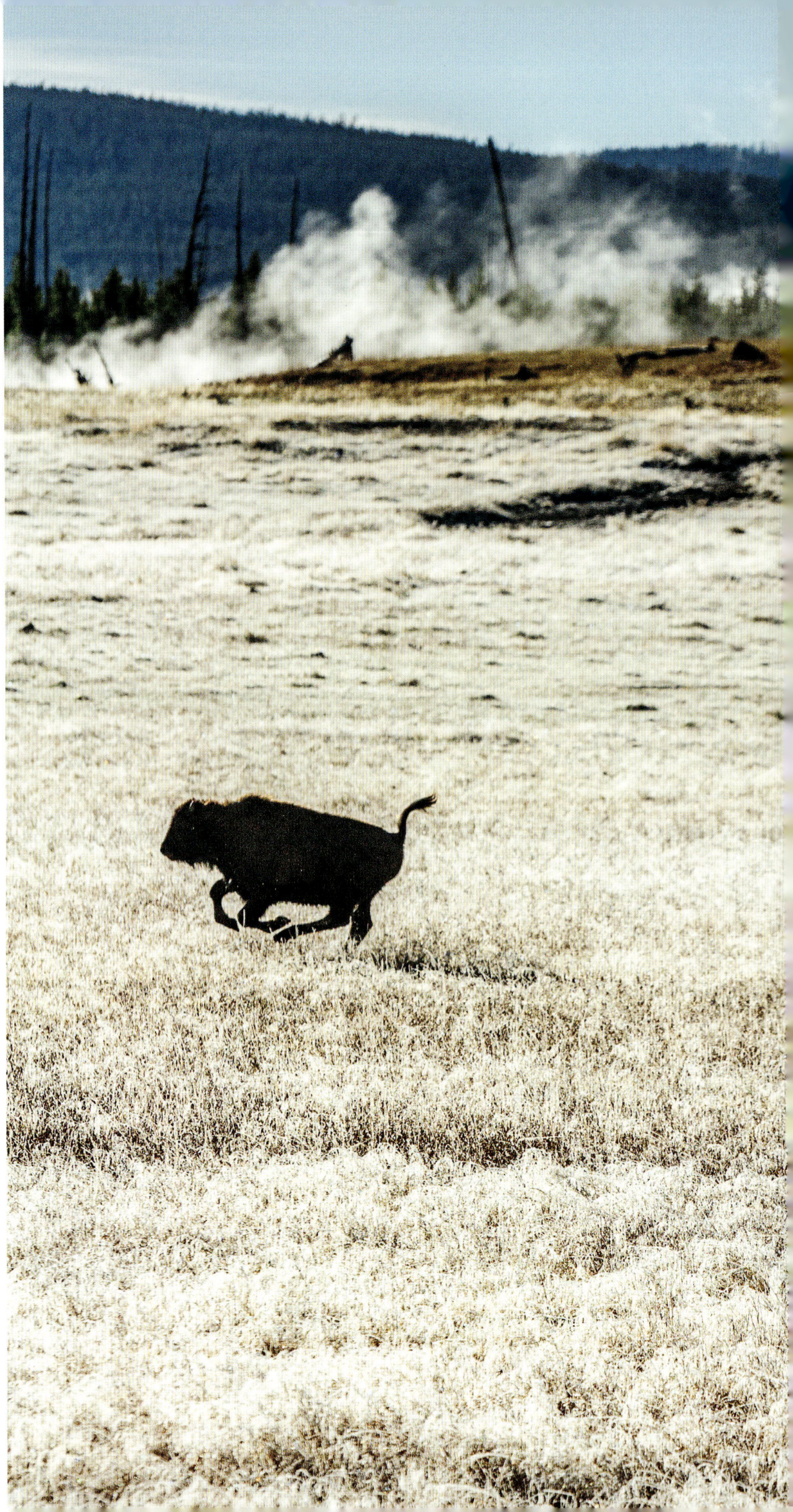

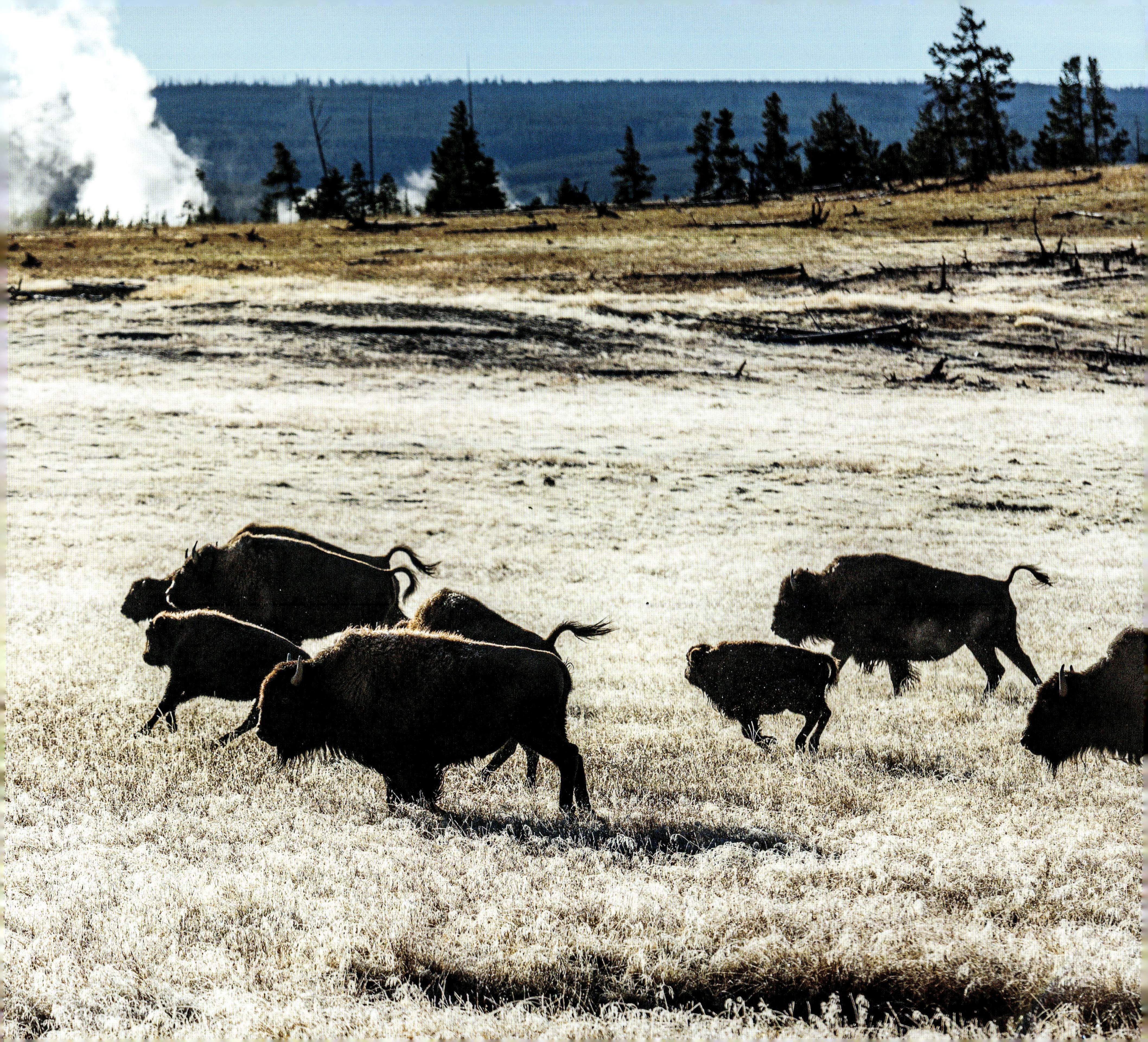

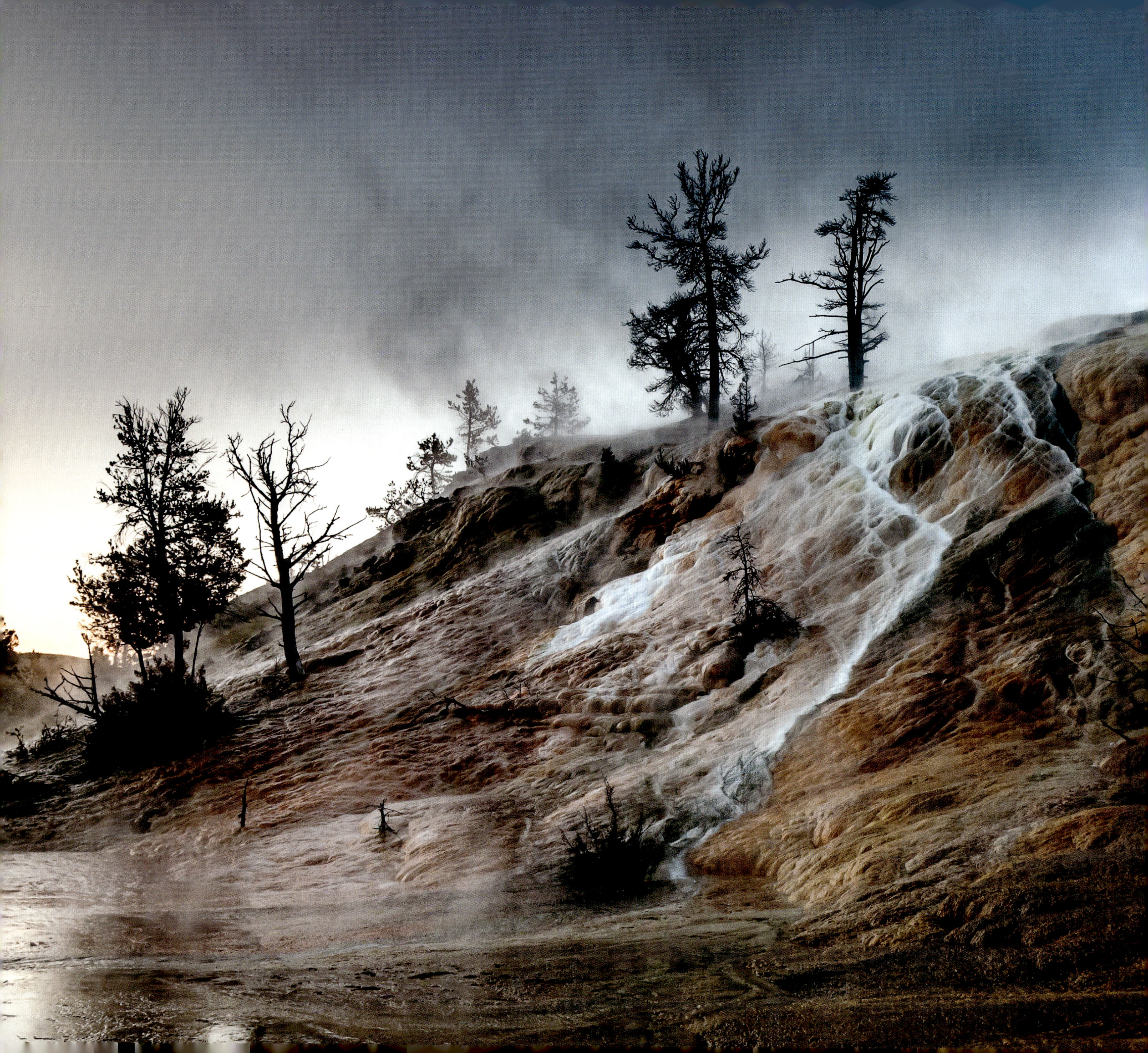

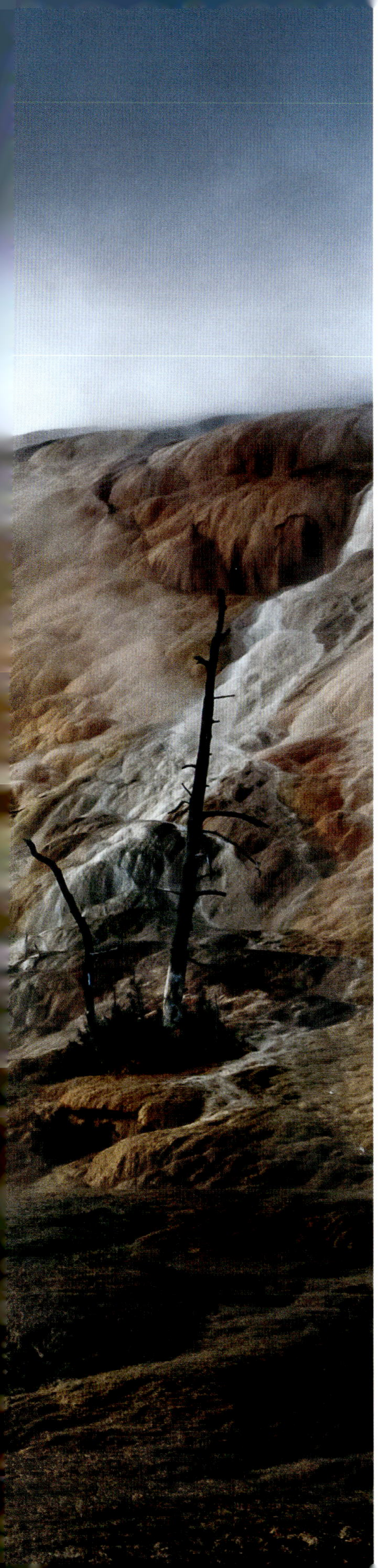

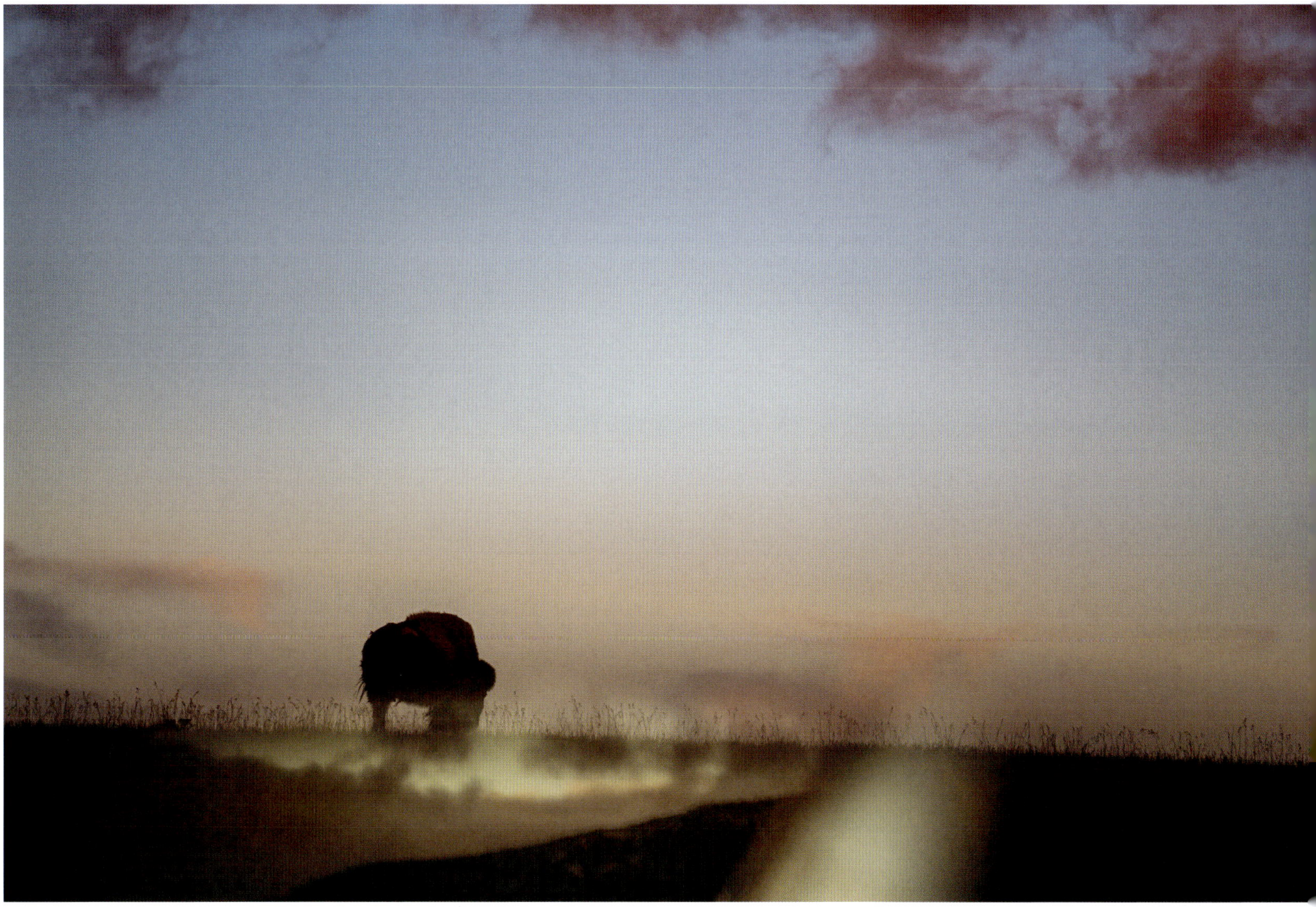

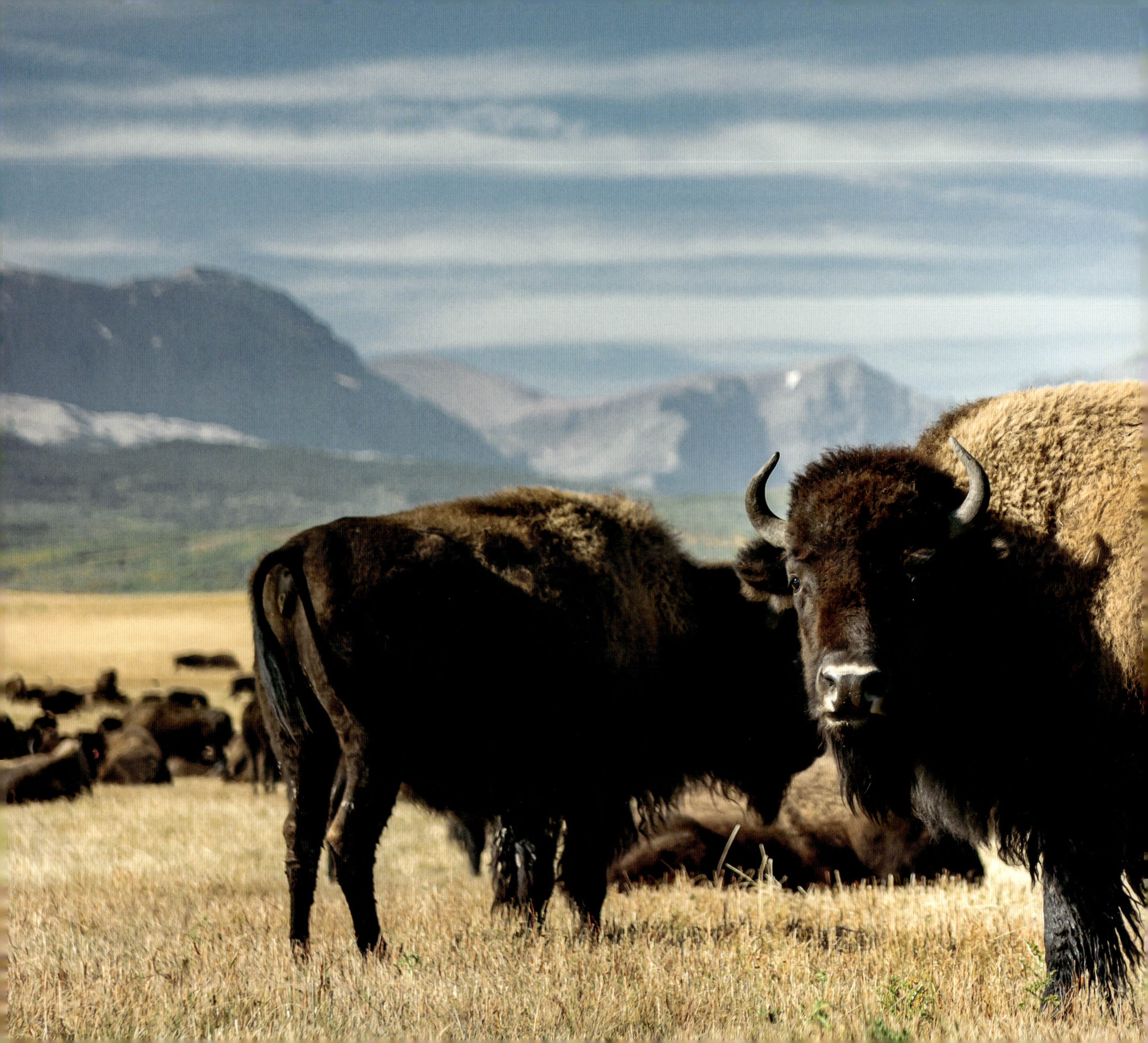

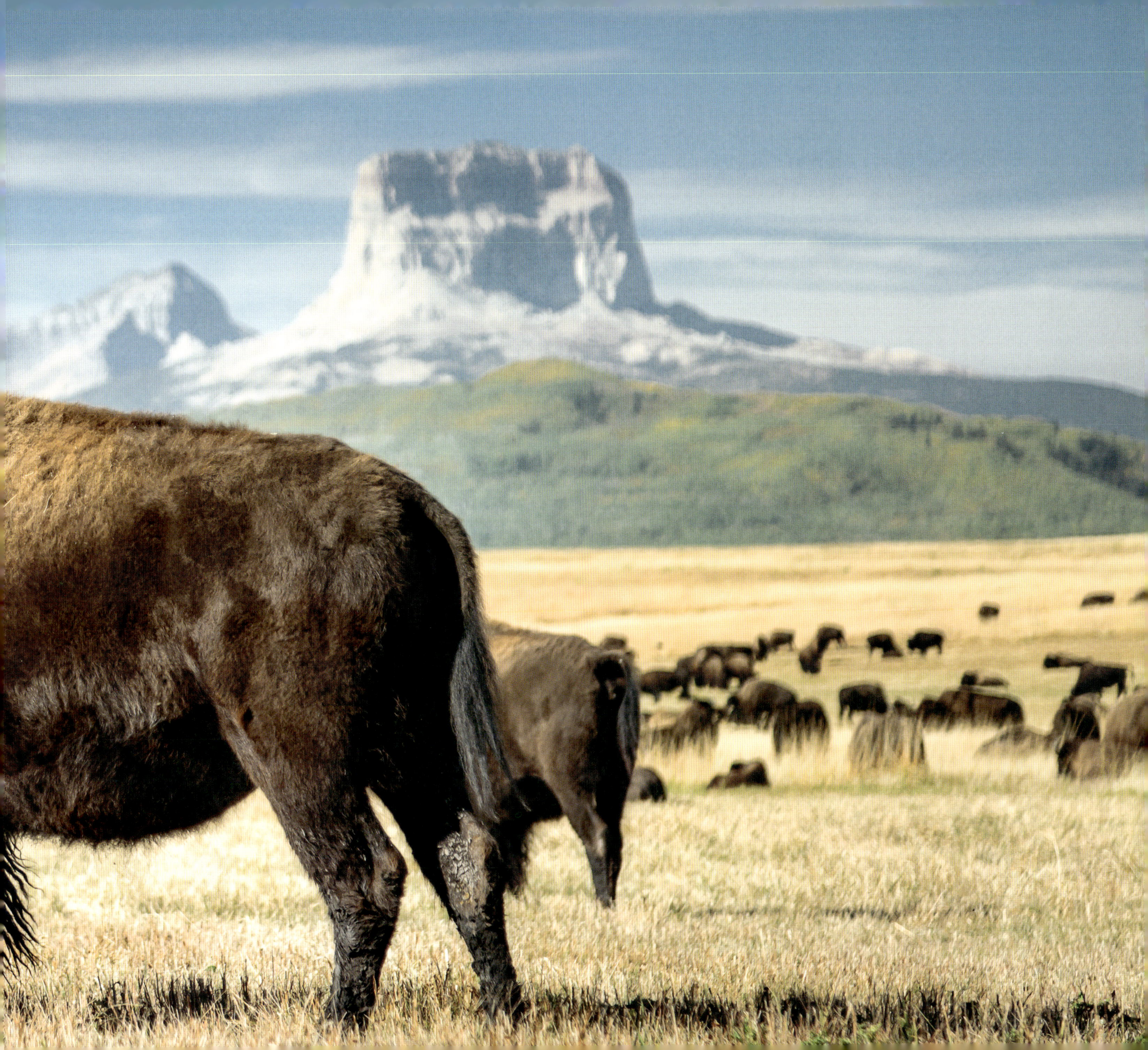

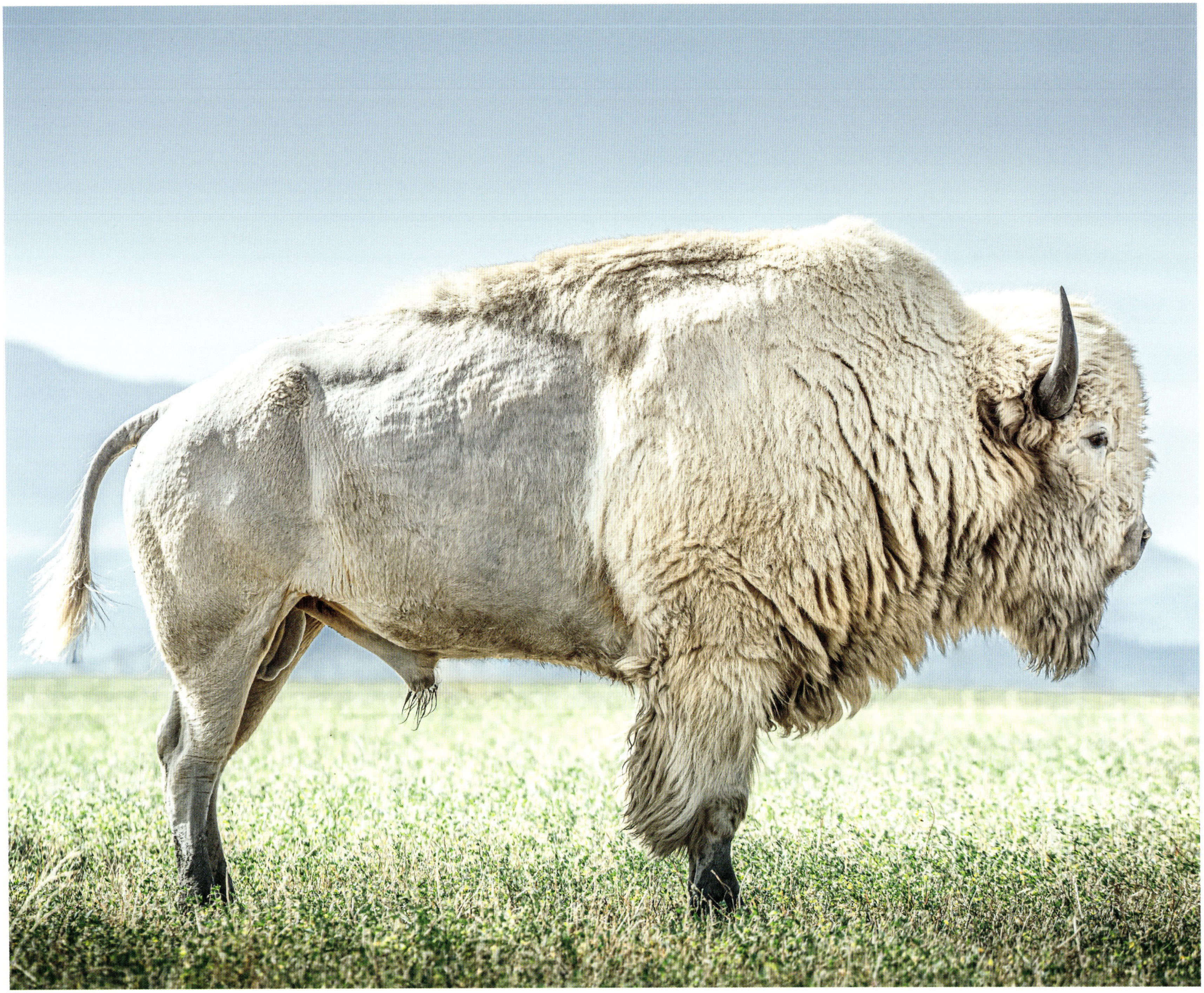

The Picturesque

The American bison, *Bison bison*, goes by many names. To scientists and park administrators, it is bison. To many native people, it's buffalo. To the Lakota, it's Tatanka.

Technically the animals are bison, but most Americans know them as buffalo. Place names like Buffalo, New York; household items like the buffalo nickel; popular bands (Buffalo Springfield); team mascots (see the University of Colorado's Ralphie, team mascot to the Buffaloes); and western mythology (thanks, Buffalo Bill!) have resulted in buffalo being the dominant name used throughout the country. In Yellowstone Park literature and at scientific conferences, however, it's bison.

Common wisdom indicates that the moniker buffalo was attached to the animals by Easterners and Europeans familiar with Asian water buffalo and African Cape buffalo. But some historians point to early nineteenth-century French trappers; they believe buffalo was an evolution of *boeuf*, the French word for beef. The conundrum was summed up most succinctly by one rancher with a strong scientific background. When he's conducting business, it's bison. When he suggests to his kids they go check on the herd, it's buffalo.

One needn't have picnicked in Yellowstone nor studied Plains Indian culture to have some idea of what a buffalo looks like—despite the fact that most Americans have probably never seen one in the flesh other than at a zoo, and it was only in 2016 that the bison was declared America's National Mammal. The animal has nonetheless pervaded our collective consciousness through film, through art, and through tropes found in Western literature and design. Its likeness has been widely disseminated since the earliest days of westward expansion and has reached into schools and homes from coast to coast, to lifestyles and regions far removed from their historic range. Despite their exotic nature and their scarcity, the bison as an emblem is American as the eagle.

The prehistoric bison painted on the walls of the caves at Lascaux around 15,000 B.C. ultimately evolved to their more modern-day relatives depicted on North American teepees and in reservation-era ledger art. During the nineteenth century, as Easterners and Europeans began discovering the wonders of the

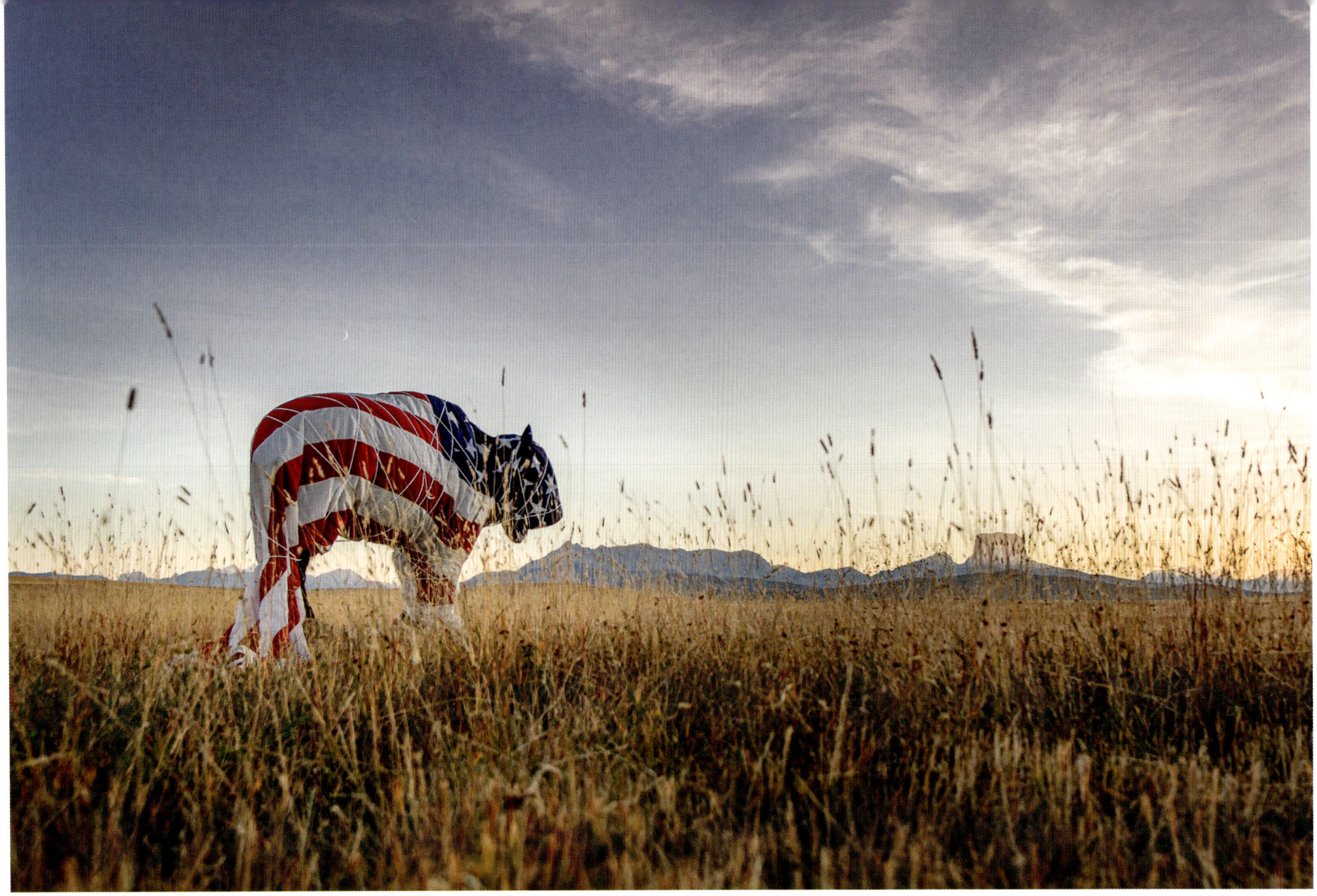

West, bison were frequent subjects in artwork by George Catlin, in the 1830s, the first white man to depict Plains Indians on their own ground; Titian Ramsay Peale, a scientific illustrator who accompanied an 1819 expedition to the Rocky Mountains; and Karl Ferdinand Wimar, who traveled up the Missouri in the 1850s to document Plains tribes.

The widely influential artist Charlie Russell grew up devouring books about the Wild West. He left home at sixteen to go work on a Montana sheep ranch, but he was a bit late to the party, not arriving in the Rockies until the 1880s. "Buffalo Hunt," Russell's first bronze of bison, wasn't sculpted until 1905, but by the time of his death in 1926, he'd produced more than fifty paintings and sculptures featuring buffalo. Frederick Remington was an Easterner and artist of the Hudson River School, but in 1891, when he took his first trip west, he was sketching what were by then merely remnants of the great herds.

Buffalo were a popular subject for those who chronicled the disappearing West and remained so even after most of the animals were gone. During the heyday of Western style, from the 1920s through the '50s, bison imagery was employed widely in decor, particularly on ranches owned by Easterners and Europeans and on dude ranches that catered to guests seeking an immersive Western experience. In 1904, Buffalo Bill Cody—famous for killing bison, then later for introducing

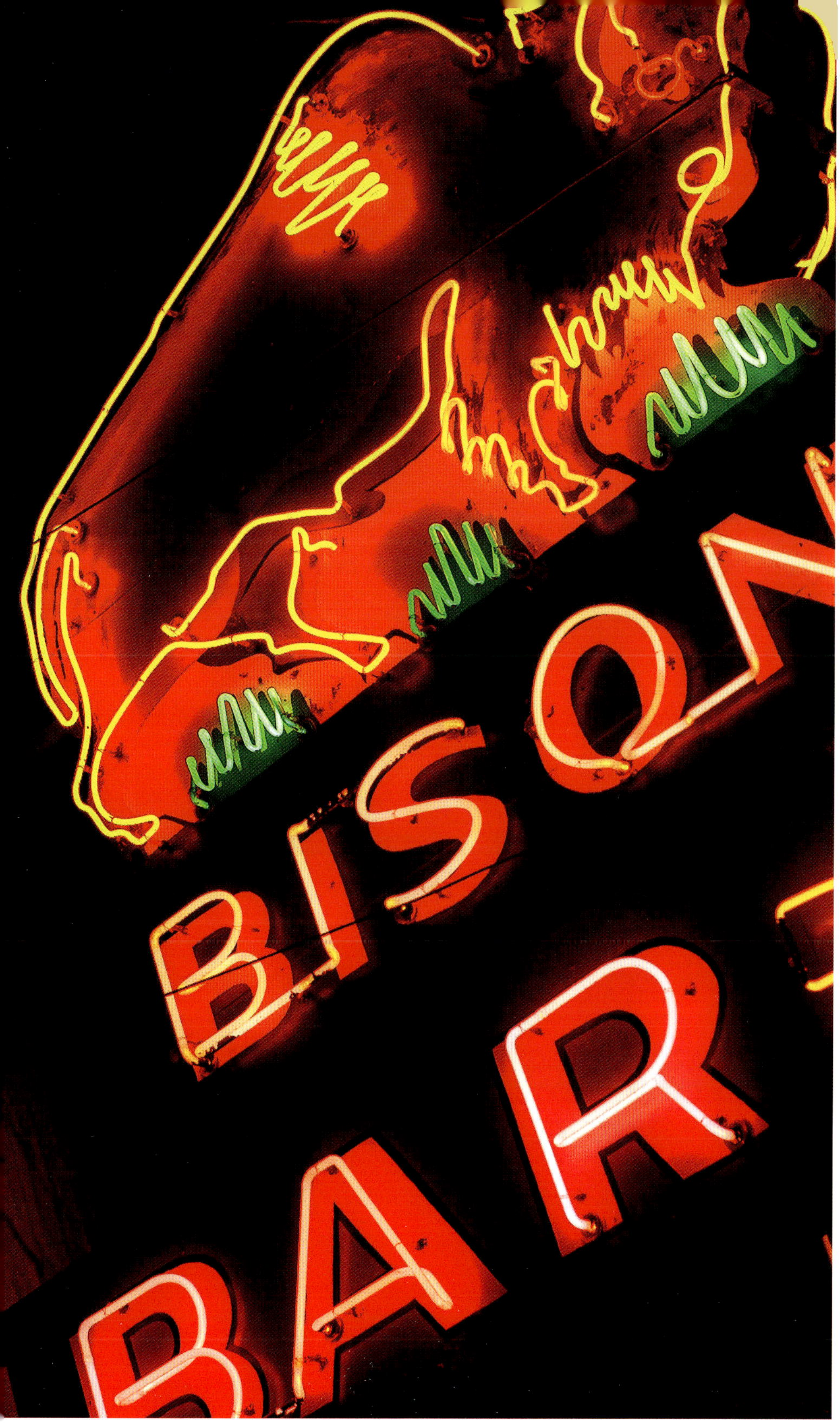

them to audiences around the world through his Wild West Shows—opened a log lodge for tourists on the Shoshone River at the eastern edge of Yellowstone National Park. At Pahaska Teepee, which is still open to Yellowstone travelers today, the bison reigns supreme. It's represented in taxidermy mounts, in a decorative buffalo-hunt chandelier by the bar, and in a life-size bison carving by the gift shop. Visitors to Pahaska will often see live bison as well. These animals, having wandered out of the park, might be walking down the road or grazing placidly on the lawn.

As a motif buffalo are just as popular today, appearing not only in fine art but employed on everything from fine woodwork to decorative metalwork on gates, chandeliers, stair rails, and fireplace screens; these can be mass manufactured and sold in western Walmarts or custom hand-crafted by master artisans for multimillion-dollar mountain homes. Linens, toys, pottery, and children's pajamas decorated with buffalo hold timeless appeal. Bison appear frequently in contemporary art as well, in traditional historic paintings of the old school, in pop works found on Santa Fe's famed Canyon Road, and in Native American art.

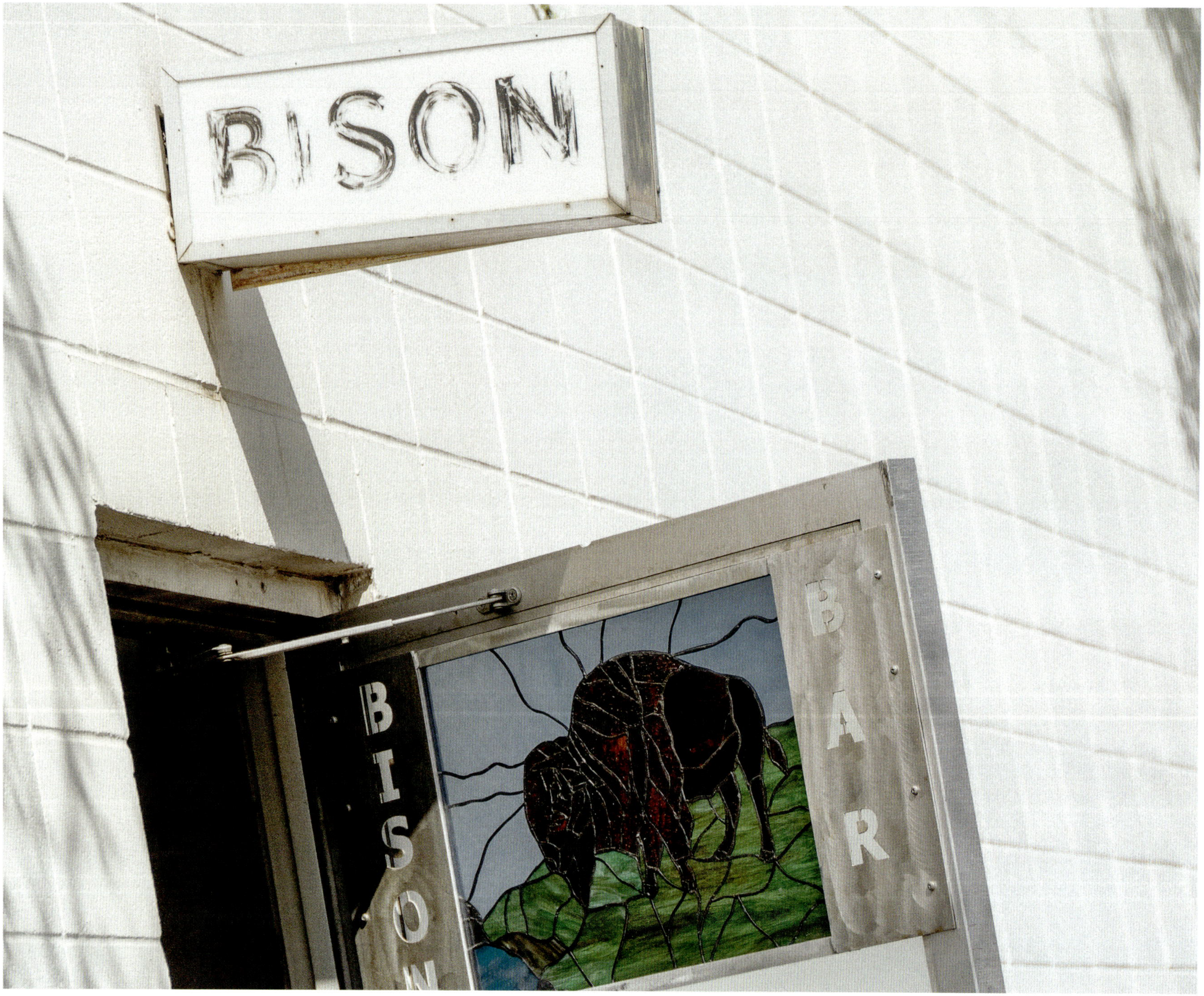
BISON
BISON
BAR

Terrance Guardipee studied at the Institute of American Indian Arts in Santa Fe. His works can be found in homes and museums around the world, including the Smithsonian. Guardipee creates art as an expression of his Blackfeet heritage and as a way of staying connected to his culture. His meticulously composed award-winning works depict tribal leaders, sacred ceremonies, and historic events, including the 1870 Marias massacre on the Bear River in northern Montana, during which his great-great-grandfather Chief Heavy Runner was killed. Guardipee's art references ancient ways of life, and his use of antique documents such as maps, ledger paper, and receipts speaks to the period when his people, having been forced onto reservations, made art using whatever materials were at hand. But he brings an immediacy to his work through bold use of color and a dynamic, contemporary style.

The buffalo is integral to his personal story and often appears in his work, he explains.

> *The buffalo was everything to my tribe: food, shelter, clothing. Almost our whole world revolved around it. To me it has greater meaning. If you believe in our old-time belief systems, a significant dream will come to you. I had a powerful dream about buffalo in 1996 and that's what set me on my path to do art for real. My whole story revolves around the buffalo being very powerful for me, because it gave me the courage to go out and do what I'm doing.*

When he's creating an artwork featuring buffalo, he says, "I'm honoring the animal that has sustained my tribe and family for thousands of years. That is what makes me who I am as an artist. I honor my ancestors as truthfully as I can. Hopefully they see me doing it and look favorably on me."

The most ubiquitous representation of the animal can sometimes still be found jangling in the pockets of Americans today. The iconic buffalo nickel, designed by sculptor James Earle Fraser (and modeled upon a resident of either the Bronx Zoo or Central Park Menagerie, depending on whether the artist's memory of an animal called Black Diamond was accurate), was minted from 1913 to 1938. Some still randomly circulate today, and whole rolls of nickels can be procured on Ebay for about forty dollars, where they're sought by collectors, or might be purchased for decorative use. The design was put back in production in 2001 for a minting of 500,000 commemorative silver dollars; the coins sold out immediately and the project, spearheaded by Colorado senator Ben Nighthorse Campbell, raised five million dollars toward the construction of the Smithsonian's Museum of the American Indian. In 1912, when Fraser was casting about for appropriate subject matter, there were as few as 2,000 bison left in existence. Even so, he later recalled in a 1947 radio interview, "When I was asked to do a nickel I felt I wanted to do something totally American—a coin that could not be mistaken for any other country's coin. It occurred to me that the buffalo, as part of our western background, was one-hundred percent American."

That's a remarkable statement—and an apt illustration of the bison's dogged perseverance—when one considers its history. Originally the bison was, quite literally, the national mammal. It thrived throughout the continent, from sea to sea, and from Mexico to Canada, numbering in the tens of millions, its herds so vast that nineteenth-century reports described them as taking two days to pass by. Bison are thought to be descended from

European steppe bison, the ones with seven-foot horn spans seen on European cave art; it's surmised that they crossed the Bering Land Bridge, eventually reaching what is now Alberta, Canada, after a corridor opened up between ice ages. By the time westward expansion began decimating the herds in the second half of the nineteenth century, bison had evolved to an approximation of what we see in the parks and on ranches today: dark brown 2,000-pound animals (the females average 1,100 pounds), with a horn span of three feet from tip to tip.

Although at one time bison were everywhere on the continent, the place they truly thrived was on the Great Plains. And no lives were as intertwined with the buffalo as those of the Plains Indian tribes.

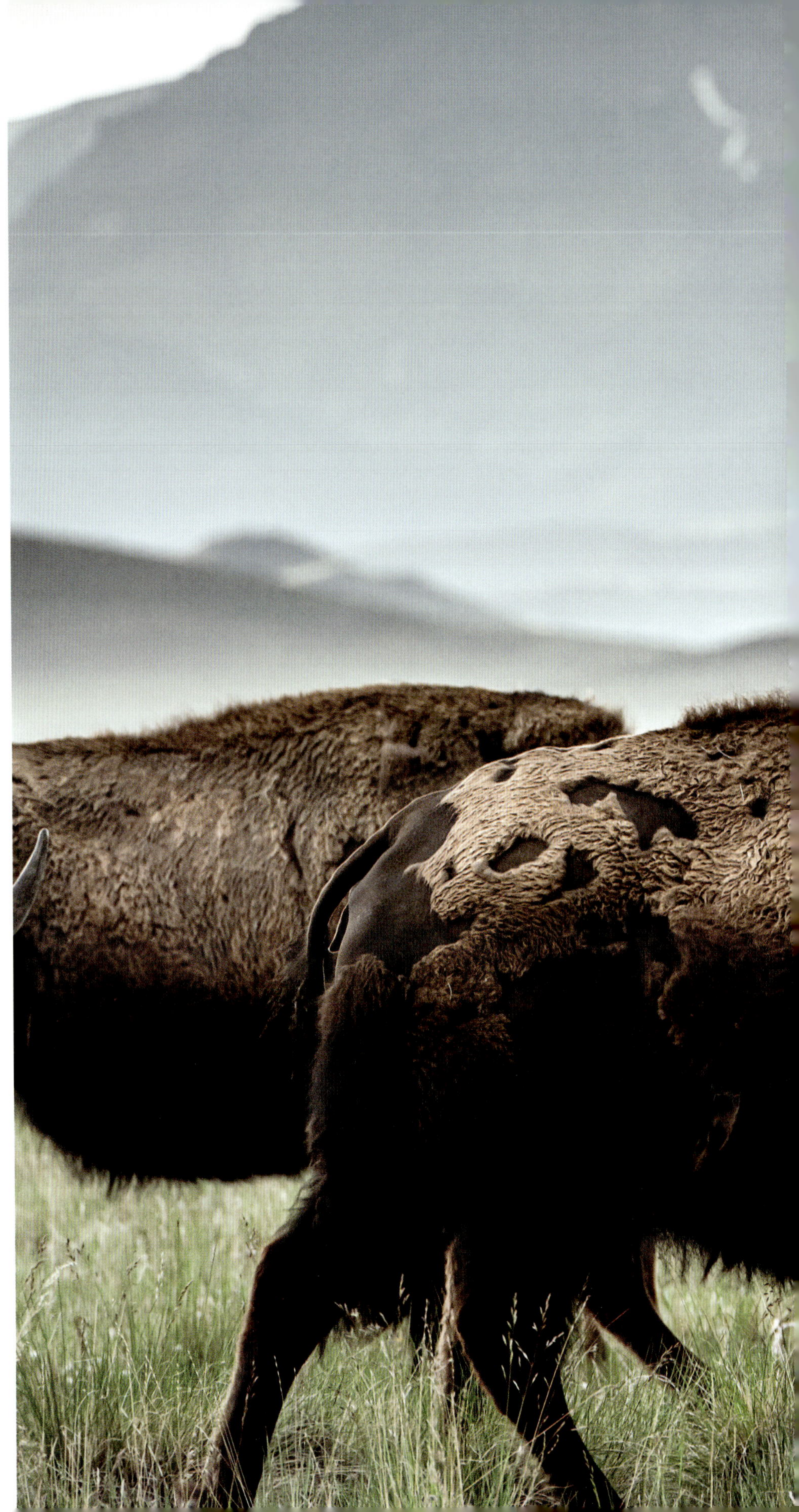

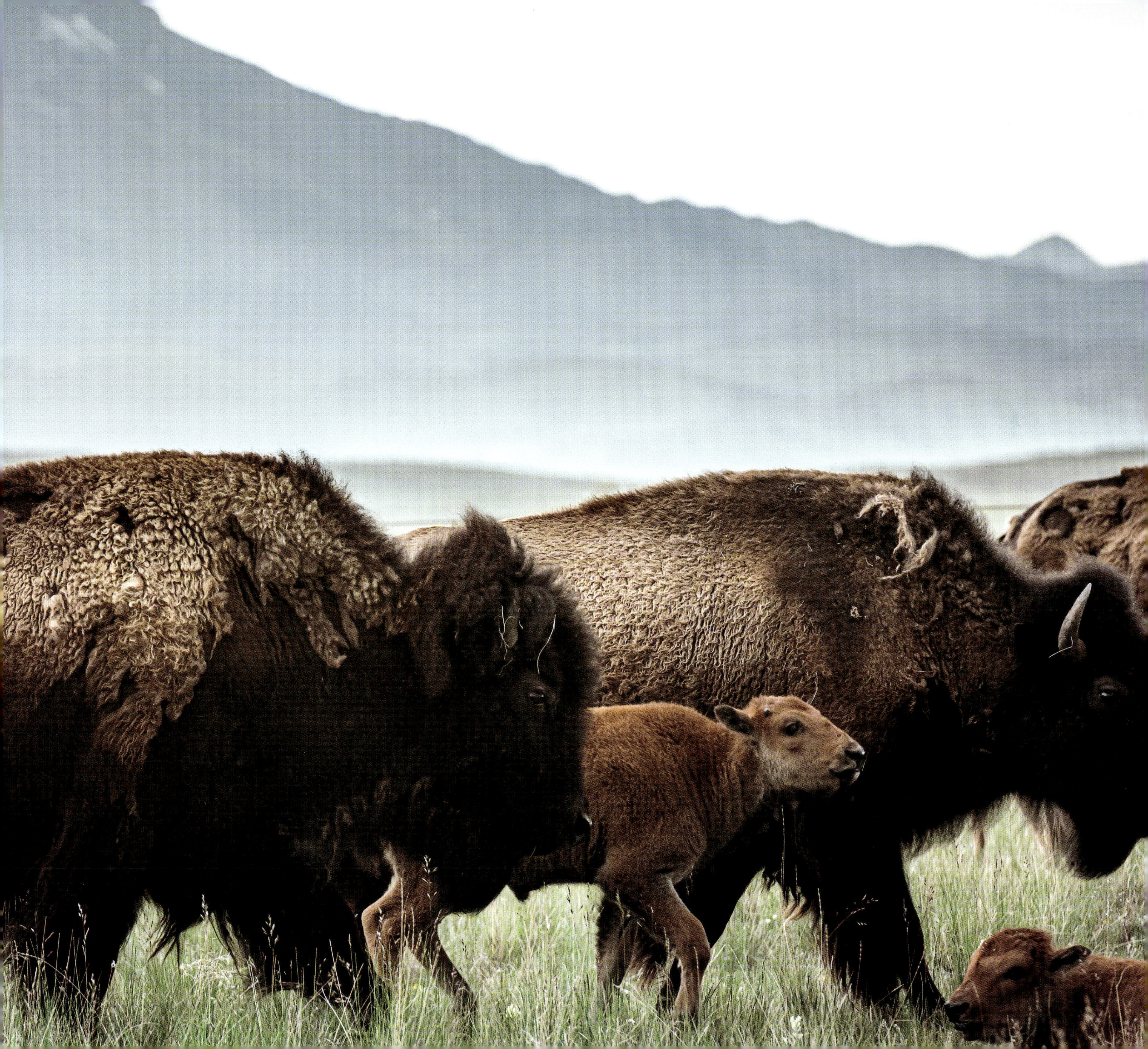

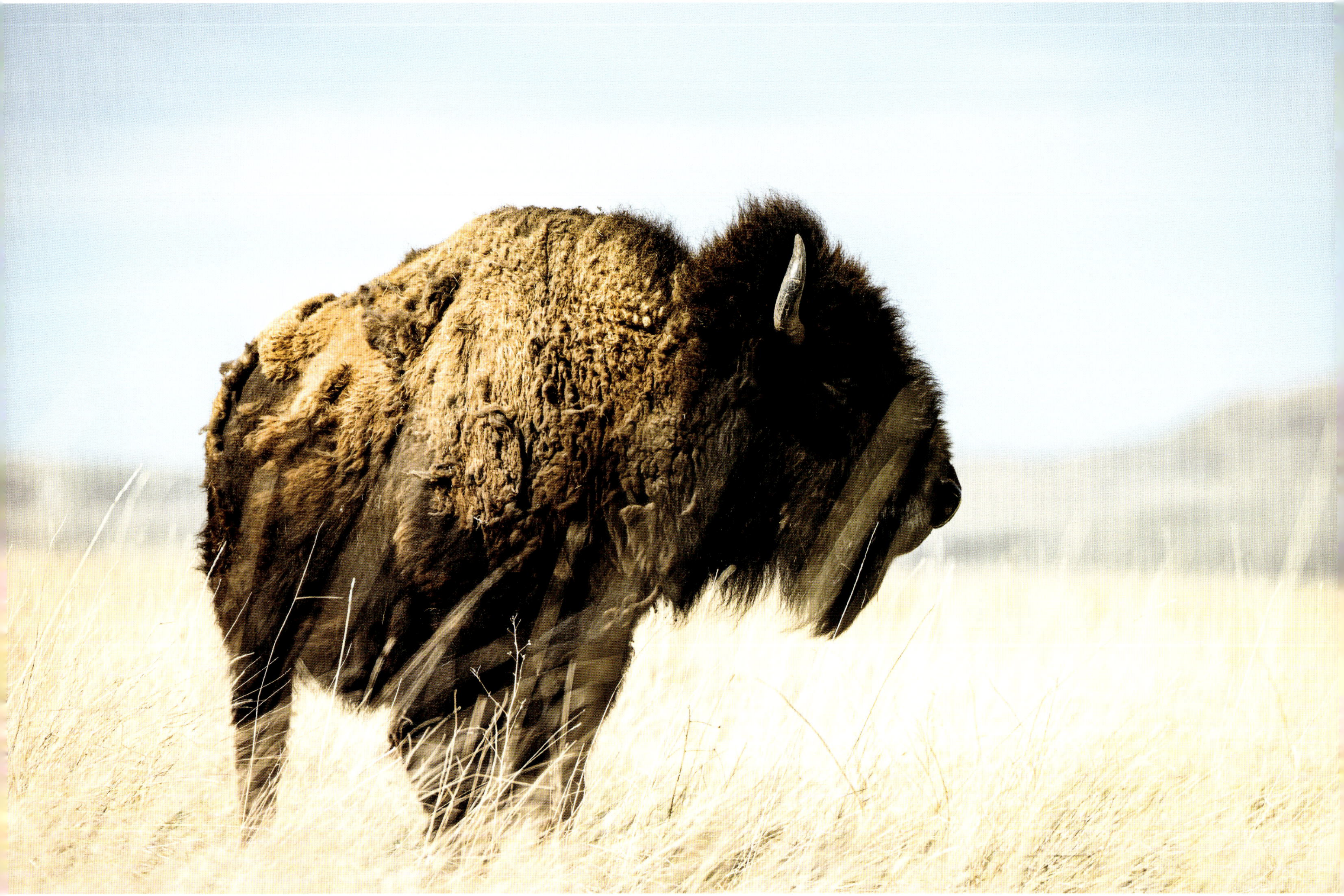

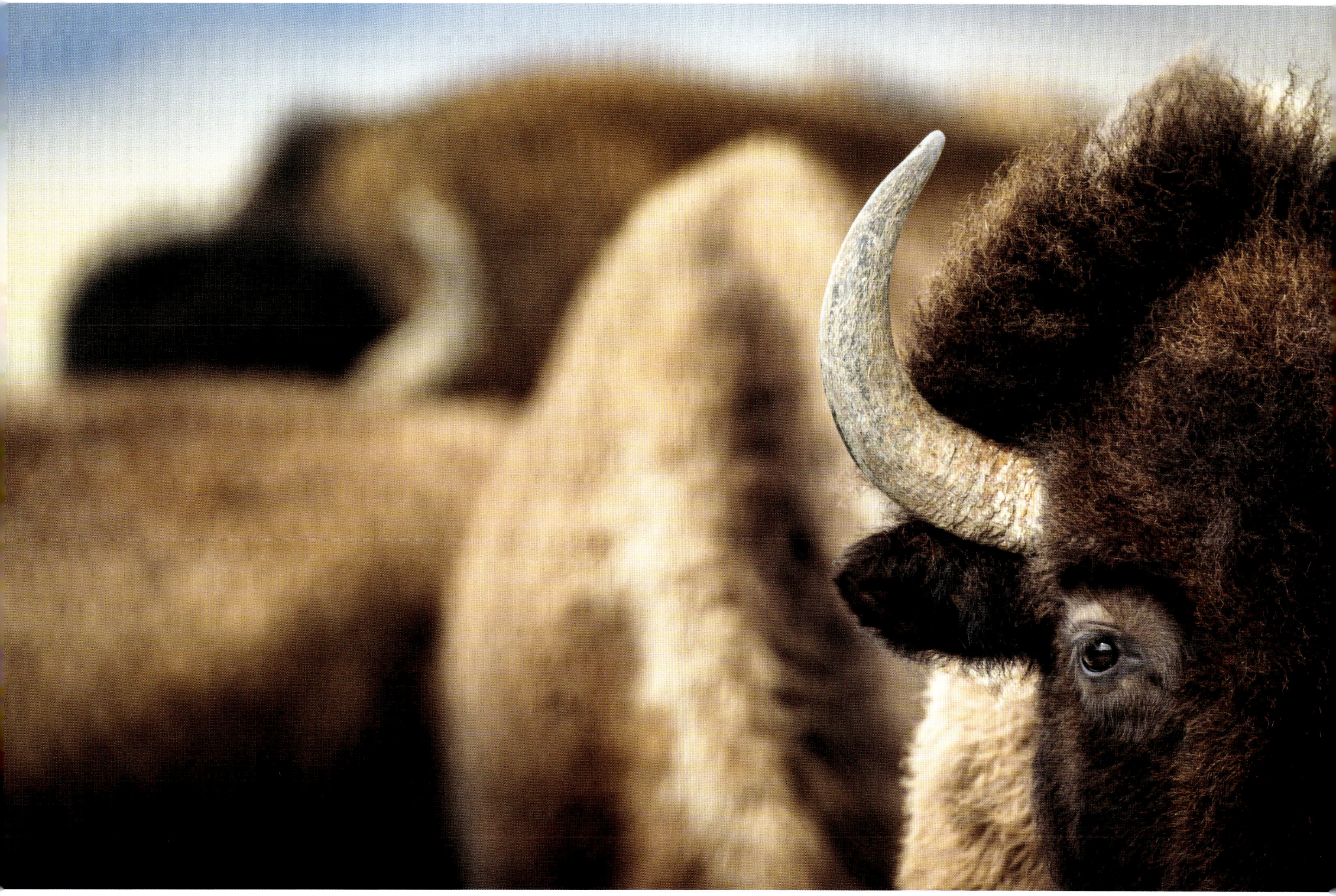

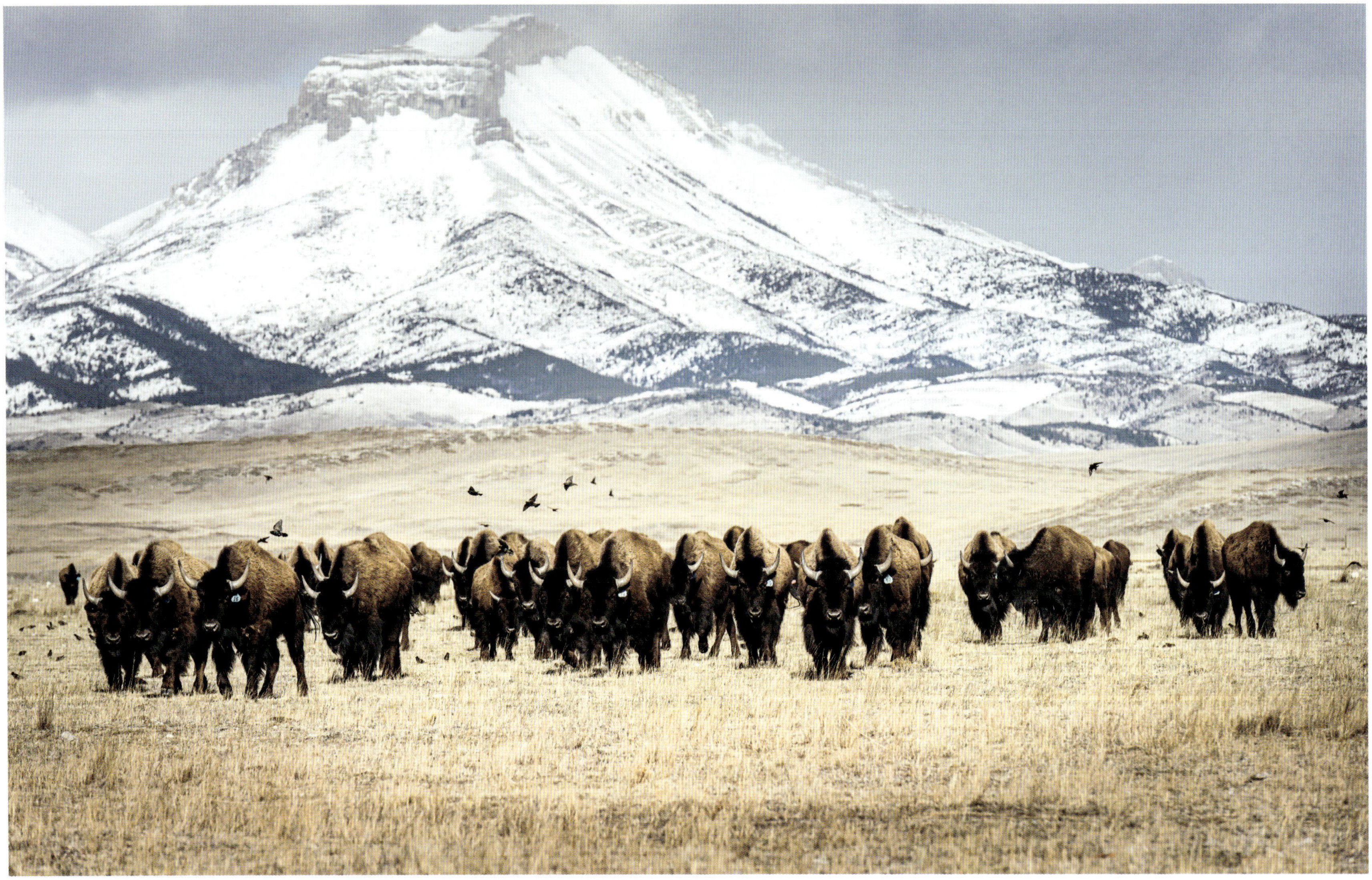

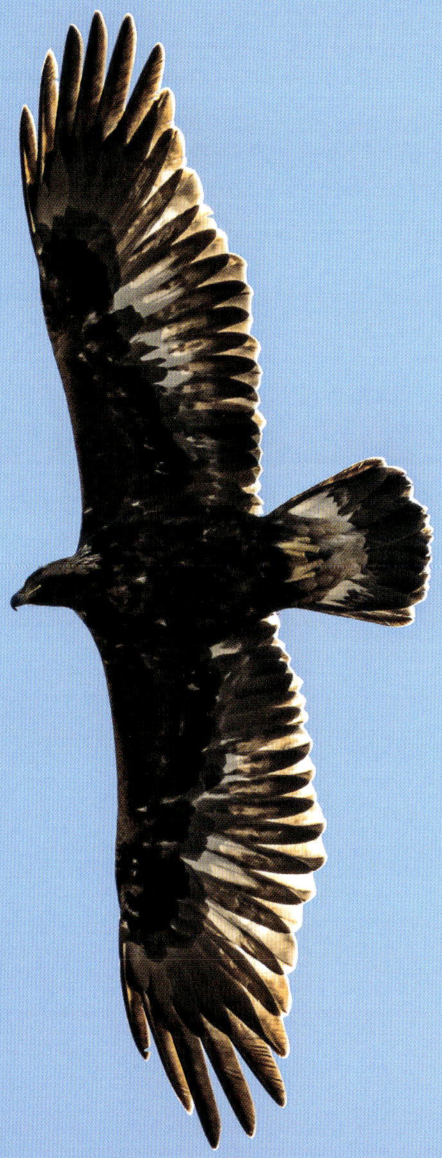

Despite their exotic nature and
their scarcity, the bison as an emblem
is as American as the eagle.

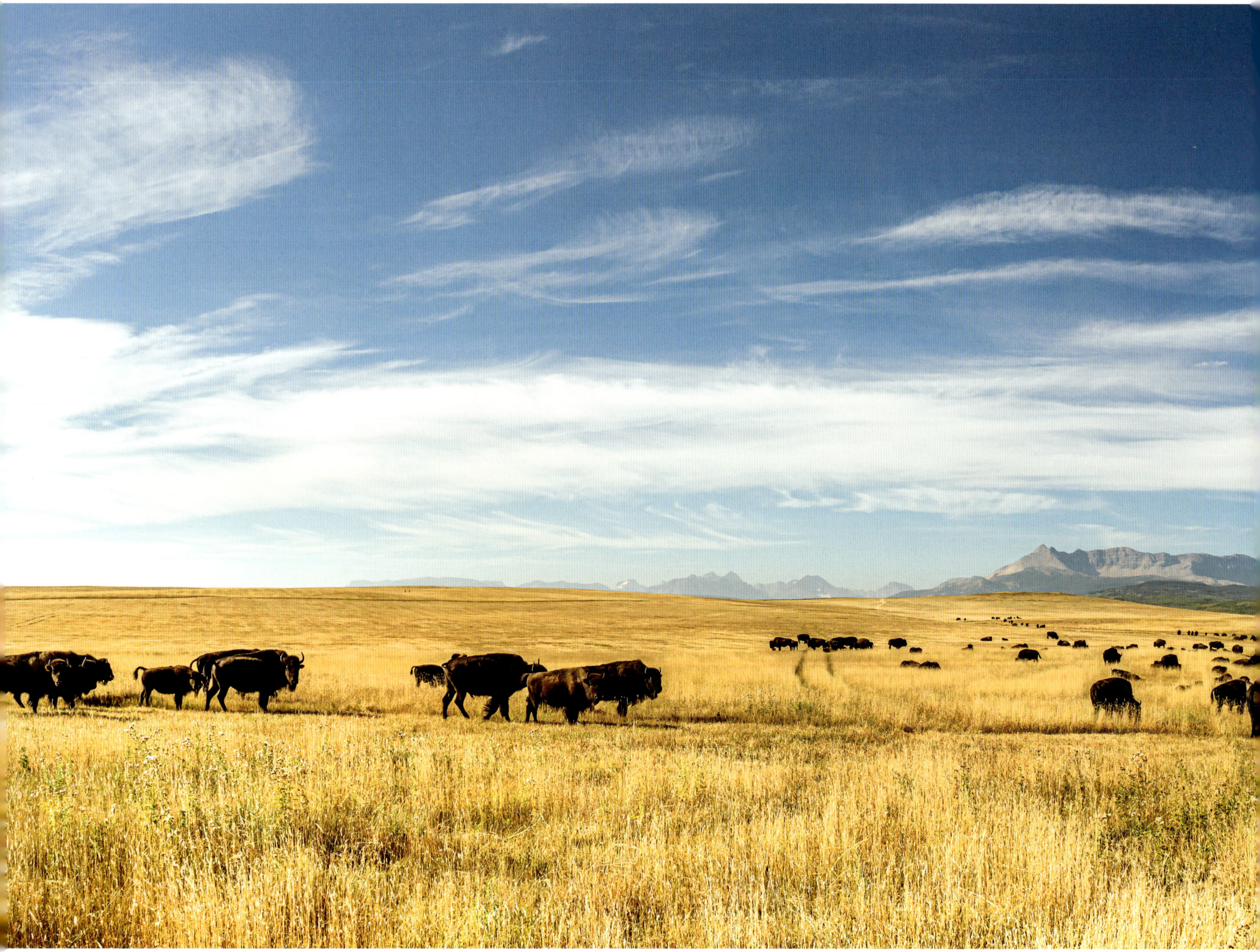

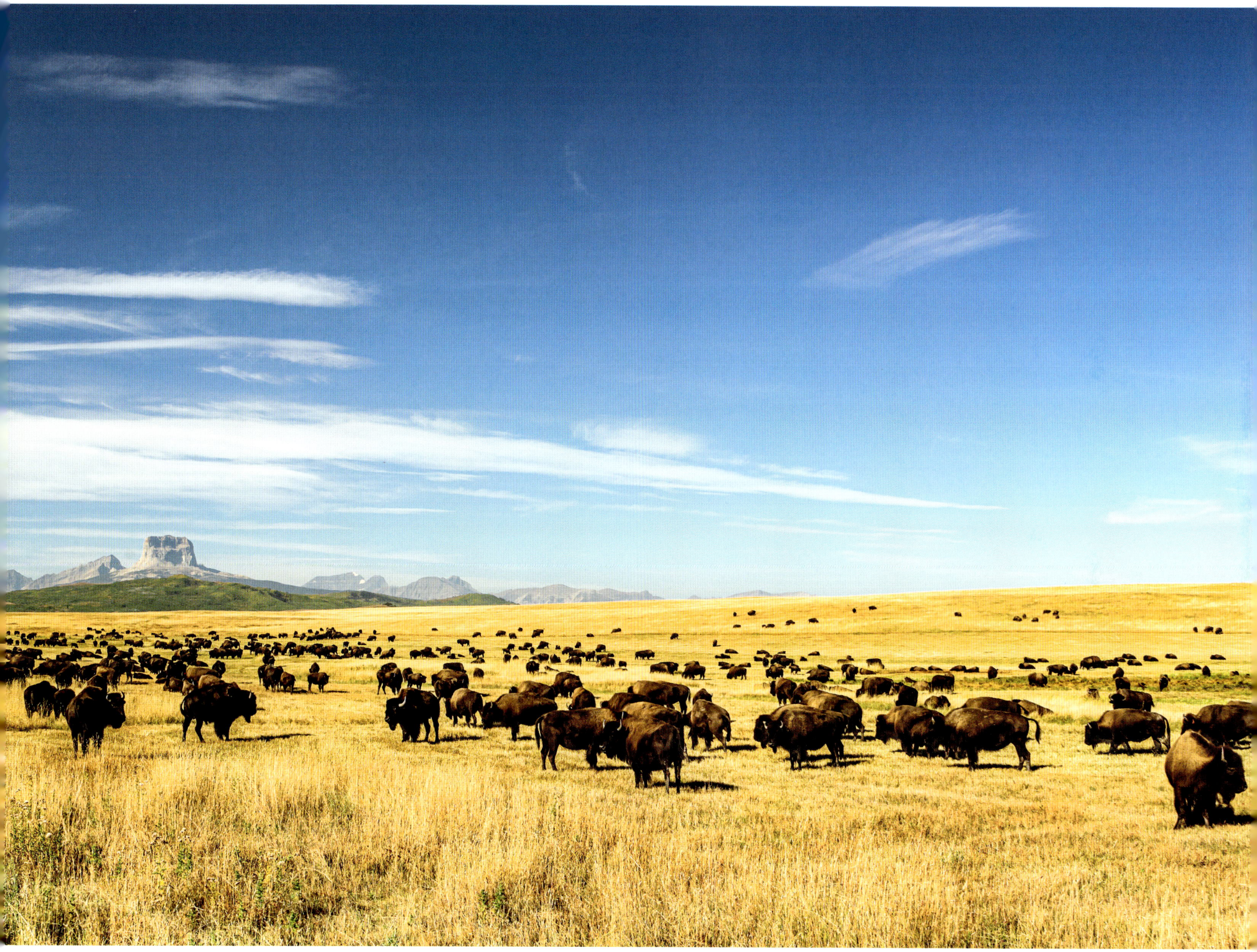

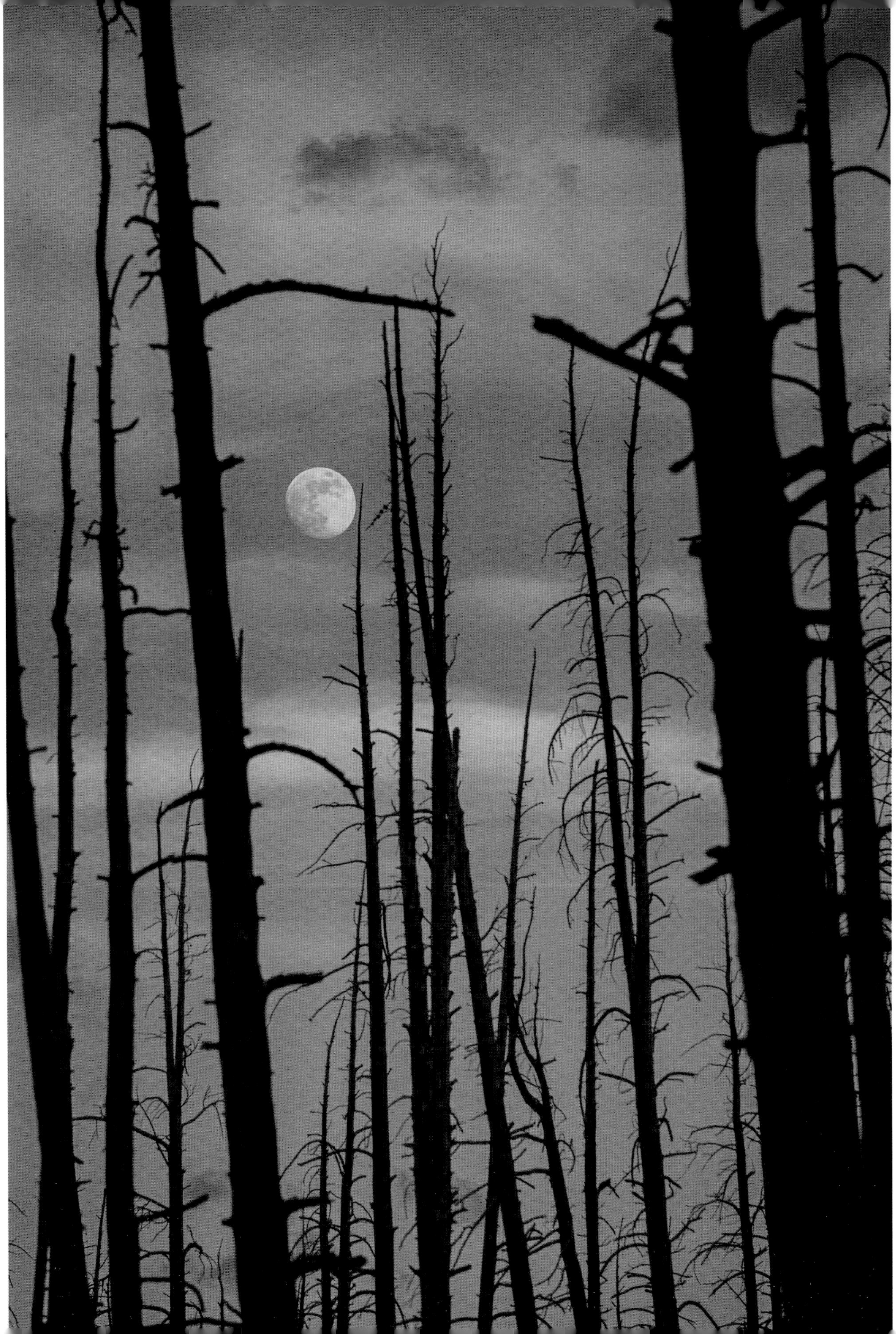

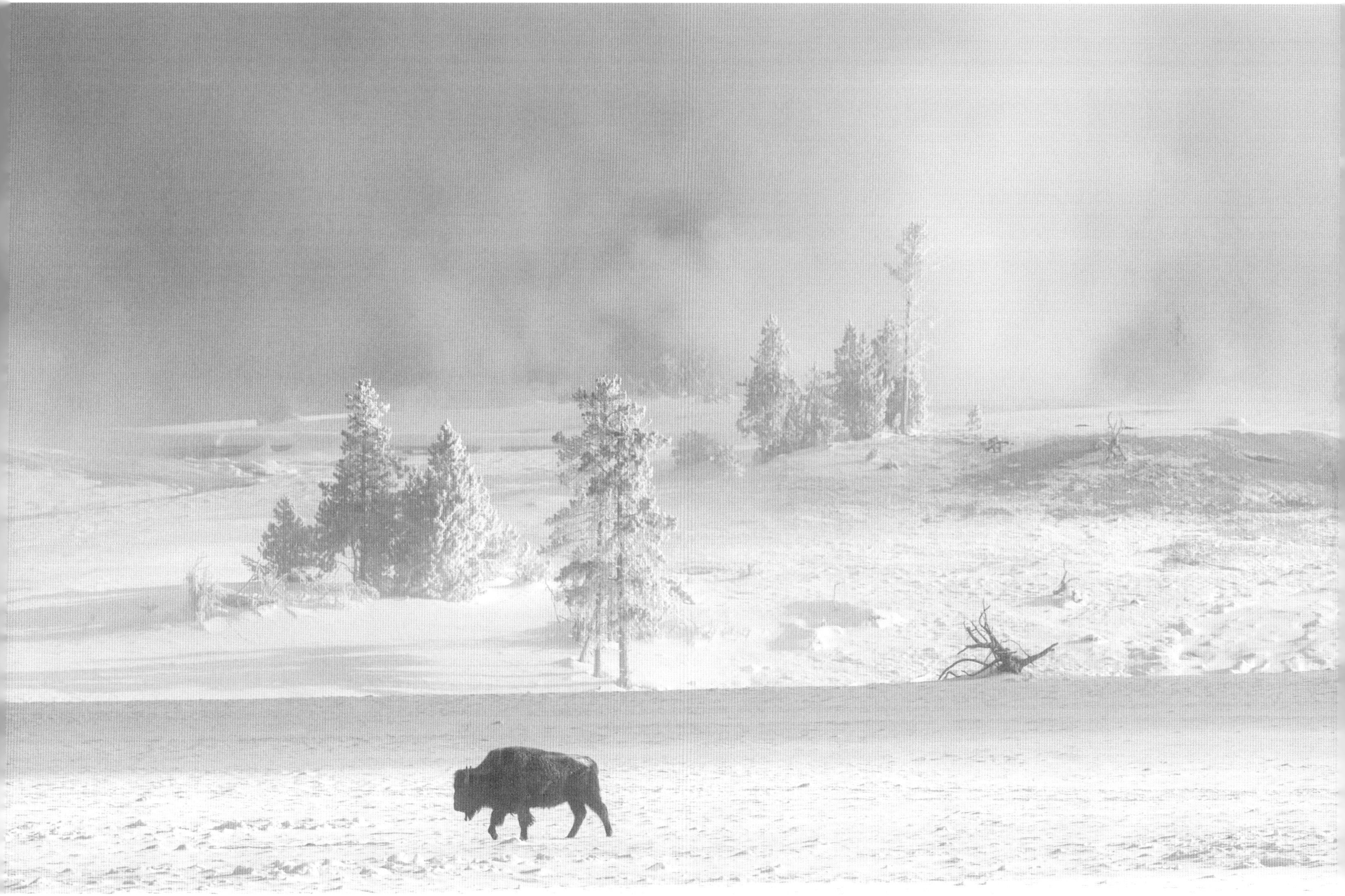

The Seven Buffalo Bulls Become the Big Dipper

At one time, a long time ago, there were some buffalo bulls, known as the Seven Buffalo Bulls, and they were wandering around and wondering how they could be forever upon sacred Mother Earth. At first, they thought they could become a large tree to be forever. Then they thought, No. A large tree can someday die, become dry, and fall to the ground to break into many pieces. *So they decided to not be a great, big tree upon sacred Mother Earth.*

The second idea was to become a huge rock, a boulder up in the mountains, so the Seven Buffalo Bulls could be forever upon sacred Mother Earth. But they thought, No. A huge rock boulder would someday crumble from the elements, like cold ice and wind. It would break up to become sand blowing in the wind or be carried away by a big, snow runoff flood. *So they decided to not become a huge rock, a boulder up in the mountains, to be forever upon sacred Mother Earth.*

The Seven Buffalo Bulls' third heart's way was to become the wind, so they could be forever upon sacred Mother Earth. The wind is all over Mother Earth and truly is forever, so they were happy to become the wind—then one buffalo bull said, "But no, bad beings like the snake can breathe in the wind, and we would be inside of a snake." That would be awful, *the Seven Buffalo Bulls thought, and they decided to not become the wind to be forever upon sacred Mother Earth.*

So they again wondered what they could be to become forever upon sacred Mother Earth. This time, they wanted to become the sacred fire, and they thought hard as to how it would be, then one buffalo bull said, "It would not be good because people throw bad things into the fire." And the Seven Buffalo Bulls also decided against being the sacred fire to be forever upon sacred Mother Earth.

One last time, they thought of how they could be forever upon sacred Mother Earth. After much thought, they decided to become the seven stars of the Big Dipper. (Also, when they went to become the Big Dipper, it is said that Eewilikkooloo, *one of the buffalo's wolfdogs, wanted to go too, and close to one of the handle stars is the dim star he became.)*

The Seven Buffalo Bulls wanted to be at the place where the Upsauloeka *(Crow Indians) point the sacred pipe to pray to help them forever upon sacred Mother Earth. So when life does not go well, and the Upsauloeka need help, we can smoke with the Seven Buffalo Bulls and ask for it.*

—HENRY REAL BIRD

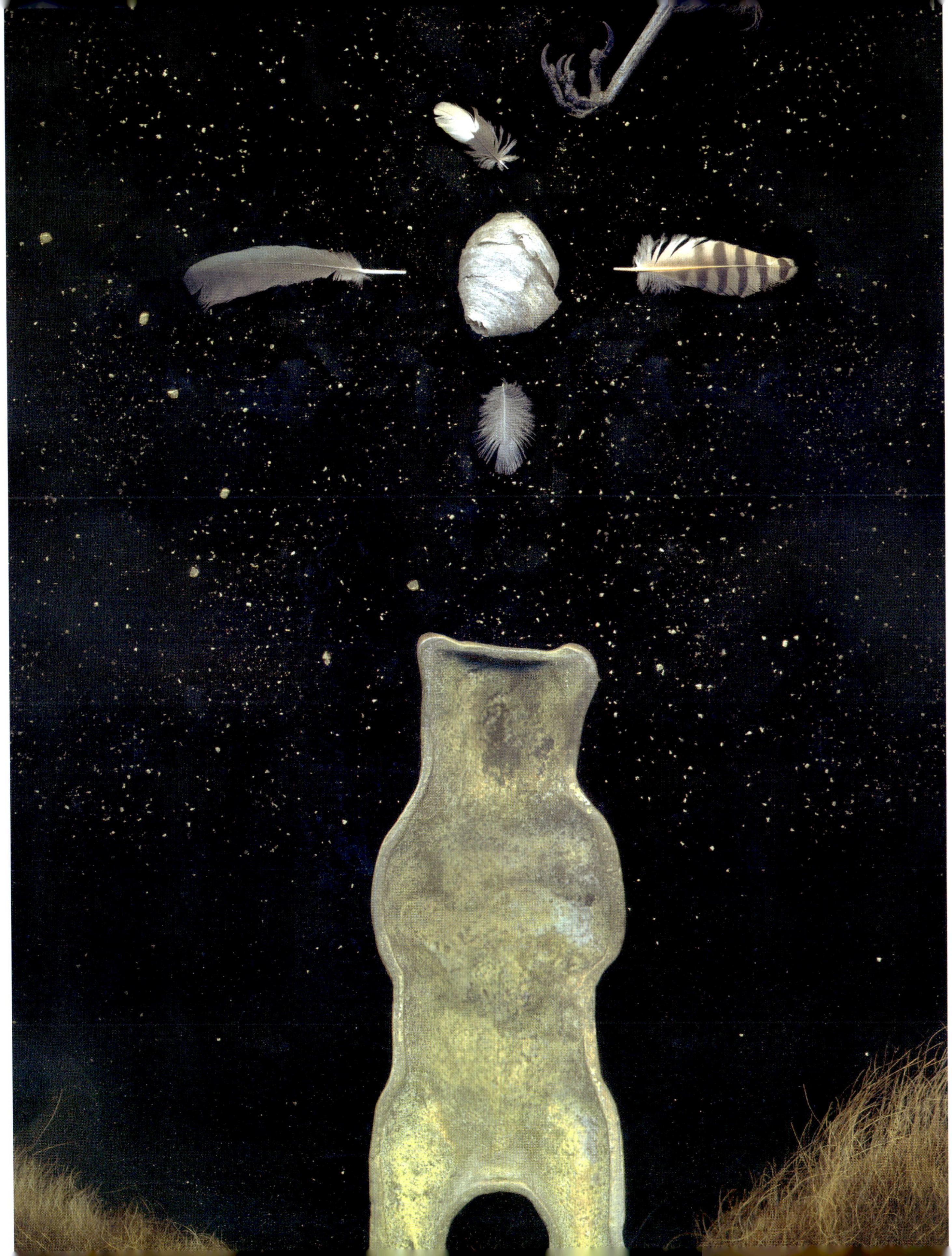

Like many North Americans, bison are the descendants of immigrants. Somewhere around 200,000 years ago, small herds of bison grazed their way east across Beringia, the broad grassy steppe that connected eastern Asia to Alaska when the sea level was three hundred feet lower than it is today. Without knowing it, they became the first American bison. Then, for tens of thousands of years, they lived in a land full of other large mammals like mammoths, mastodon, camels, giant ground sloths, horses, stag moose, helmeted muskox, short-faced bears, saber-tooth cats, and bear-size beavers. It was only about 12,000 years ago that these other species departed or became extinct and the bison inherited the Great Plains and the surrounding woodlands. By this time, people had arrived, traveling the same route from Asia, and a whole new chapter of the American bison had begun.

—KIRK JOHNSON, SANT DIRECTOR, NATIONAL MUSEUM OF NATURAL HISTORY, SMITHSONIAN INSTITUTION

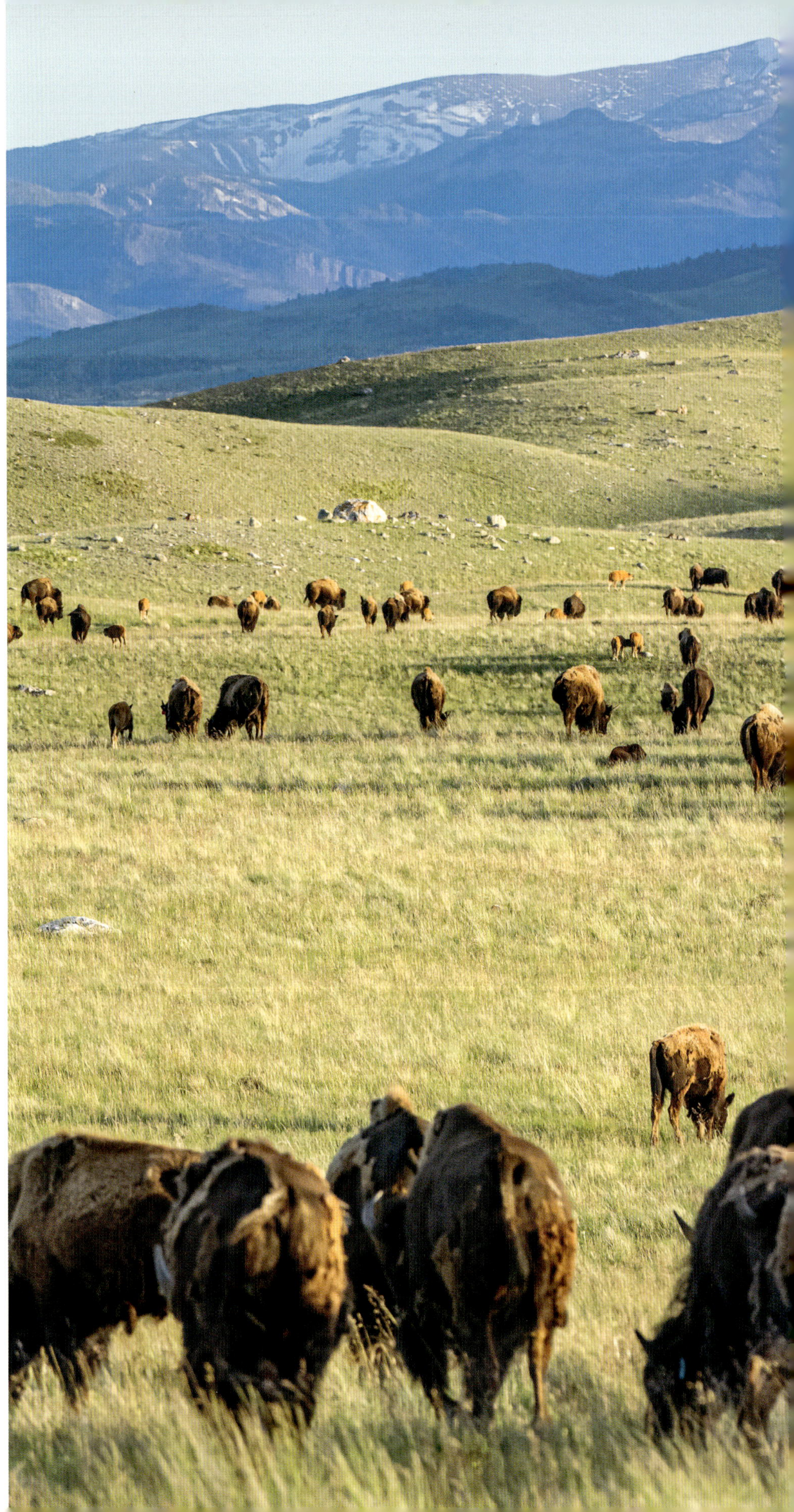

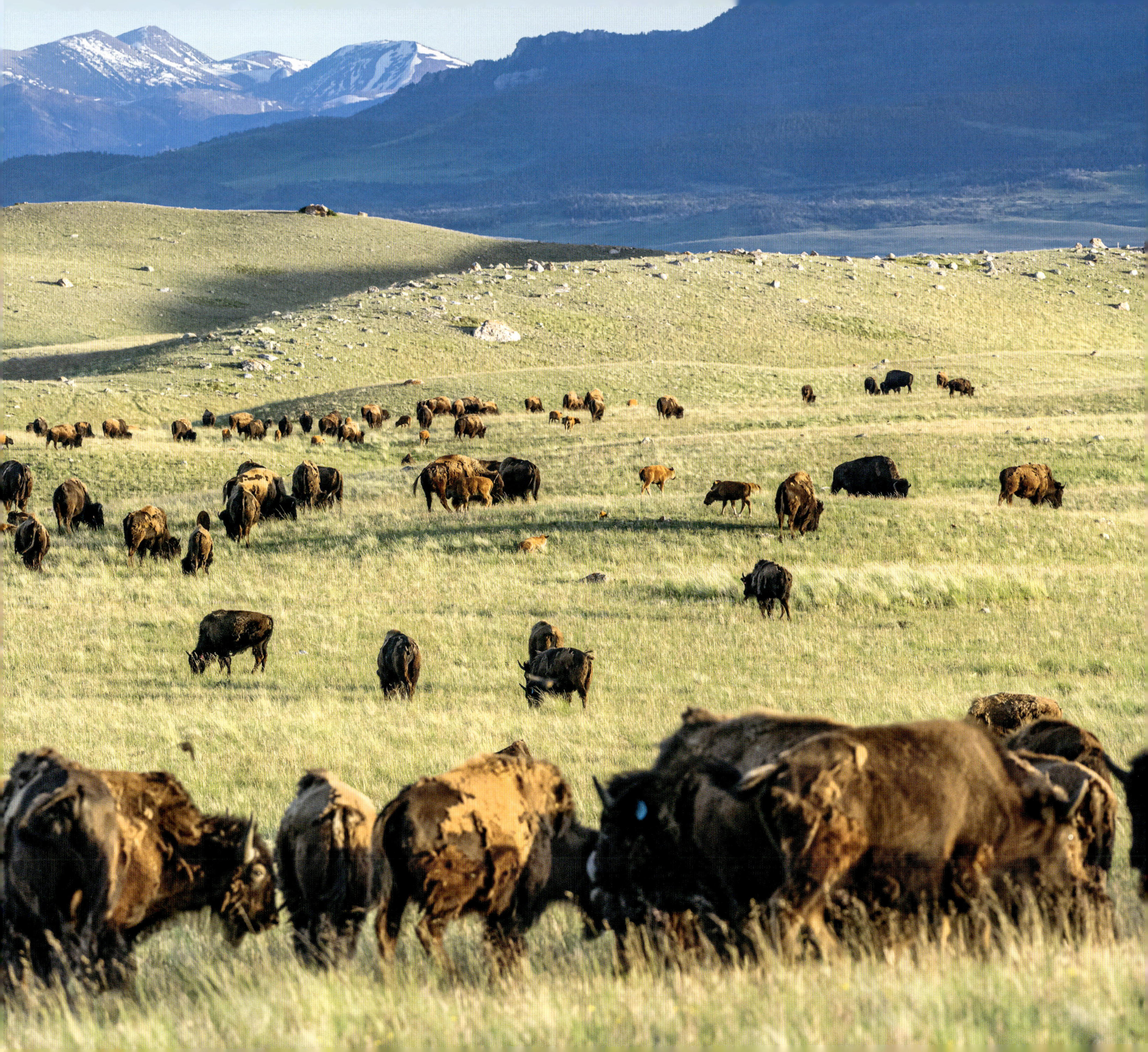

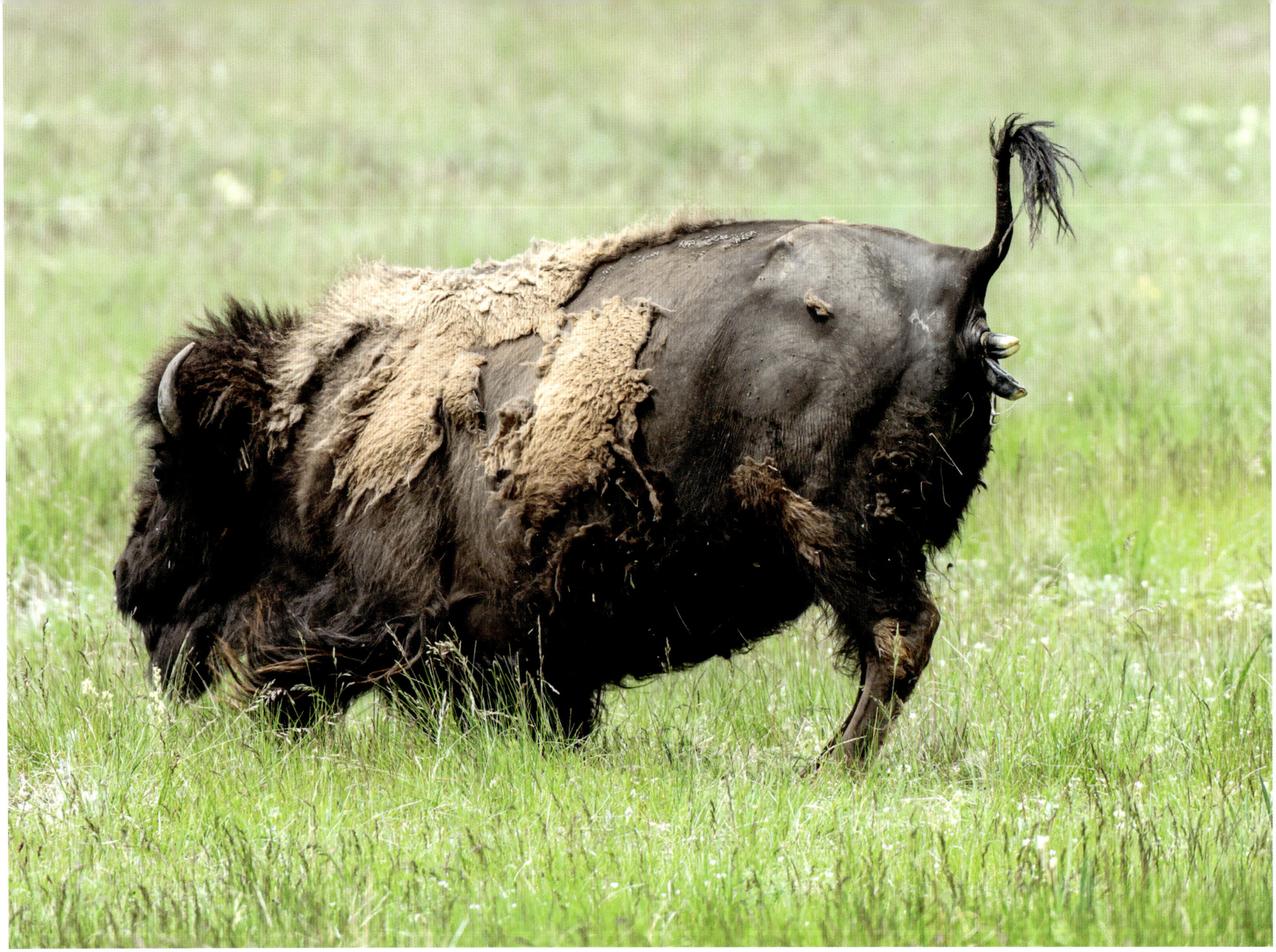

Bison manage their own calving, something that keeps most cattle ranchers up in the middle of the night in inclement weather for weeks on end.

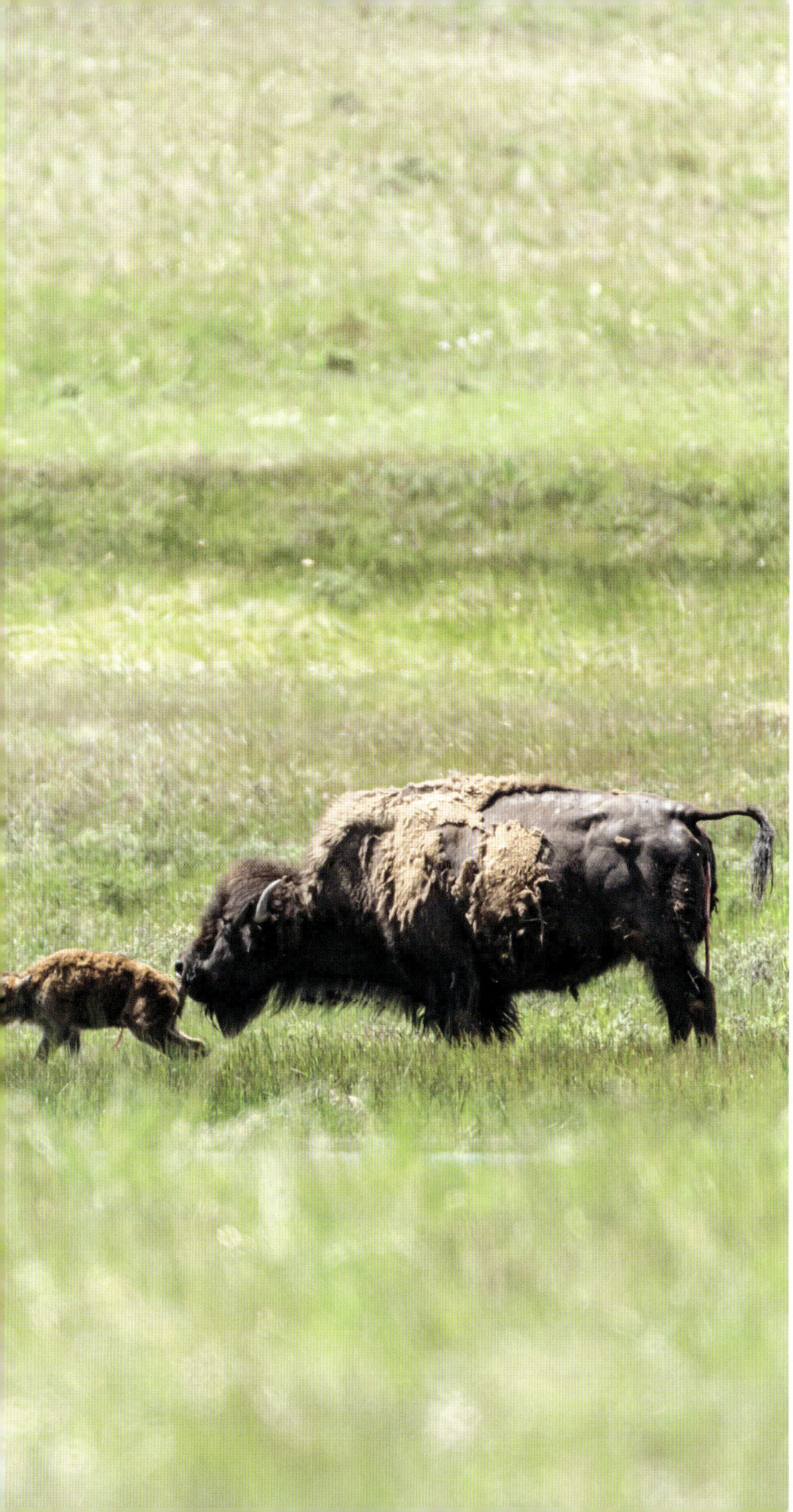
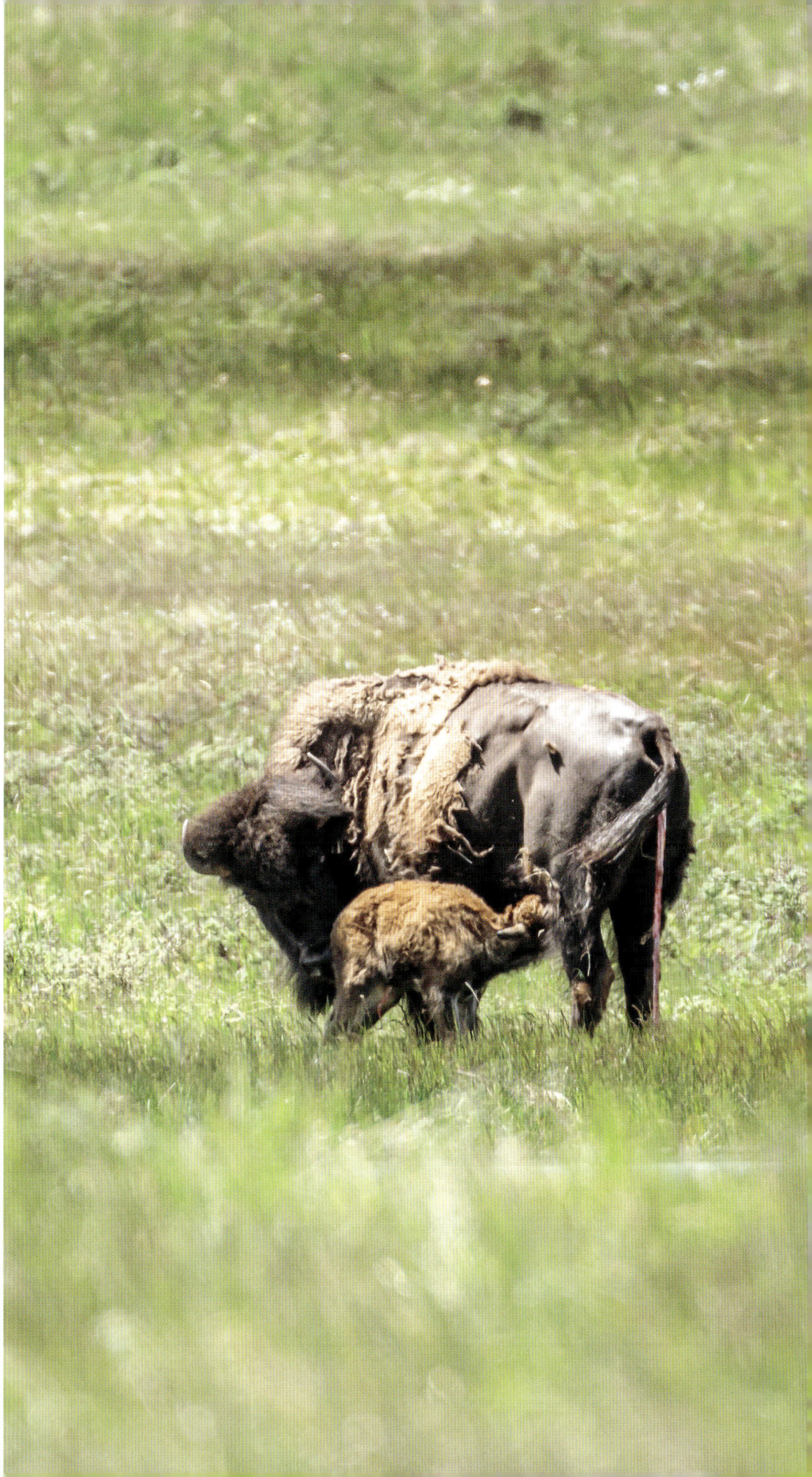

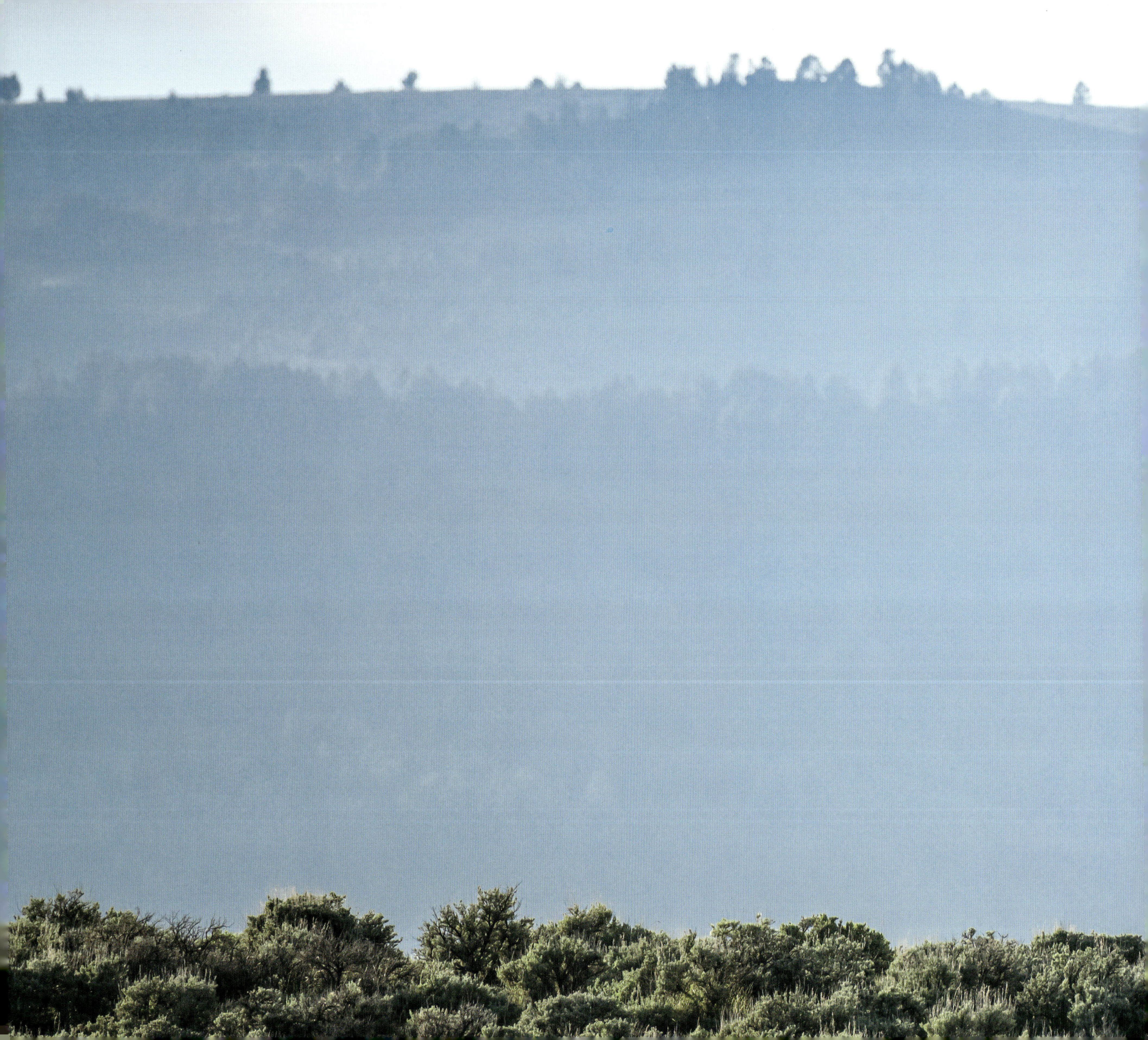

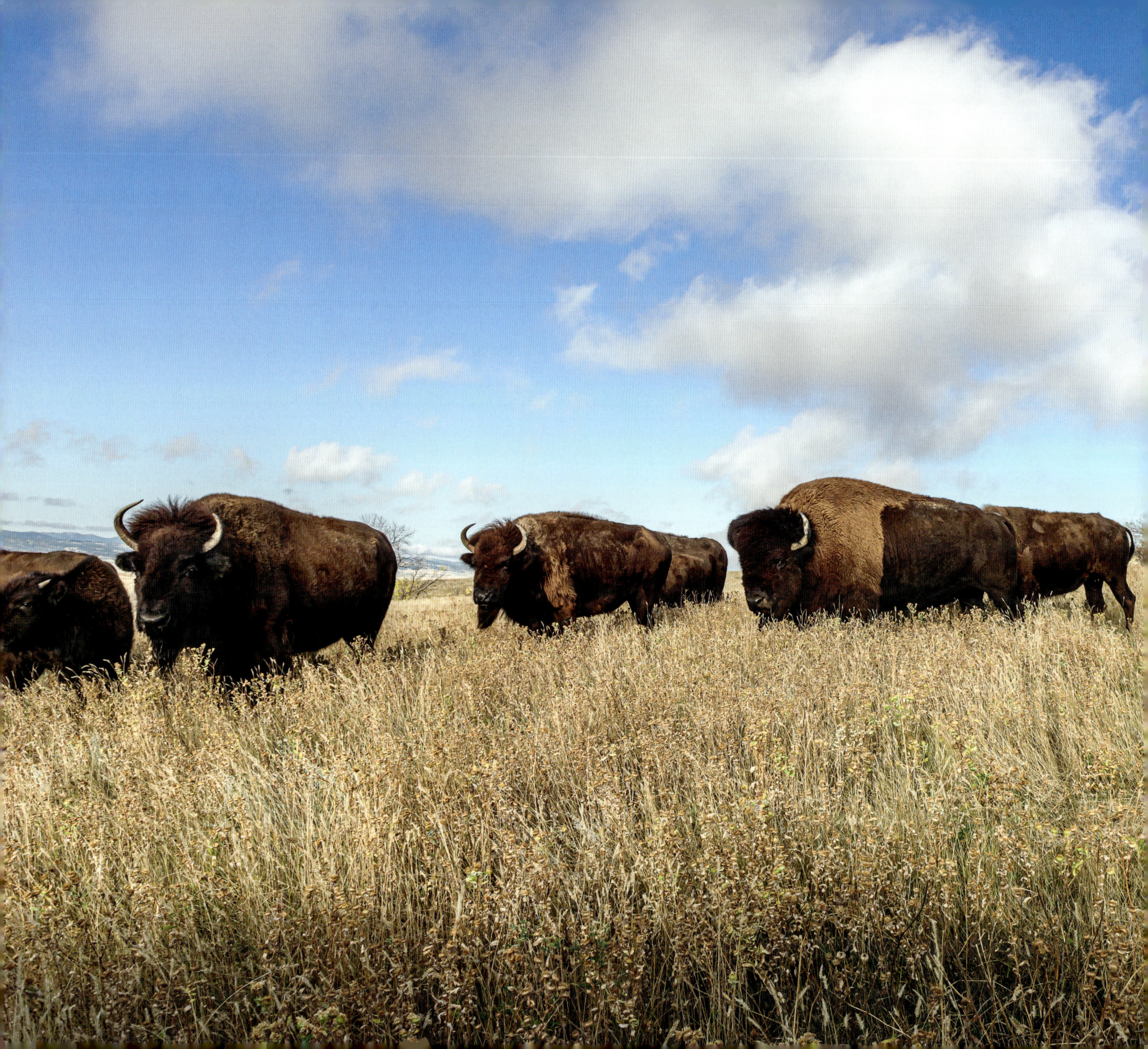

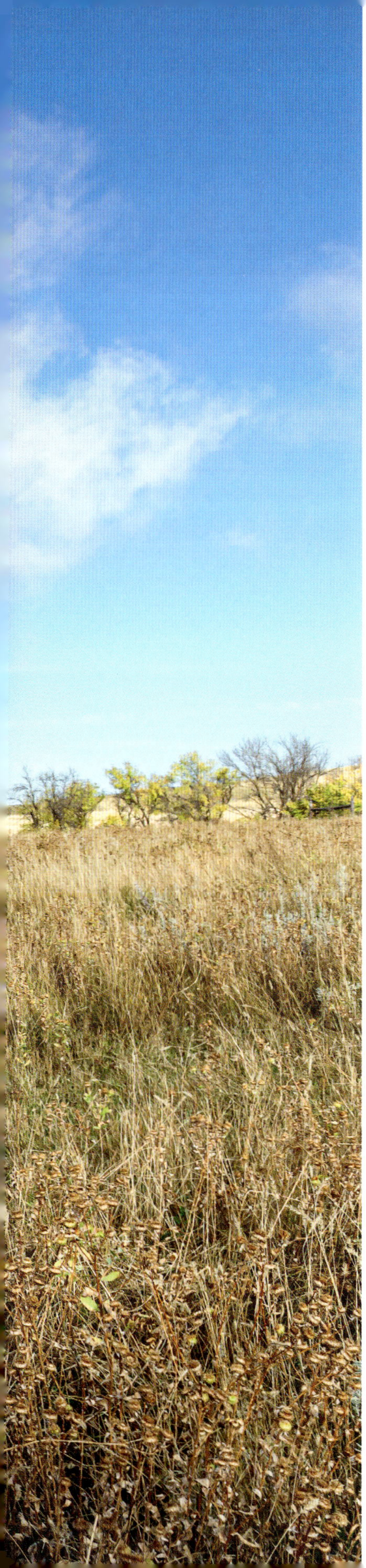

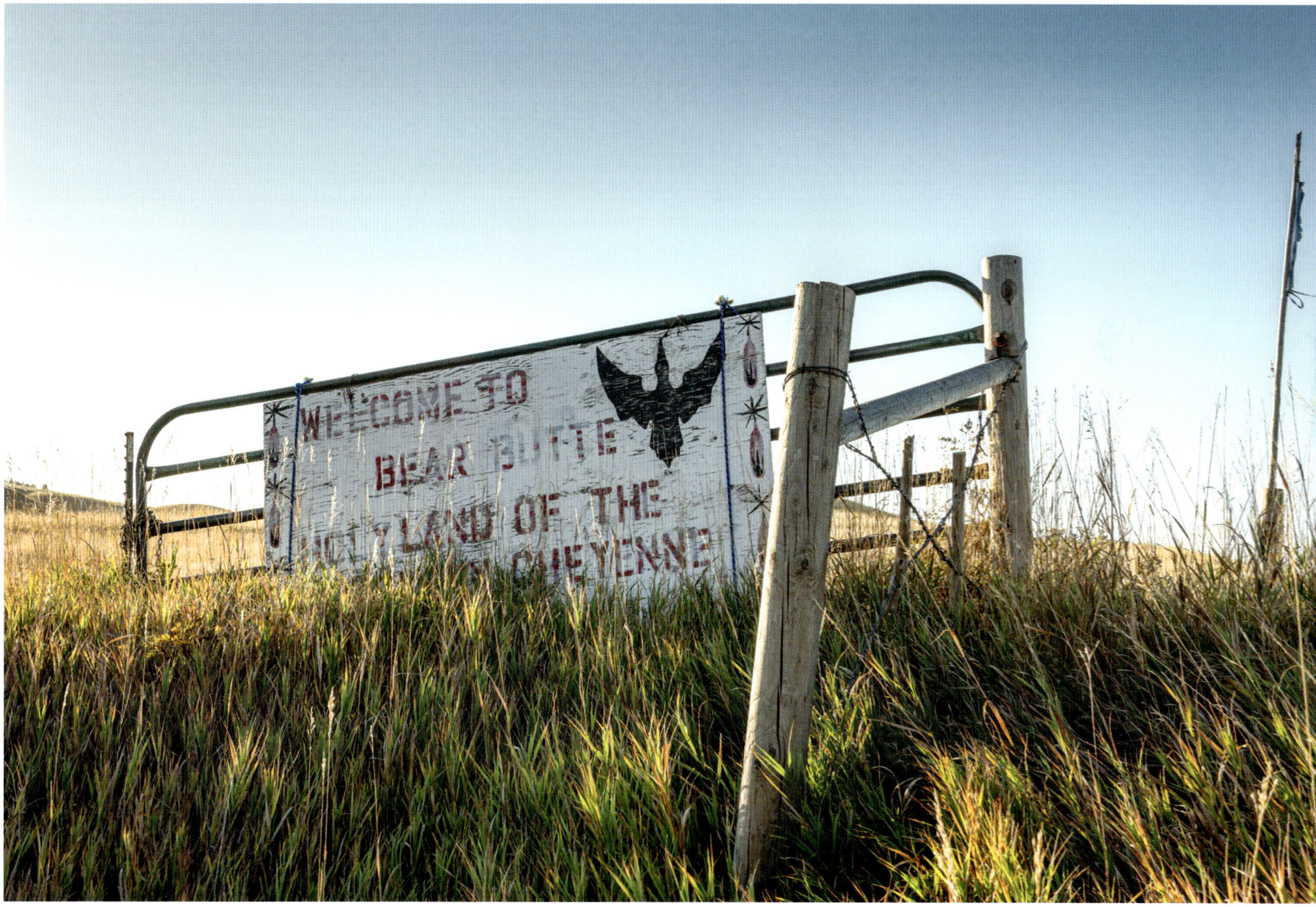
WELCOME TO
BEAR BUTTE
LAND OF THE
CHEYENNE

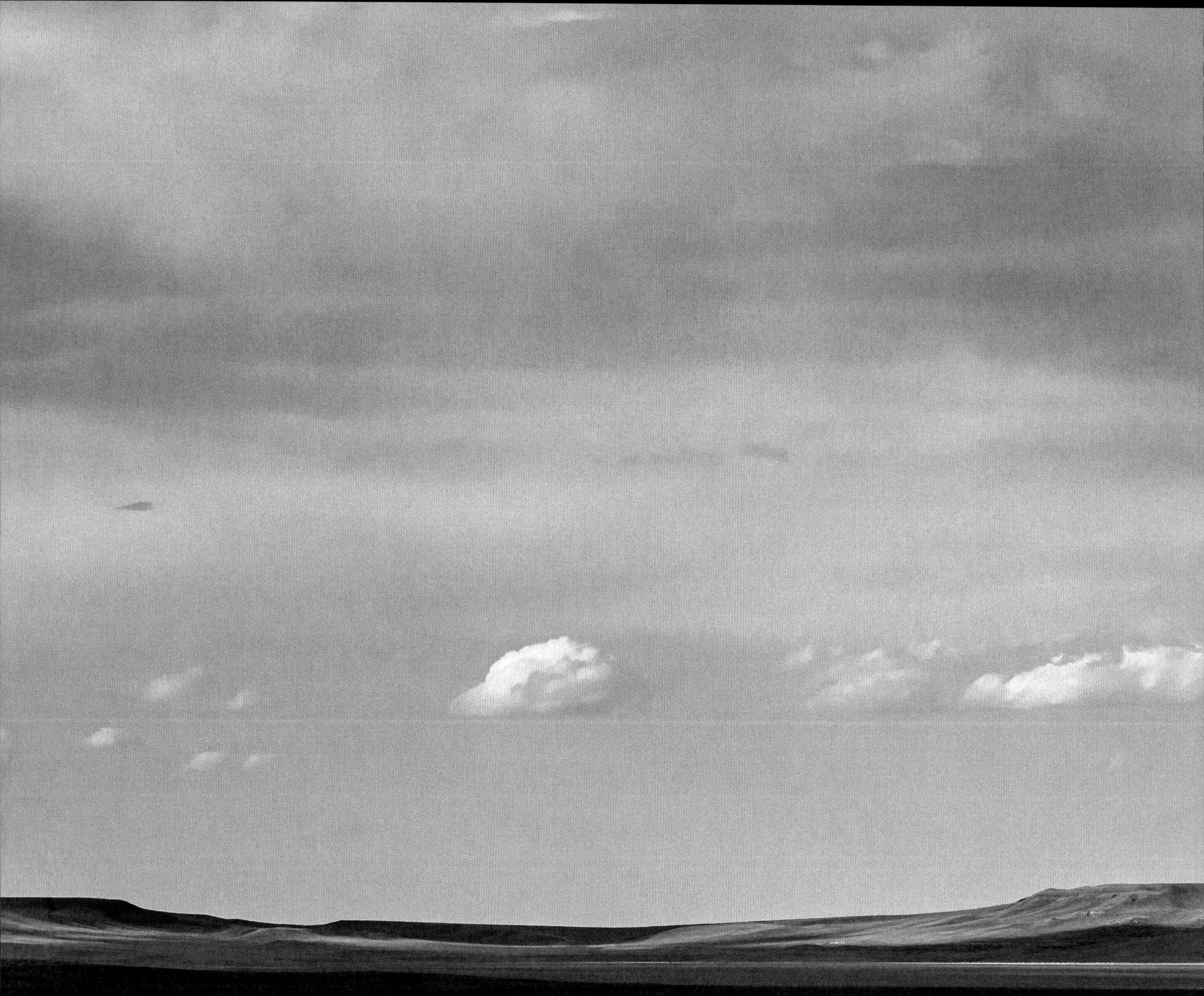

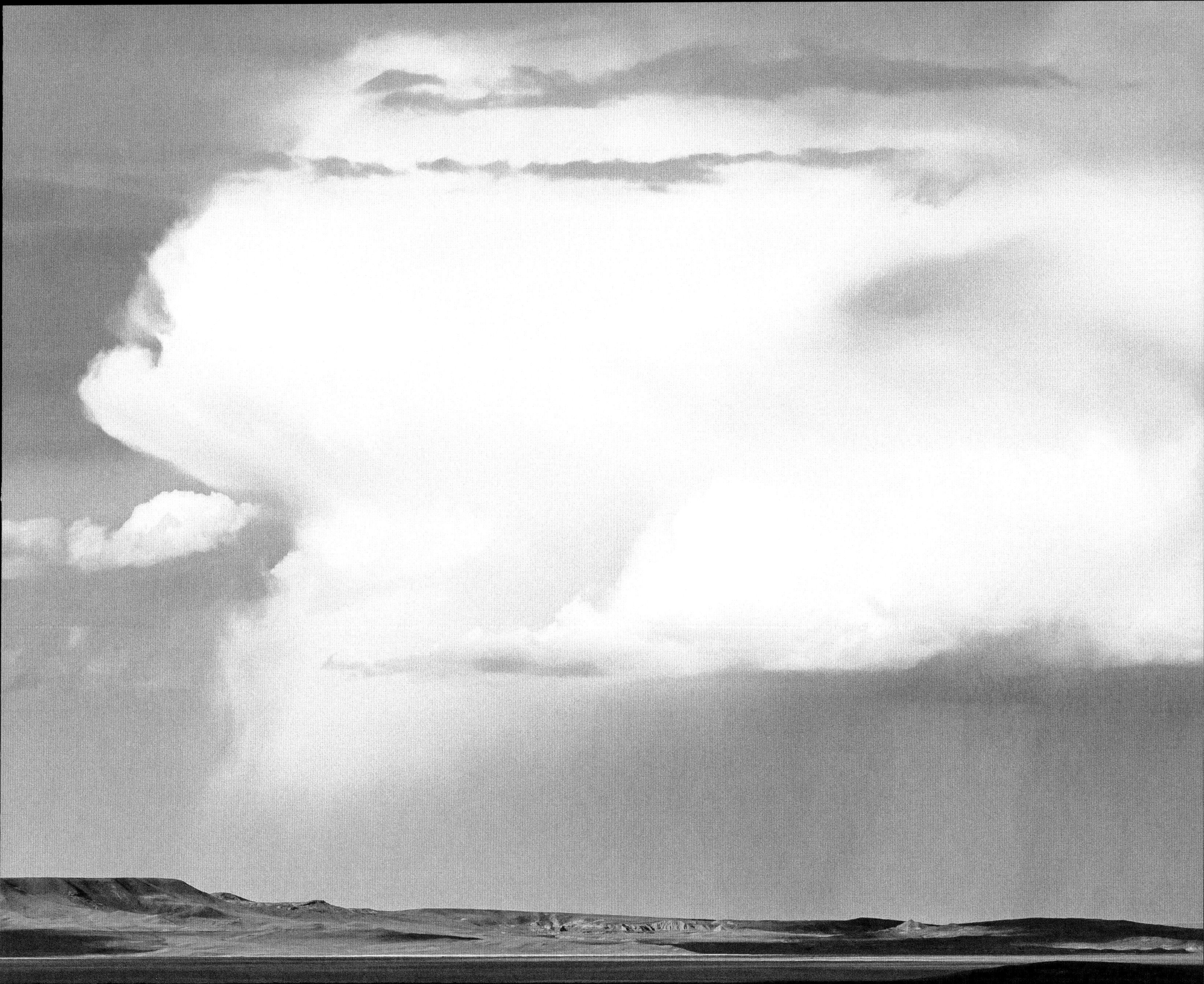

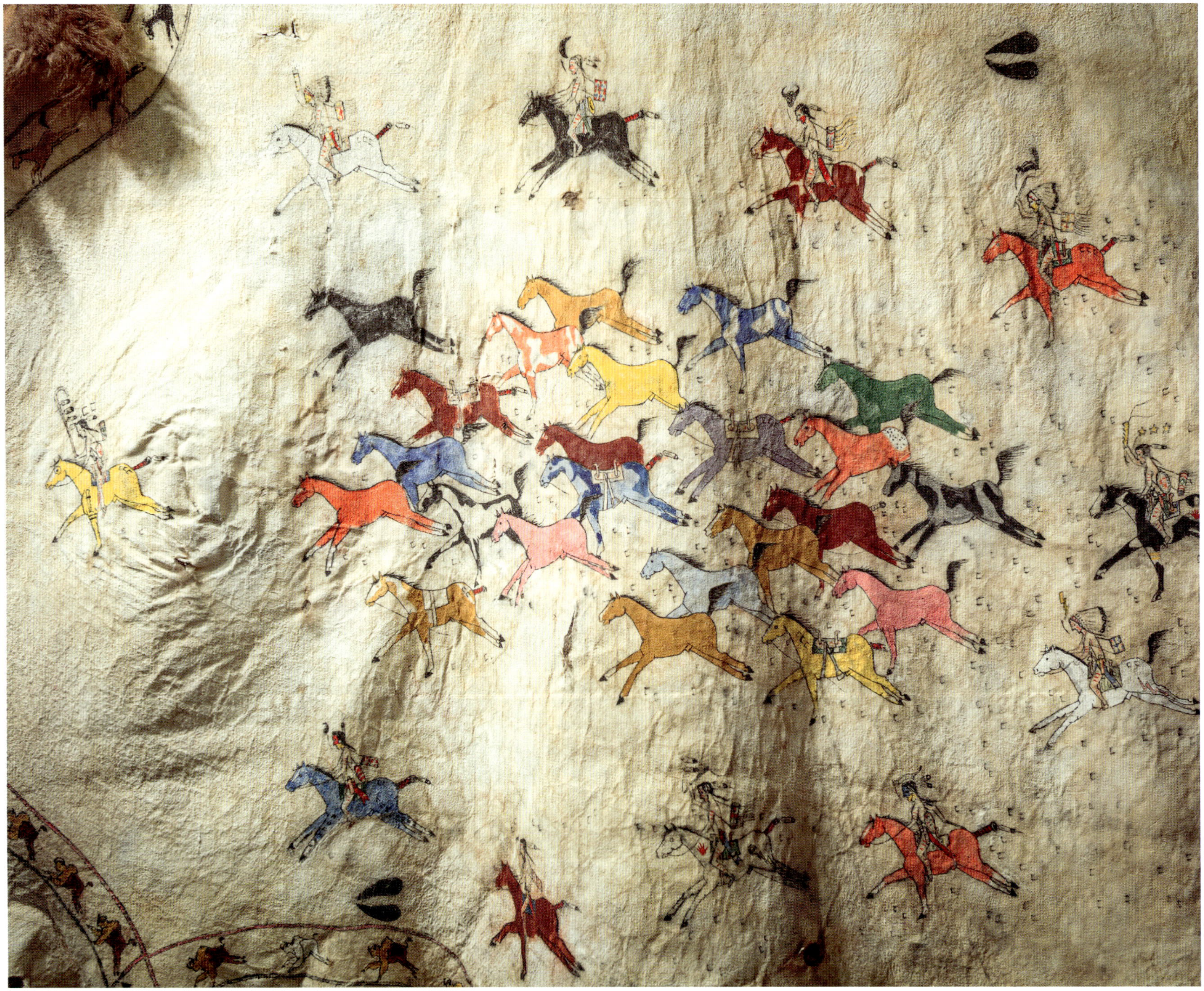

WASNA

Geraldine Goes in Center makes *wasna* the traditional way, using her grandmother's recipe, from strips of dried bison meat and local dried chokecherries. As she prepares pemmican for Native American food journalist Loretta Oden for her TV show *Seasoned with Spirit,* Goes in Center points to bison's heart-healthy qualities (seventy percent leaner than beef, it has more protein and significantly lower cholesterol than beef) and the fact that chokecherries have more antioxidants than green tea. This traditional bison jerky, she says, "is actually very healthy. It's a food as people we should go back to eating."

Her friend Cathy Smith shows Oden bow strings made from bison sinew, a buffalo wearing robe painted with details of the owner's accomplishments, and a beaded tobacco bag made from a bison bull's scrotum. Smith won an Emmy Award for costume design in the TV miniseries Son of the Morning Star and was cultural advisor to the film Dances with Wolves. She made the Lakota Sioux costumes in that film and dressed the Lakota actors every day for five months, helping imbue the film with such authenticity that it won multiple Academy Awards. A renowned expert on the clothing and accoutrements of Native Americans, fur trappers, and the people of the frontier West, Smith keeps alive ancient crafts like beadwork and porcupine quillwork, always using traditional methods and materials such as brain-tanned hides and natural dyes from plants.

As a young girl growing up among the Lakota Sioux in the Black Hills of South Dakota, Smith displayed a passion for the ancient learning, crafts, customs, and sacred ceremonies, so much so that she was adopted by a powerful medicine man (after he stopped in her driveway seeking, naturally, a buffalo skull).

She was called to the Sun Dance, a spiritual ceremony which requires that the participants dance from sunrise to sunset without food or water for four days, and to pledge to do this for four years. Just to prepare for the experience requires a vision quest or *Hanbleceya*, a four-day fast carried out alone on a high mountain—something she's now done for thirty-five years.

During her year of preparation for her first Sun Dance, she recalls, "The spirits wouldn't leave me alone. A herd of buffalo would run through the yard, there would be dust everywhere, doors slamming, but nothing there."

The recurring appearance of buffalo wasn't surprising to her or her spiritual advisor, given the Sioux's sacred myths and creation story. "The Lakota people believe they came from buffalo," Smith explains:

> *There are legends of how the buffalo people lived under the earth and came out of Wind Cave and became human beings. Buffalo are wholly sacred and central to life. Everything involves the buffalo. Buffalo horn spoons are used for sacred food. We use the sinew for sewing sacred things. A buffalo skull serves as the altar. Painted 'wearing' robes were your coats. Teepees were made from cow buffalo hides. The horns became cups, the toes were made into rattles. Every single part of the buffalo was used. And when we sun dance I bring the buffalo heart. It goes into a six-foot-deep hole dug for a cottonwood that's 60 feet tall and forked. Then a buffalo fetish, a rawhide cutout of a male buffalo, goes in the fork of the cottonwood tree. And for every kind of ceremony, you have to have pemmican.*

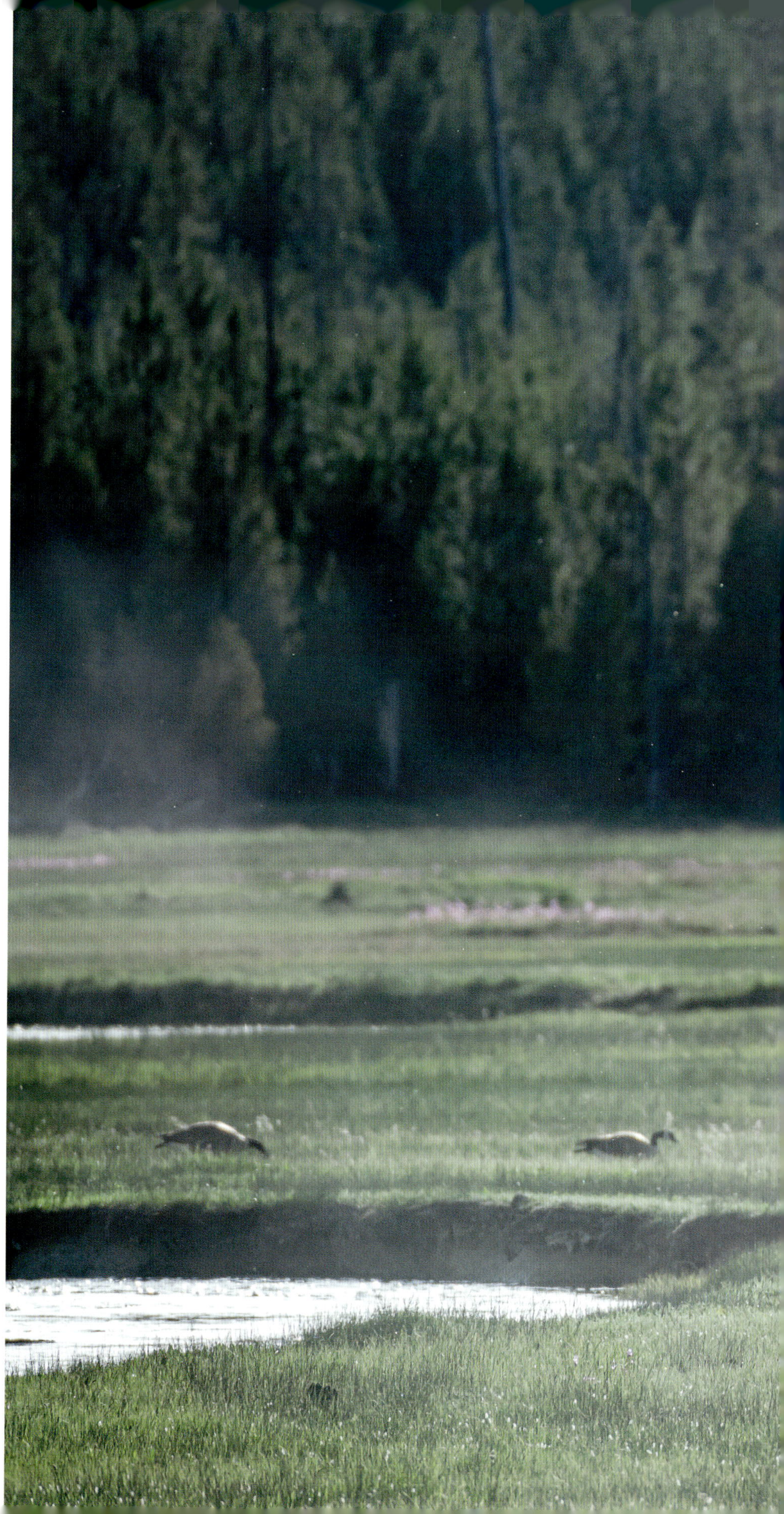

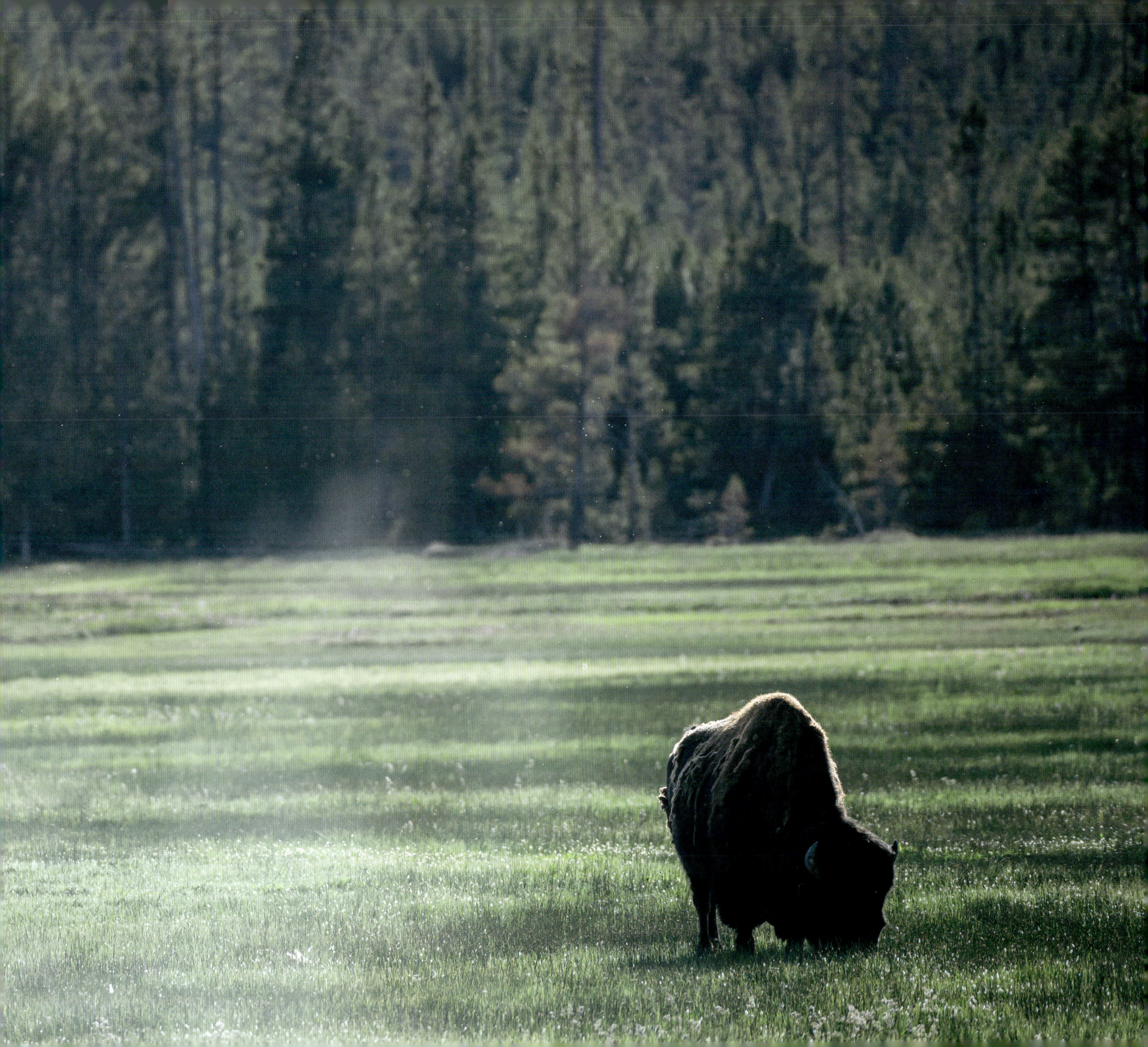

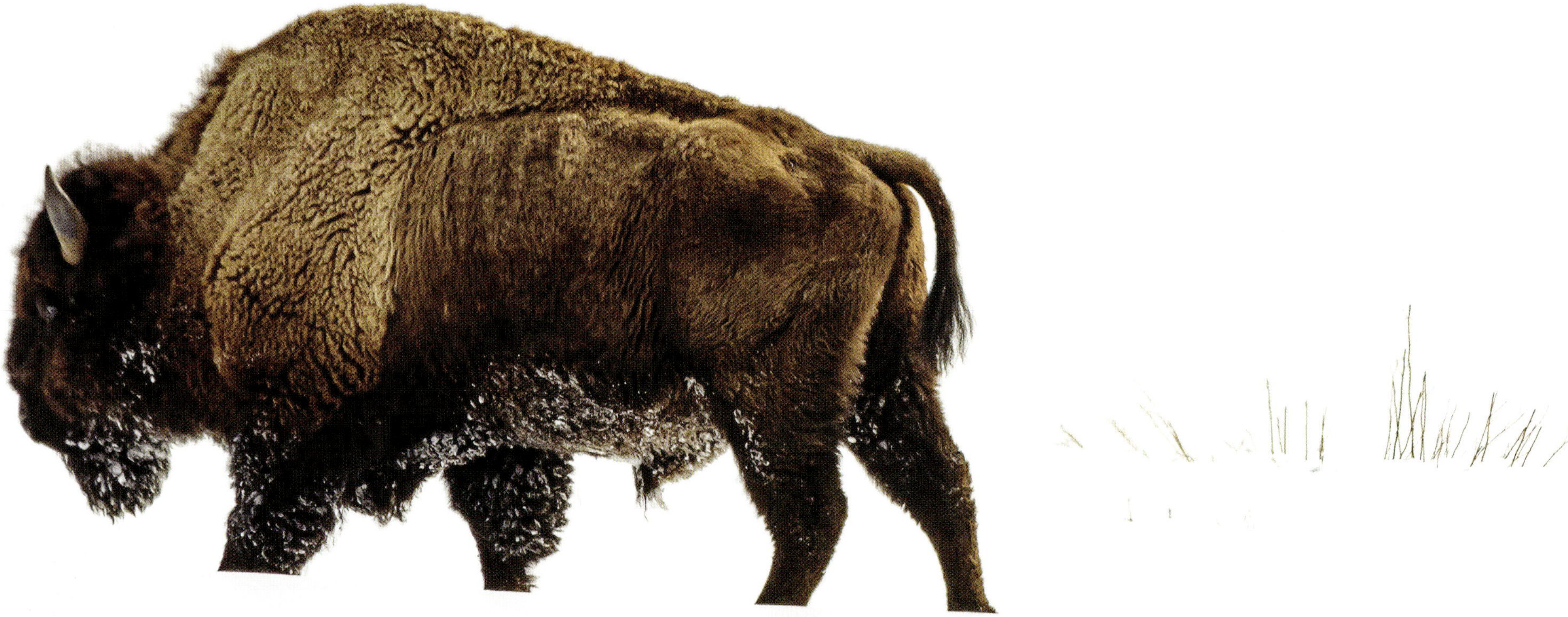

More than an icon, wild free-roaming American Bison are our family elders who teach by example—if only we will listen. Give generously. It's ok to take—just not too much. Always leave a place better than you find it. Care about each other. Stay agile. Don't complain.

—YVON CHOUINARD,
ENVIRONMENTALIST
FOUNDER OF PATAGONIA

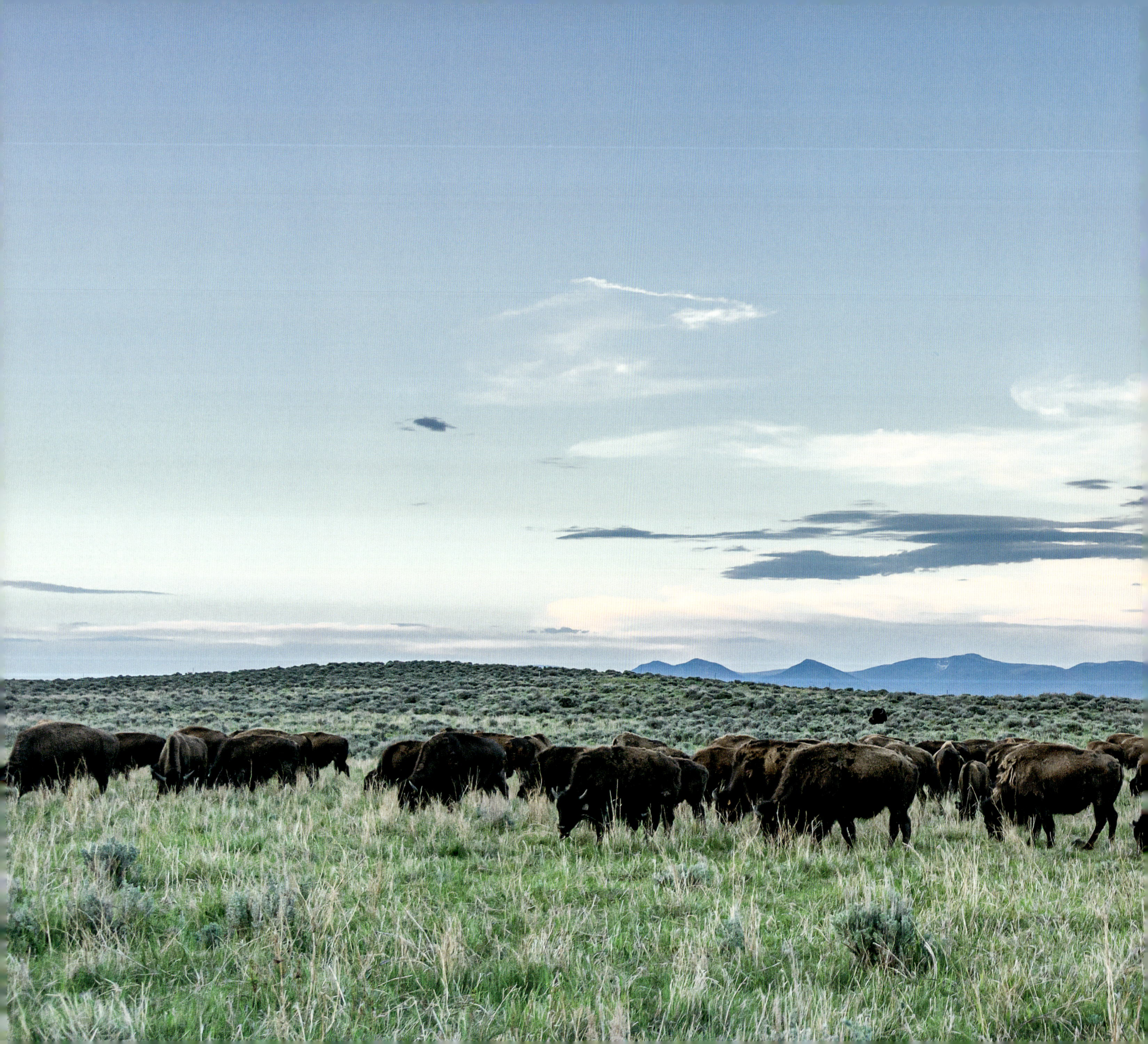

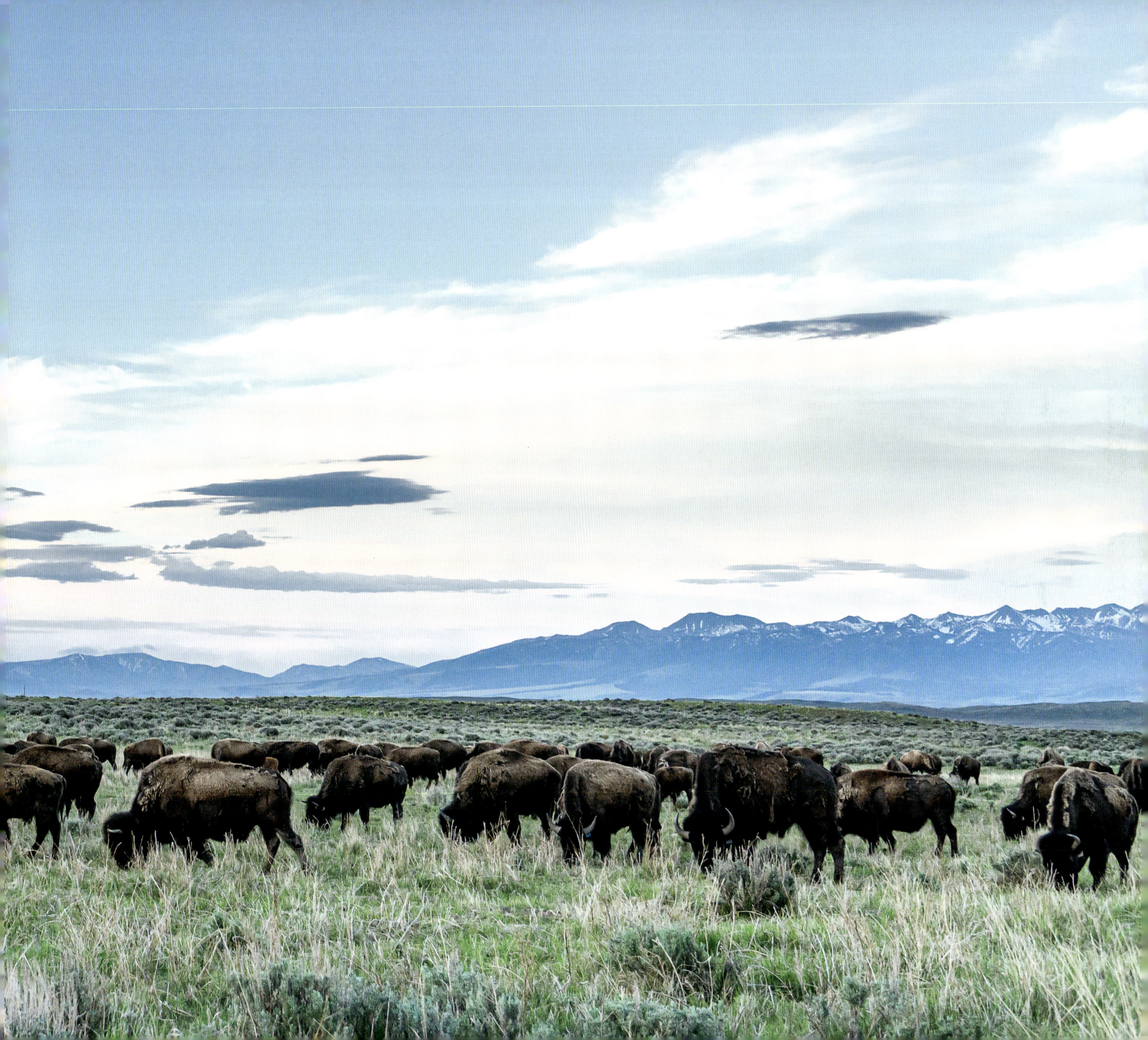

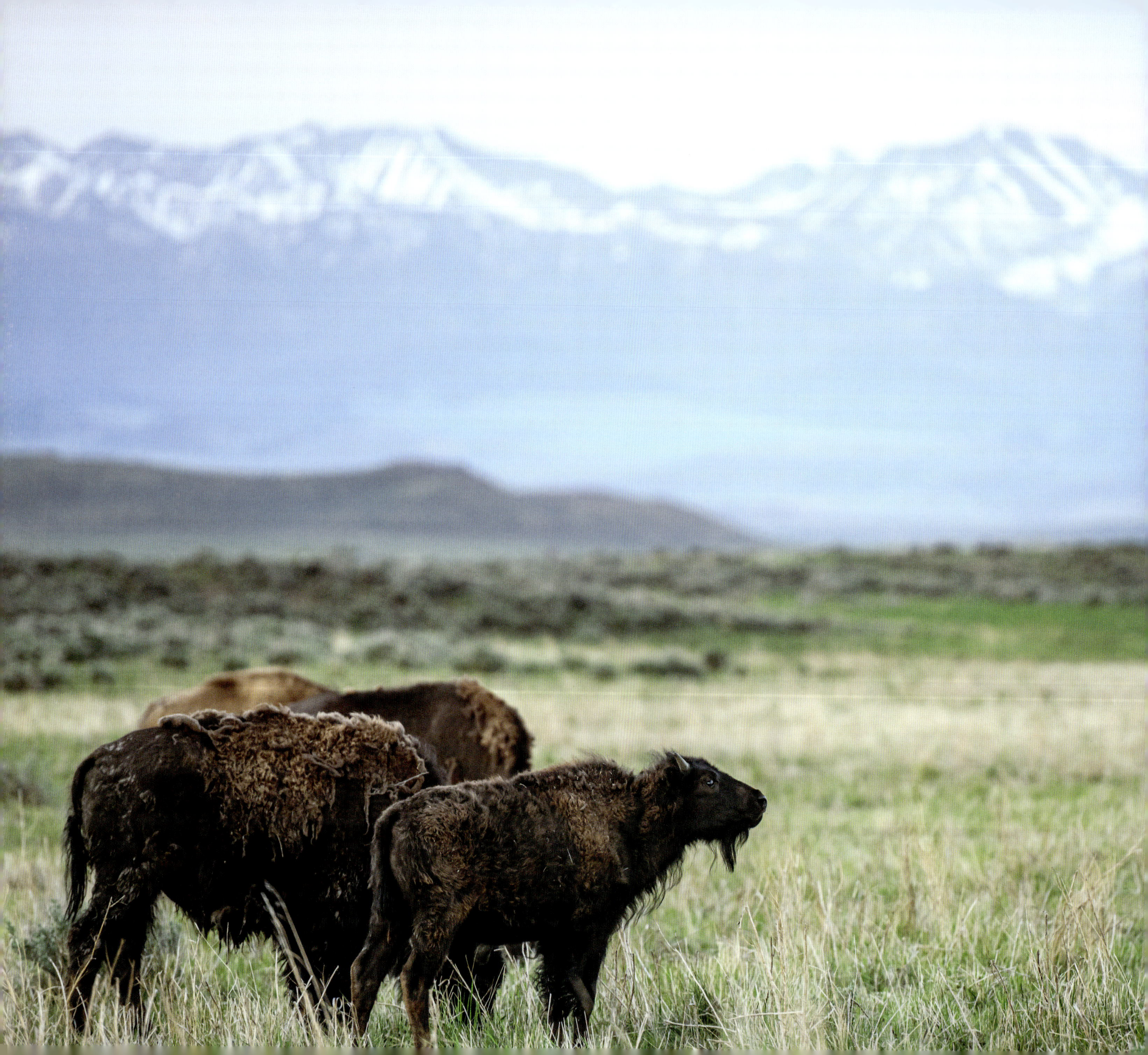

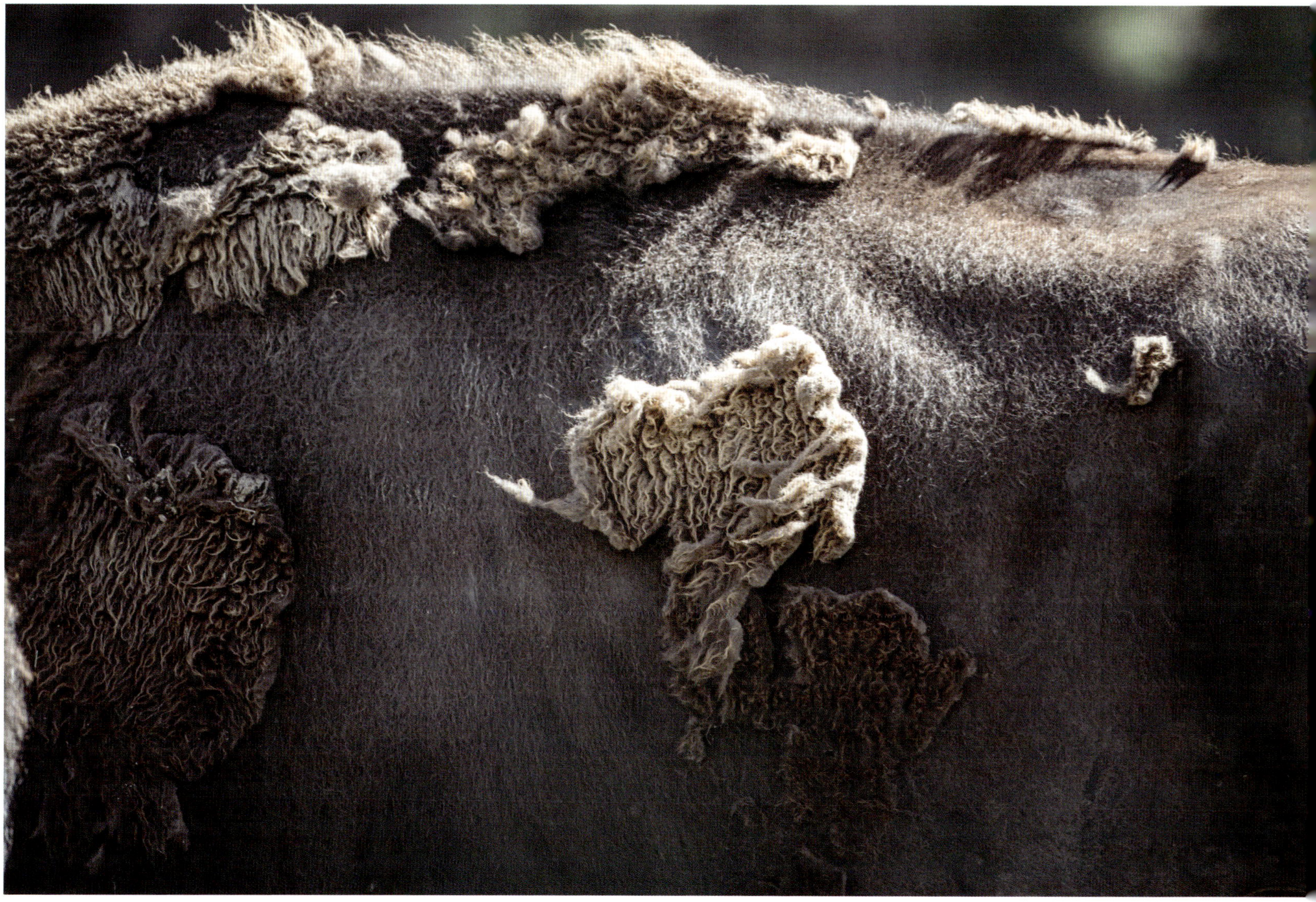

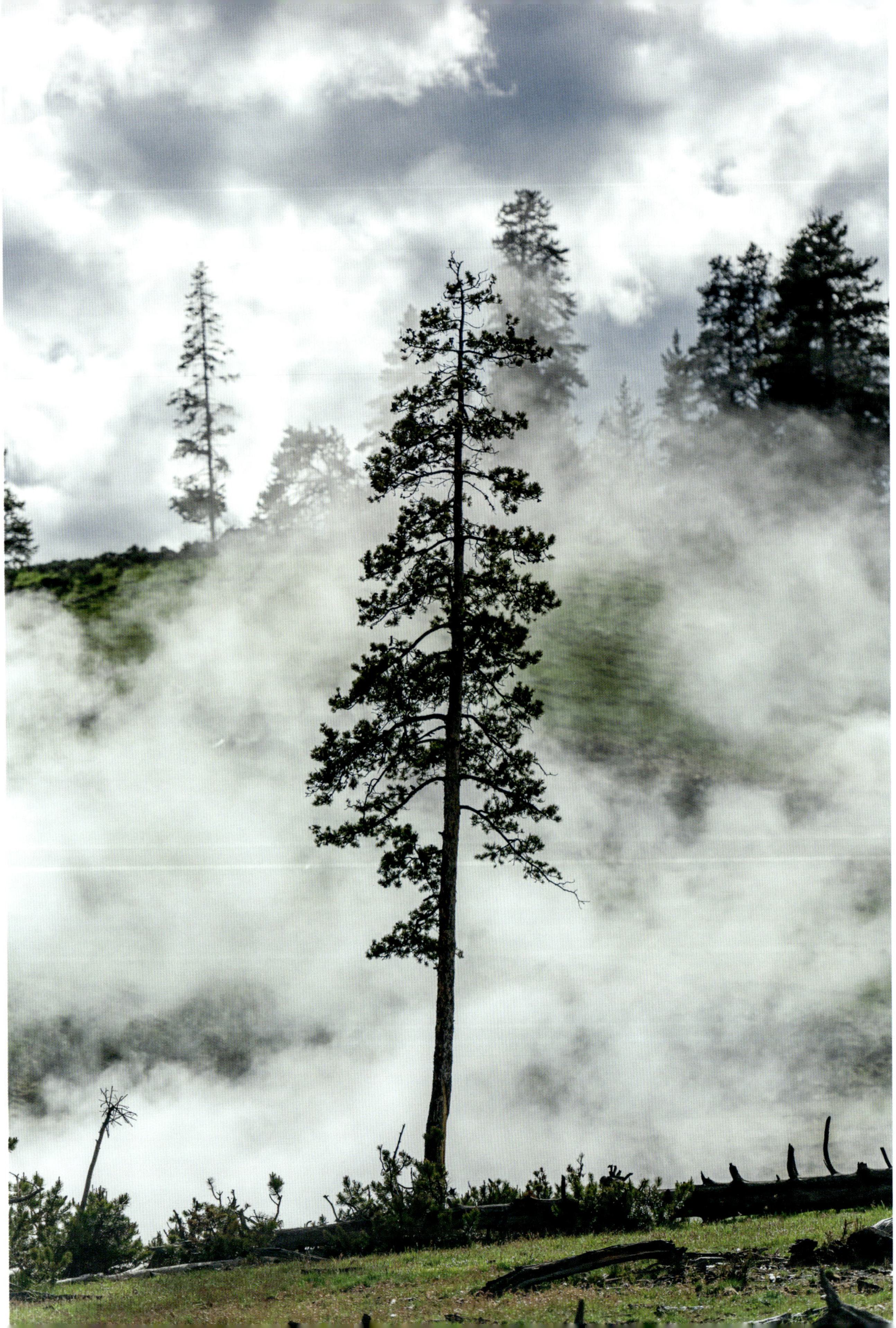

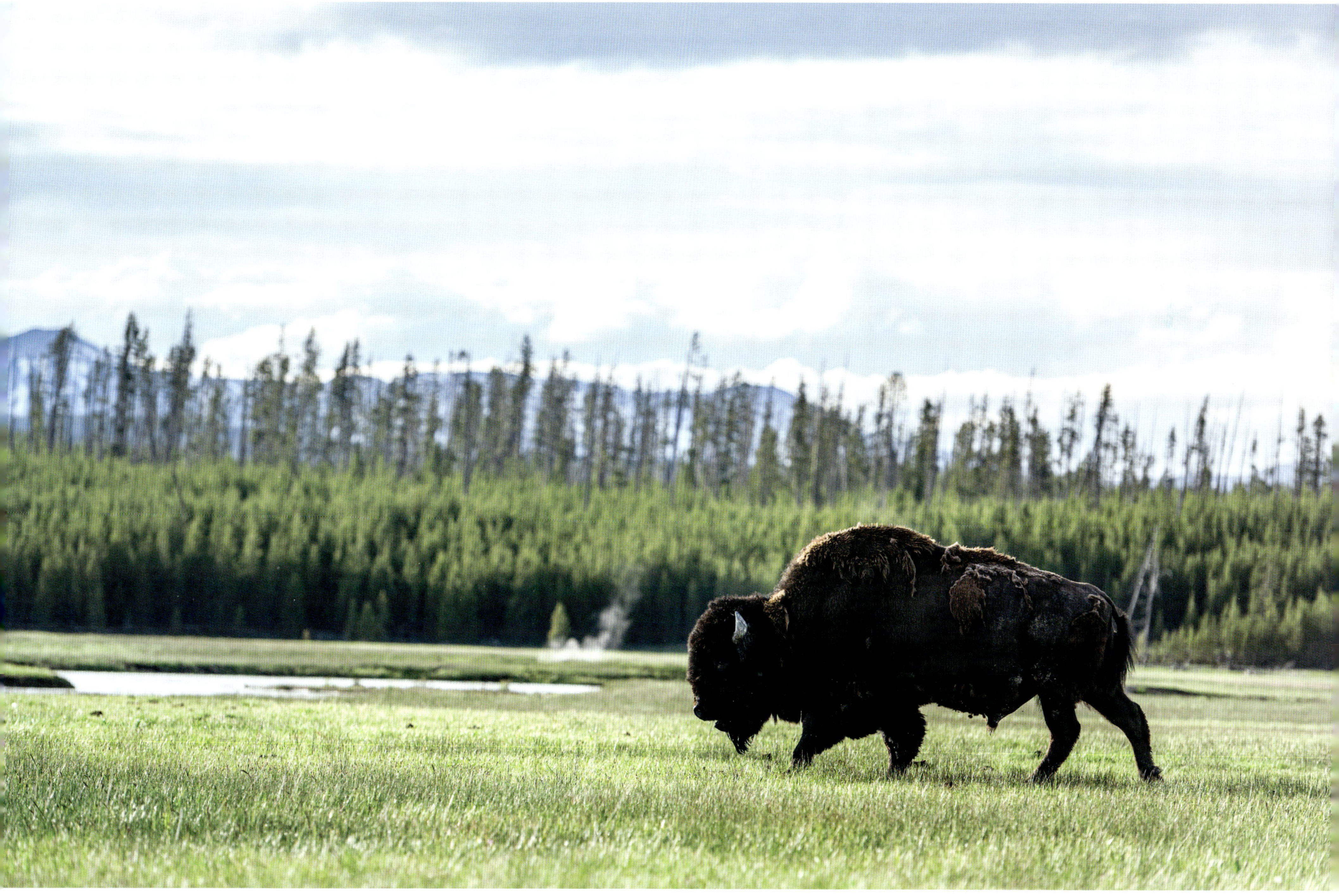

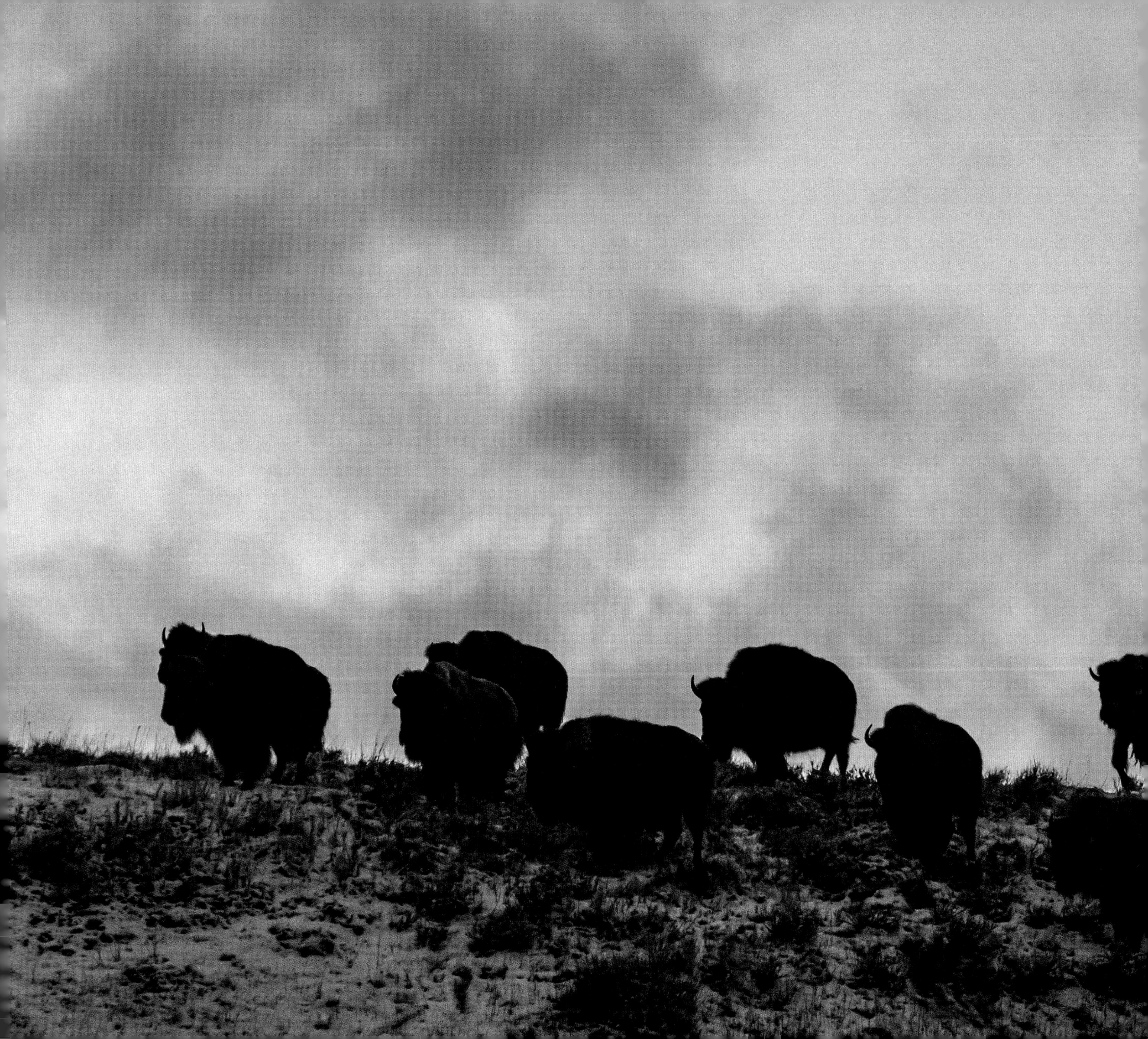

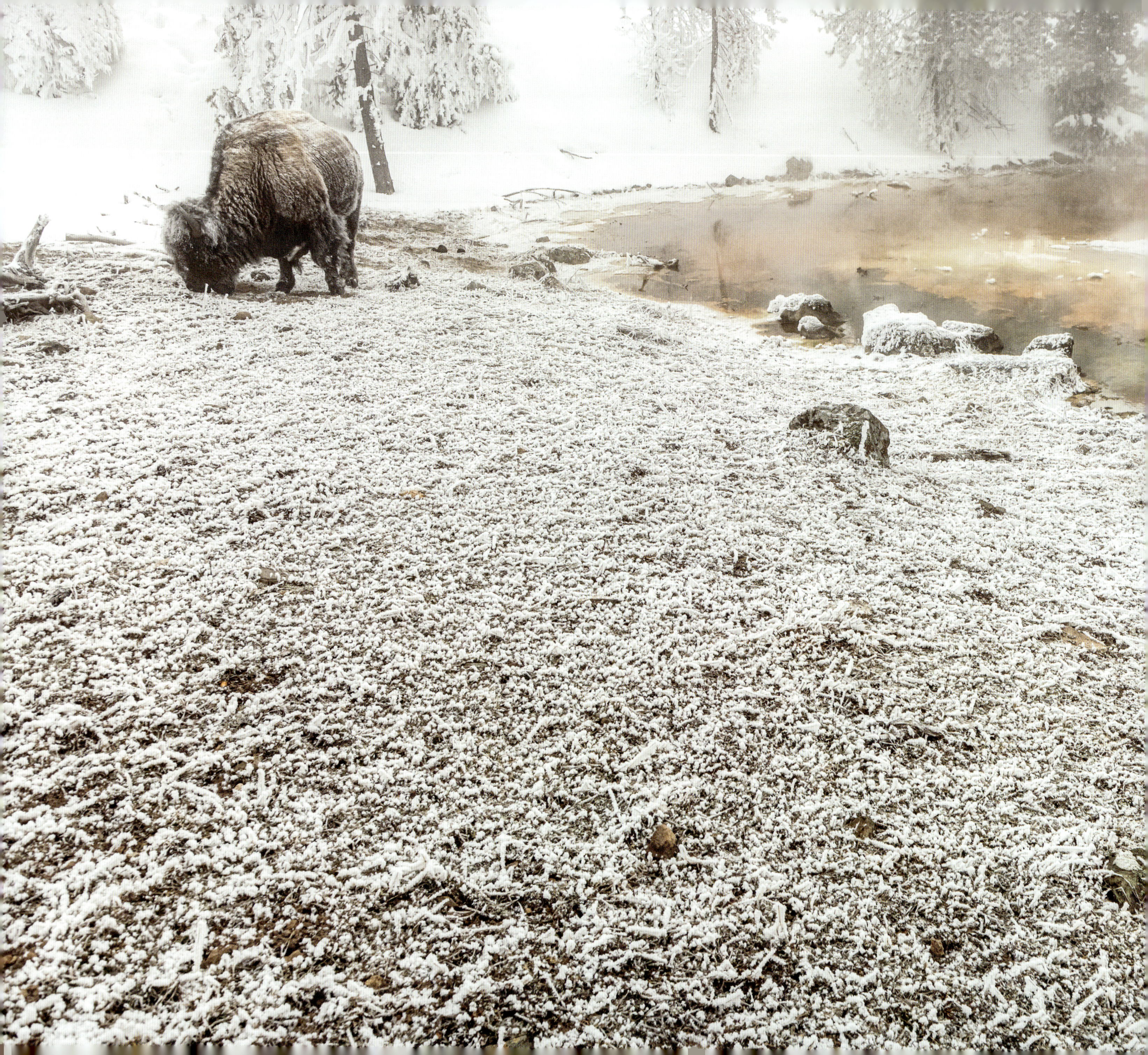

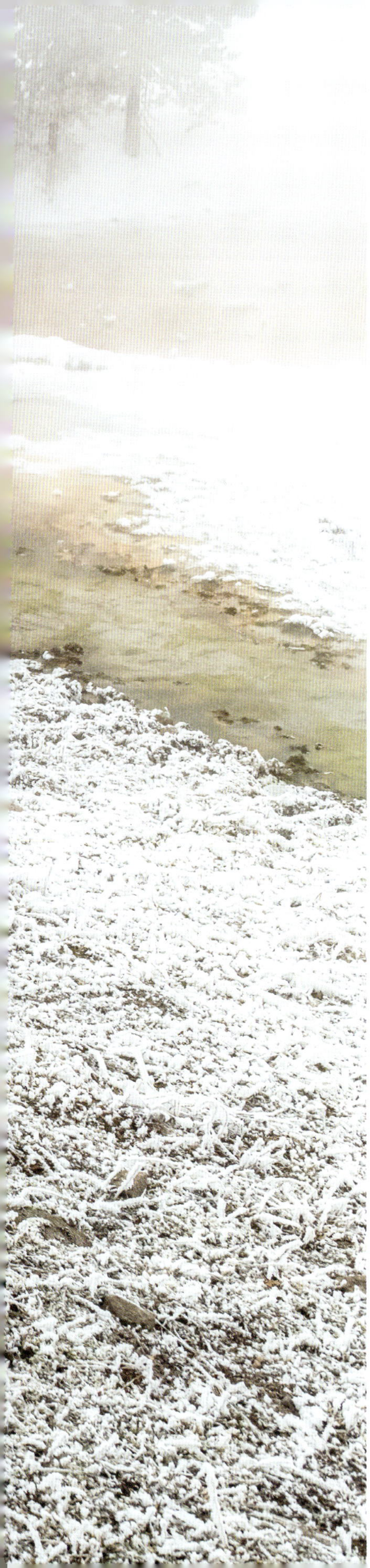

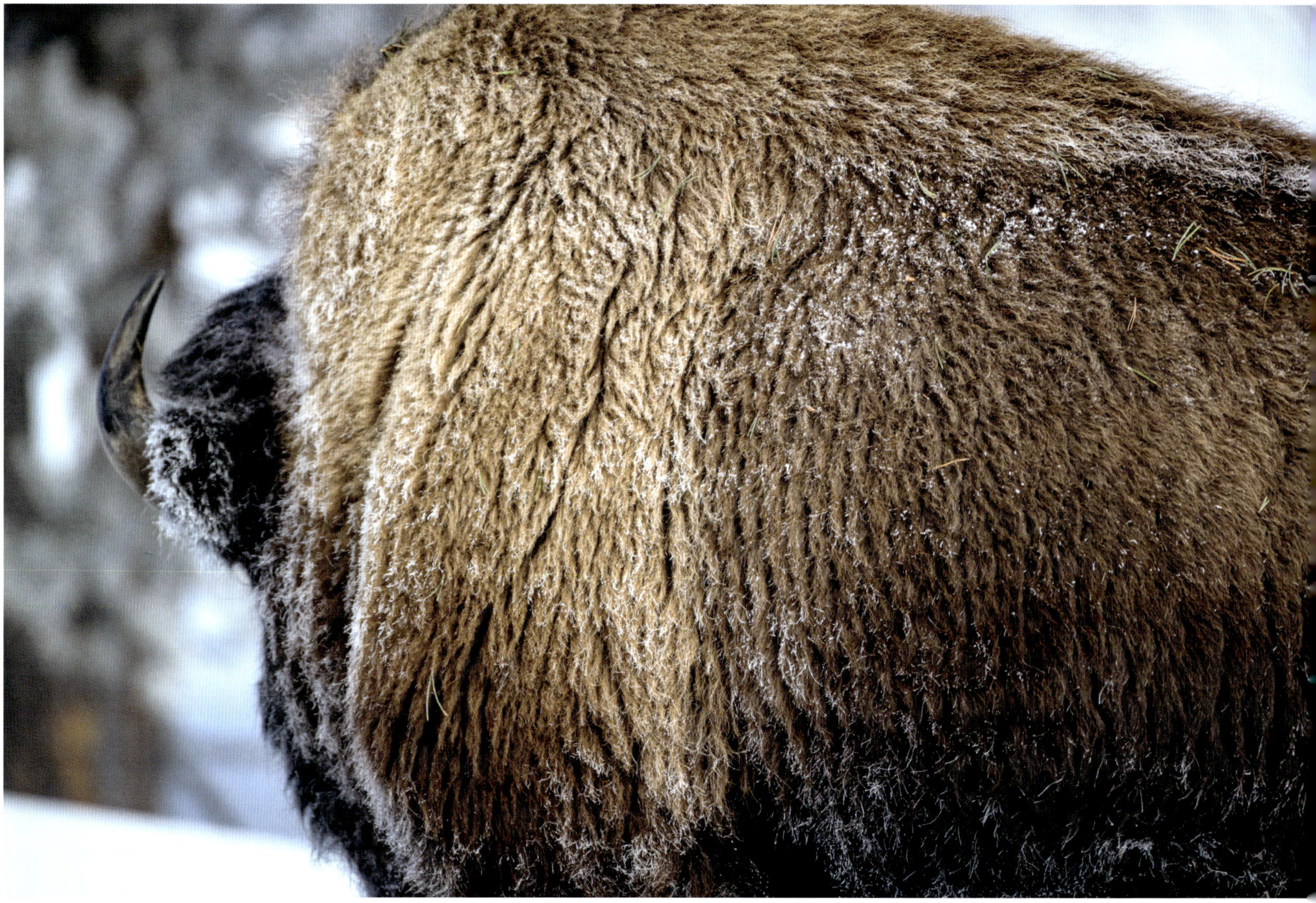

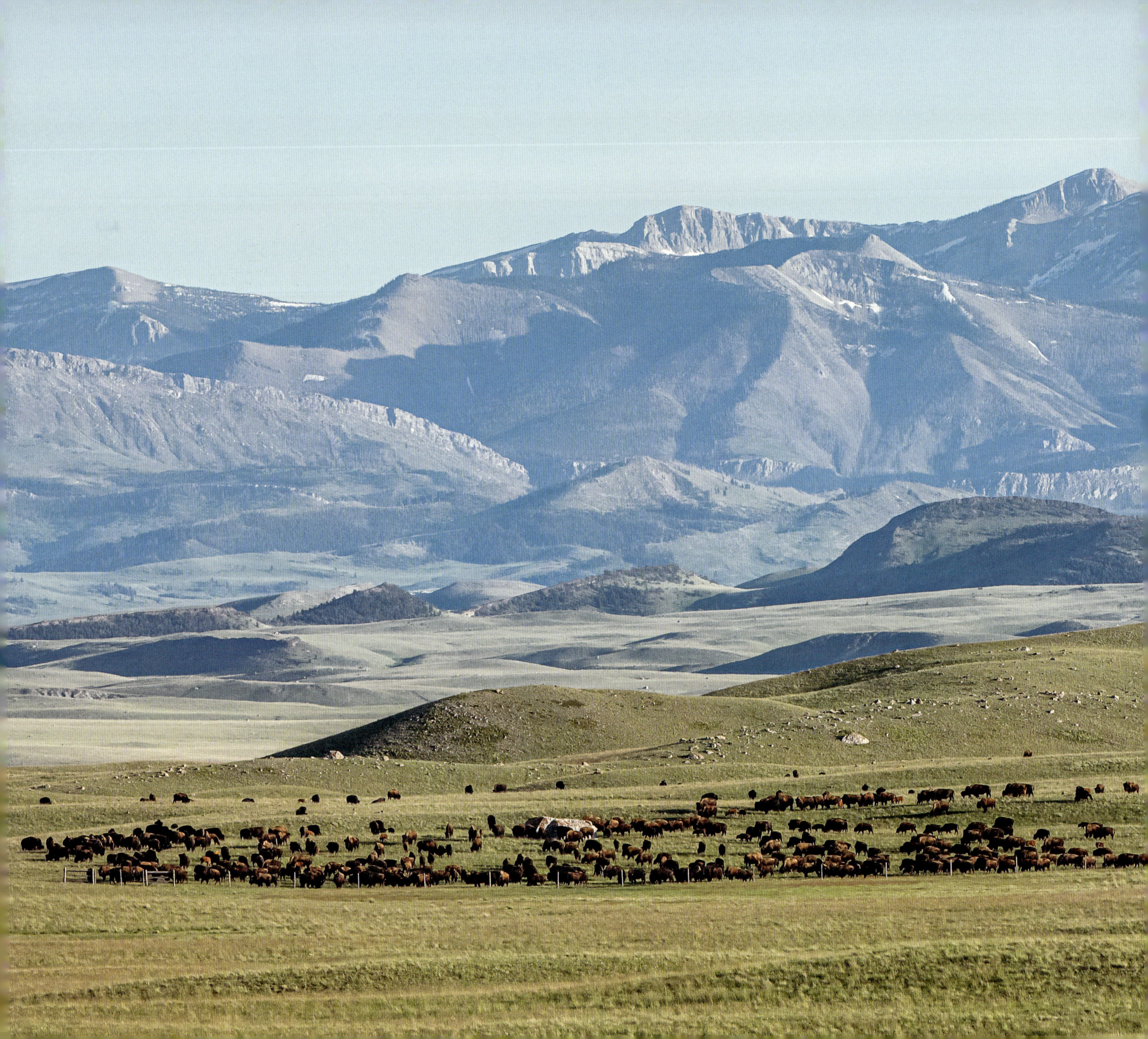

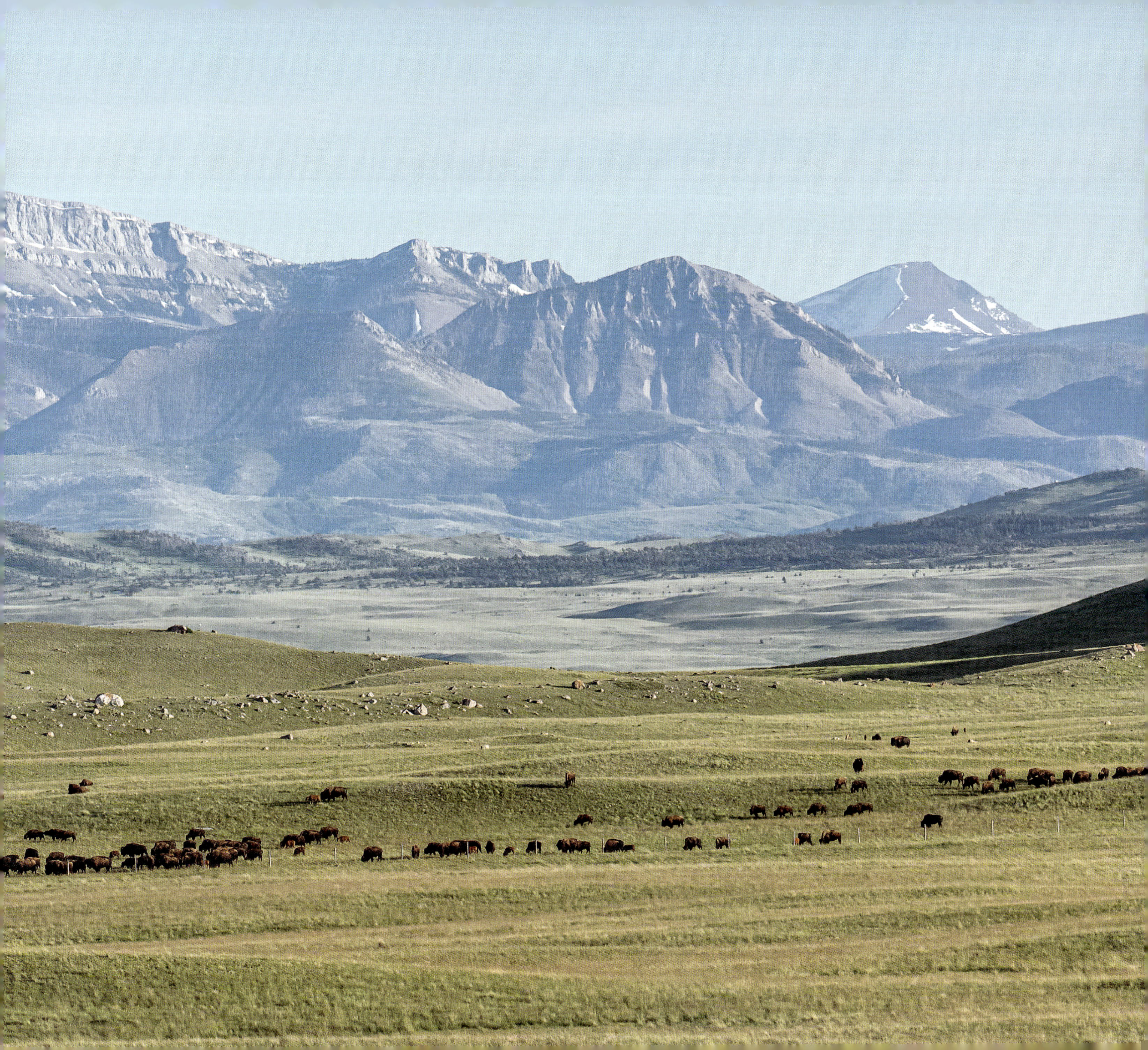

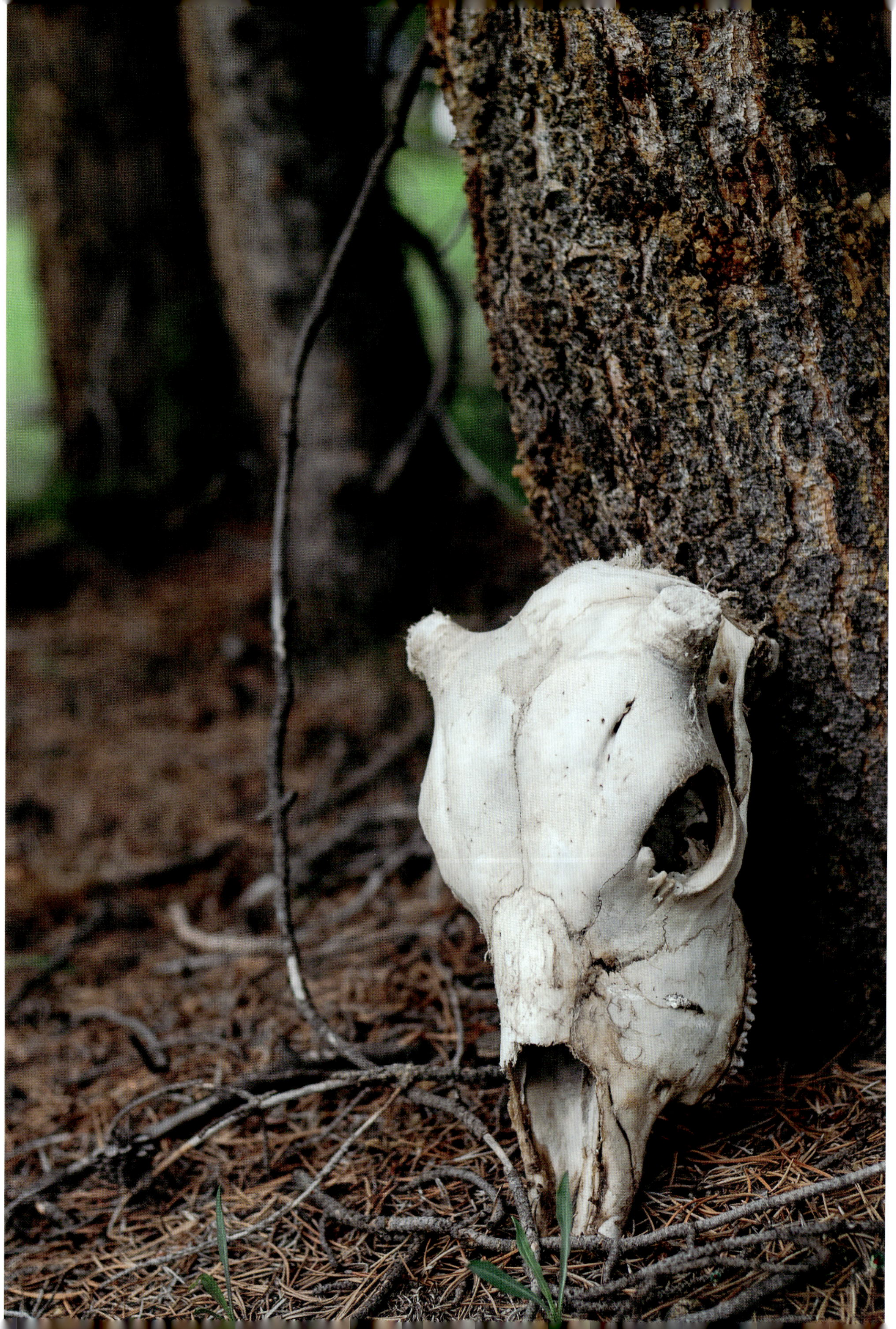

Skulls and Skeletons

Thus exists one of the existential issues that has troubled Plains Indians for the past 150 years. At least a couple of generations of tribal people grew up without pemmican.

Probably never before has an animal population been wiped out so quickly, on such a scale, and as a result a whole peoples' experience altered so drastically. And as the buffalo went, and the people became dependent on government-issued provisions, and their children were sent to boarding schools, and then were made to give up their language, teachings, and customs, so went much of the tribal learning, the self-sufficiency, the institutional knowledge, and a deep understanding of the customs and traditions.

When the Lewis and Clark Expedition ascended the Missouri River after their first winter with the Mandans, they came into open country, the true high plains of the uncharted West. It was a veritable Eden. On April 22, 1805, Lewis wrote, "I asscended to the top of the cutt bluff this morning, from whence I had a most delightful view of the country, the whole of which except the Vally formed by the Missouri is void of timber or underbrush, exposing to the first glance of the spectator immence herds of Buffaloe, Elk, deer, & Antelopes feeding in one common and boundless pasture." [sic]

The annihilation of the bison followed on the heels of the Lewis and Clark Expedition, first slowly, with trappers and fur traders taking animals for robes and tongues then, in the aftermath of the Civil War, escalating in a relentless rising tide that was dramatically hastened by the westward advance of the railroads, which not only brought more settlers and divided the herds, but provided a means to get more goods to eastern markets. And the railroad workers had to be fed. Buffalo Bill Cody, in fact, earned his moniker by reportedly killing 4,280 bison in eighteen months to help feed the crews building the Kansas Pacific Railroad.

The slaughter, first on the southern plains then spreading to the northern plains, was initially driven by commercial interests. It reached a crescendo in the 1870s, propelled by demand for leather in European and American factories. As many as 2,000 buffalo hunters flocked to the plains for work that was gruesome but lucrative.

In *We Pointed Them North*, a memoir of the early days of cattle drives on the frontier, Teddy Blue Abbott (with author Helen Huntington Smith) writes:

> *The buffalo hunters were a rough class—they had to be, to lead the life they led. That buffalo slaughter was a dirty business. They would have two skinners working with each pair of hunters, and the hunters would go out and round up a bunch of buffalo and shoot them down all they could. The skinners would follow along in their wagon and take the hides. But when it got dark they would quit, leaving maybe ten or twenty carcasses that would freeze solid, and the next spring they would just lie there on the prairie and rot, hides and all. Riding the range you would find lots of skeletons with pieces of hide still sticking to them. It was all waste.*

Andrew Garcia lived among the Indians in Montana. In his memoir, *Tough Trip Through Paradise, Montana 1878,* he recalls:

> *In regard to the buffalo hunters and trappers, no one could tell who or what they had been. Some were good and a whole lot of them were bad and as a rule stood in with the lower strata, as most of them were booze hounds who blowed in their money by gambling, and were very short on law and order. After the buffalo were all killed off, many of those gentry promptly took up horse stealing and cattle lifting as the next best job to buffalo hunting, and cost the stockmen many a bloody battle and the loss of many of thousands of dollars in stock before they got them hung up on trees, or run out of the country, and a few in the pen.*

For a time, bison hunting was a popular tourist activity, with west-bound trains stopping so that passengers could shoot through the open windows without leaving the plush Victorian comfort of their cars. In *The West*, Geoffrey C. Ward describes a church congregation in Lawrence, Kansas, which registered 300 people for a two-day shooting trip as a fundraiser; he also tells of Orlando A. Bond, who killed 300 buffalo in a single day, 5,855

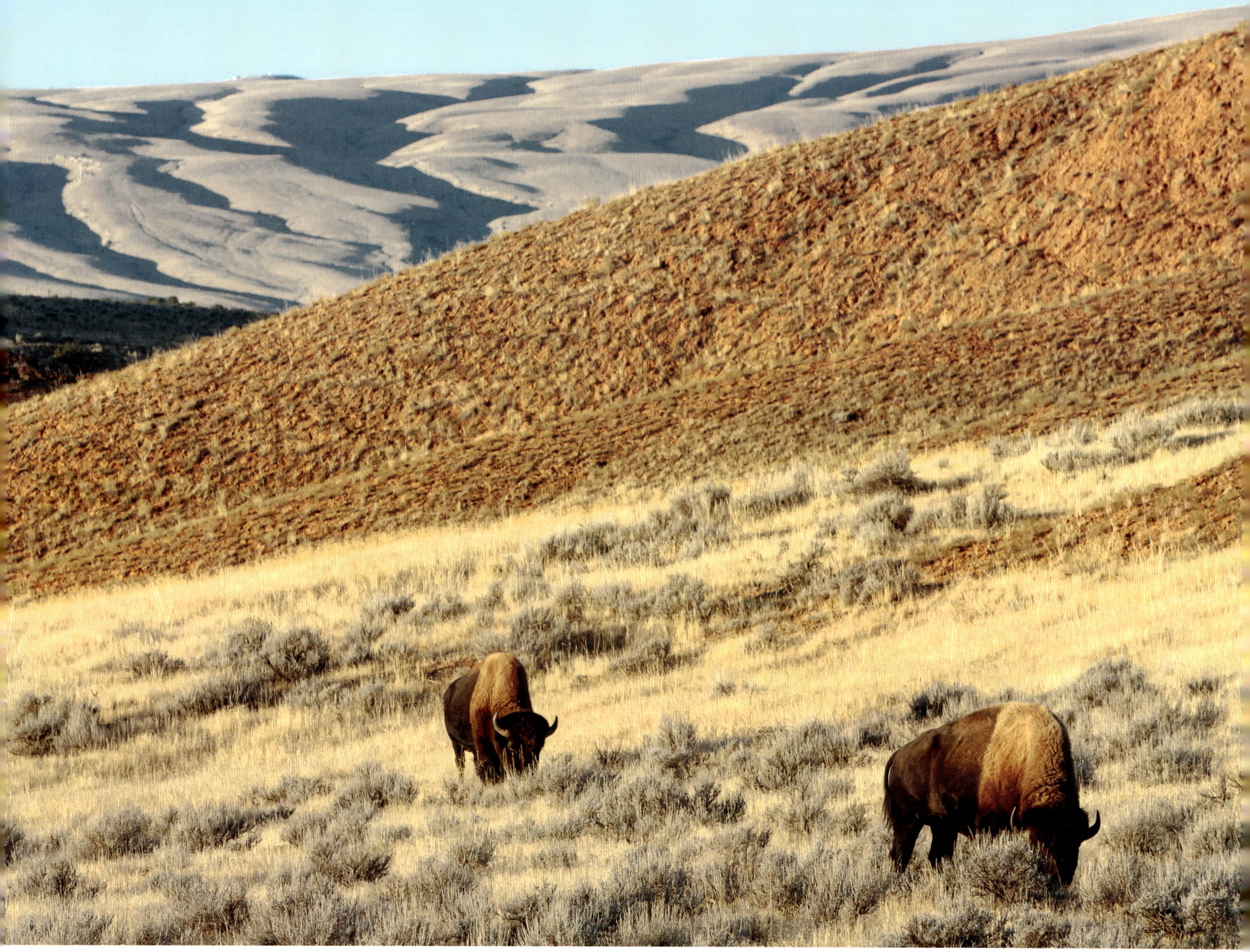

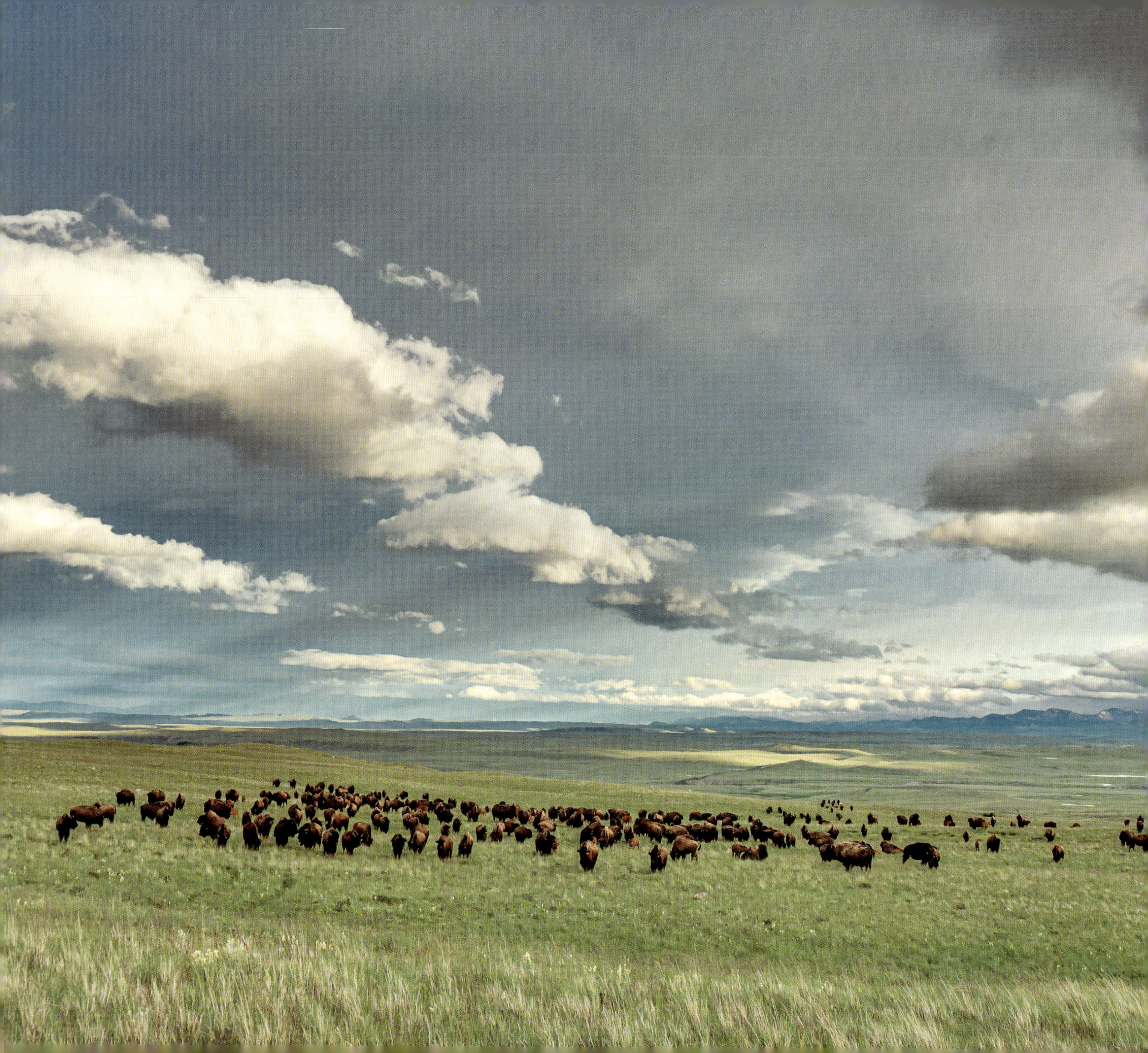

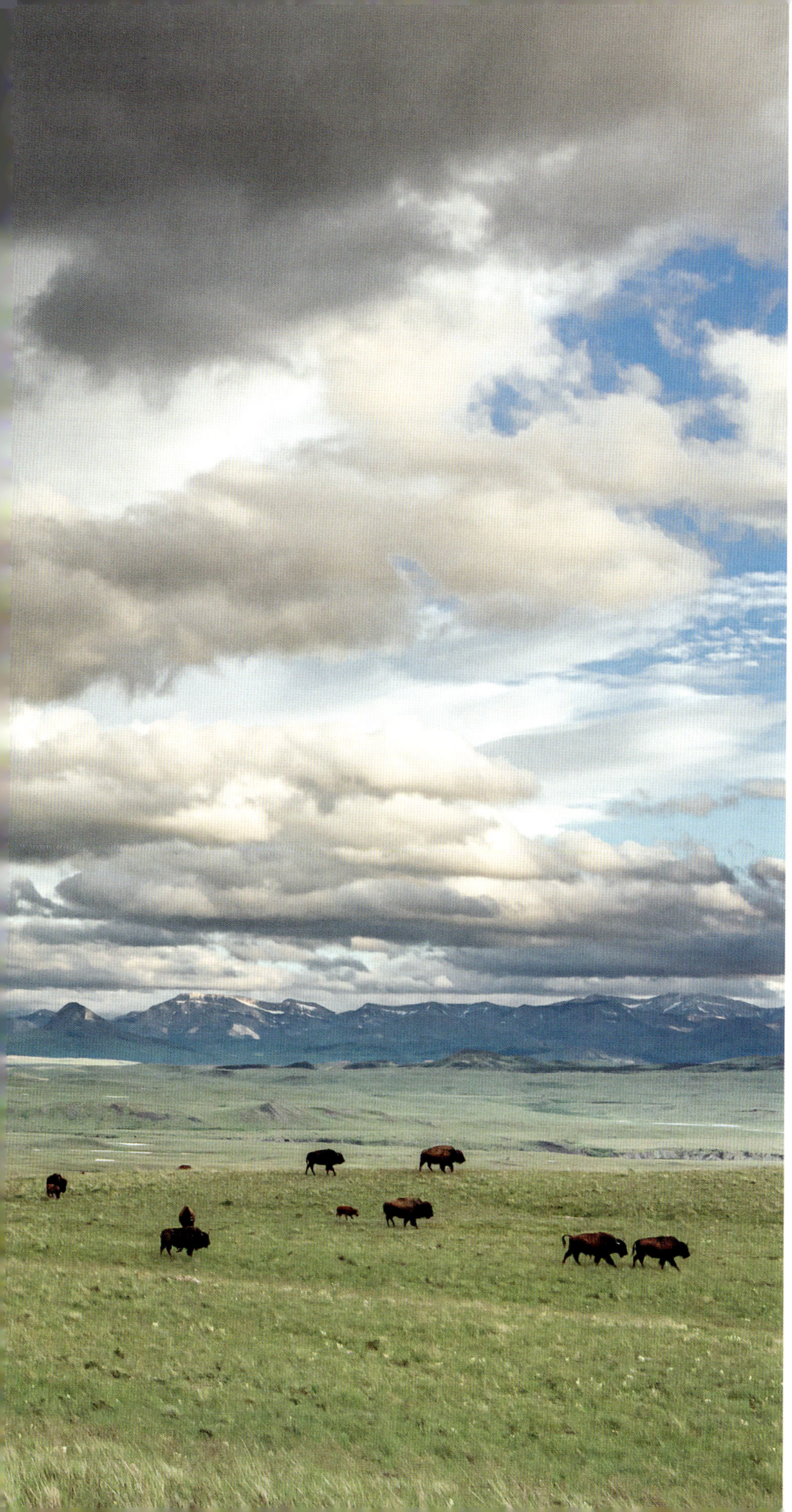

in a two-month outing, "so many that he was permanently deafened by the sound of his own rifle."

The actual extirpation was encouraged, subtly and overtly, by ranchers who wanted the bison to make way for cattle, by miners who were increasingly pushing into Indian Territory, by the government, and by the military. These authorities sought a final subjugation of the Indians in order to open the entire continent to settlement, crops, mining, and timber harvesting, and to fulfill the notion of America's manifest destiny as a settled nation stretching from sea to shining sea. In 1874, as the general population was waking up to the fact that bison were on the brink of extinction, the Texas legislature considered a bill to protect the species. In arguing against it, General Phil Sheridan blatantly connected the dots. "These [buffalo hunters] have done . . . more to settle the vexed Indian question than the entire regular army has done in the last thirty years. They are destroying the Indians' commissary."

Once the bison were gone, the bone pickers came, gathering skulls and skeletons, shaping them into veritable mountains for collection, and then shipping them to factories in the East for use in the production of sugar and as fertilizer—as many as thirty million pounds in just a few years. "Before the buffalo bones were all picked up, people . . . reported bones so thick on the ground that they looked like fallen snow," says Steve Rinella in *American Buffalo; Search for a Lost Icon.* "If you could have watched the bone pickers' progress in time-lapse photography taken from outer space, it would have looked like the snow was melting fast, pushing up river valleys and railroads first and then spreading outward until there was hardly a bone left anywhere."

By 1875, virtually all the Indians in the southern Plains had yielded to the inevitable and succumbed to the government agencies. The northern tribes held out longer, but the discovery of gold in the Black Hills provided the final nail in the bison's coffin. By 1883, when a Dakota Territory law protecting the buffalo was passed, the herds had been decimated. An 1894 survey counted only eighty-five free-ranging bison, with about 1,000 more in zoos and private hands. Remnant herds had survived in Yellowstone National Park and in Canada, but the numbers were miniscule. Congress enacted legislation to protect the remainder and a decade later, in 1905, the American Bison Society was founded at the Bronx Zoo by a group that included Frederic Remington and Andrew Carnegie, with Teddy Roosevelt as honorary president.

Still, by then the lives of native Plains inhabitants, and the ecosystem on which they depended, was irrevocably altered. "Ten years after the last wild buffalo was killed, the Great Plains was a different, certainly eerie, place," writes Dan O'Brien in *Buffalo for the Broken Heart.* "It was as if the buffalo had been expunged from the record of plains life. Nothing remained but the occasional bone, like the ones farmers might have rolled under the prairie sod with the shares of their moldboard plows."

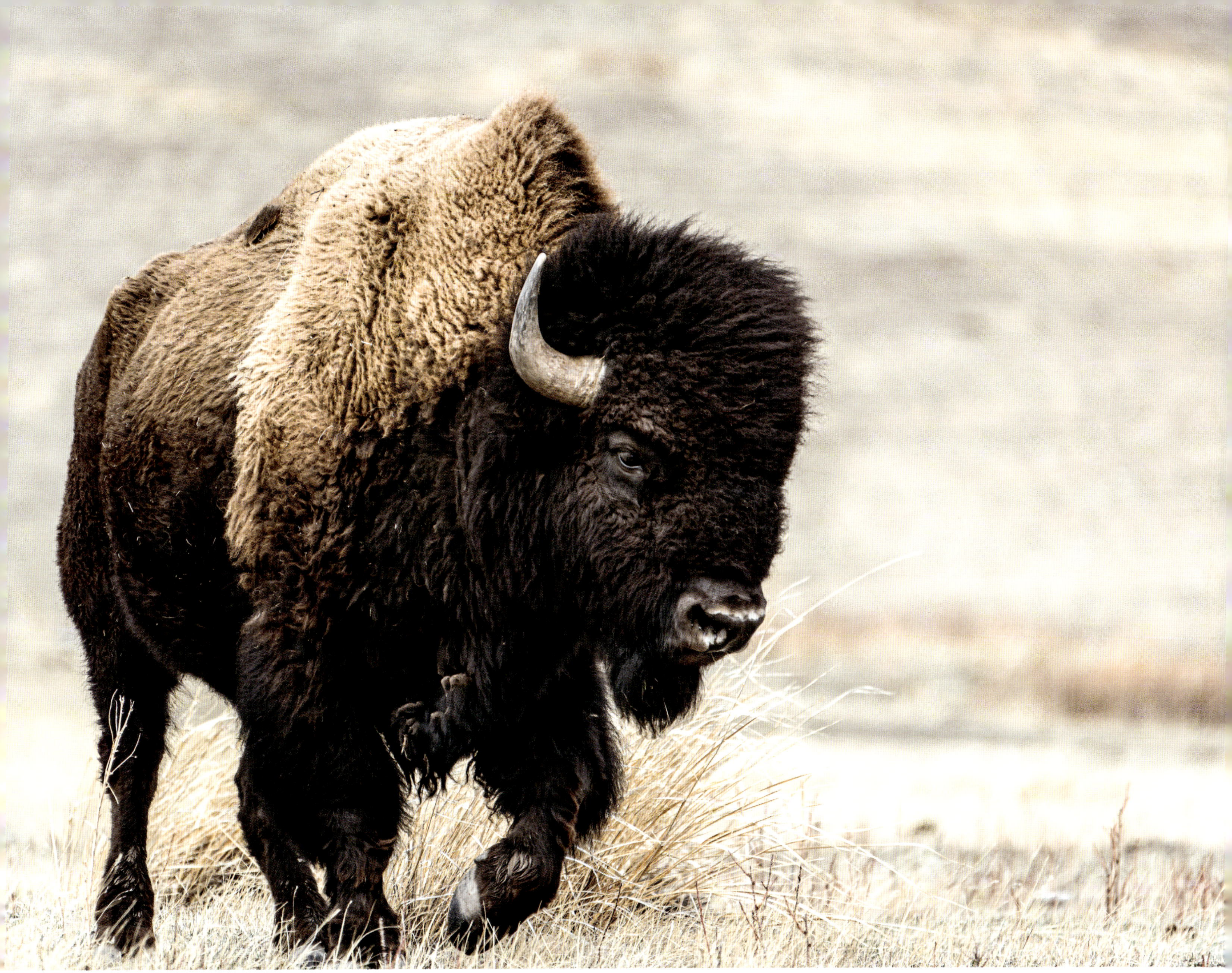

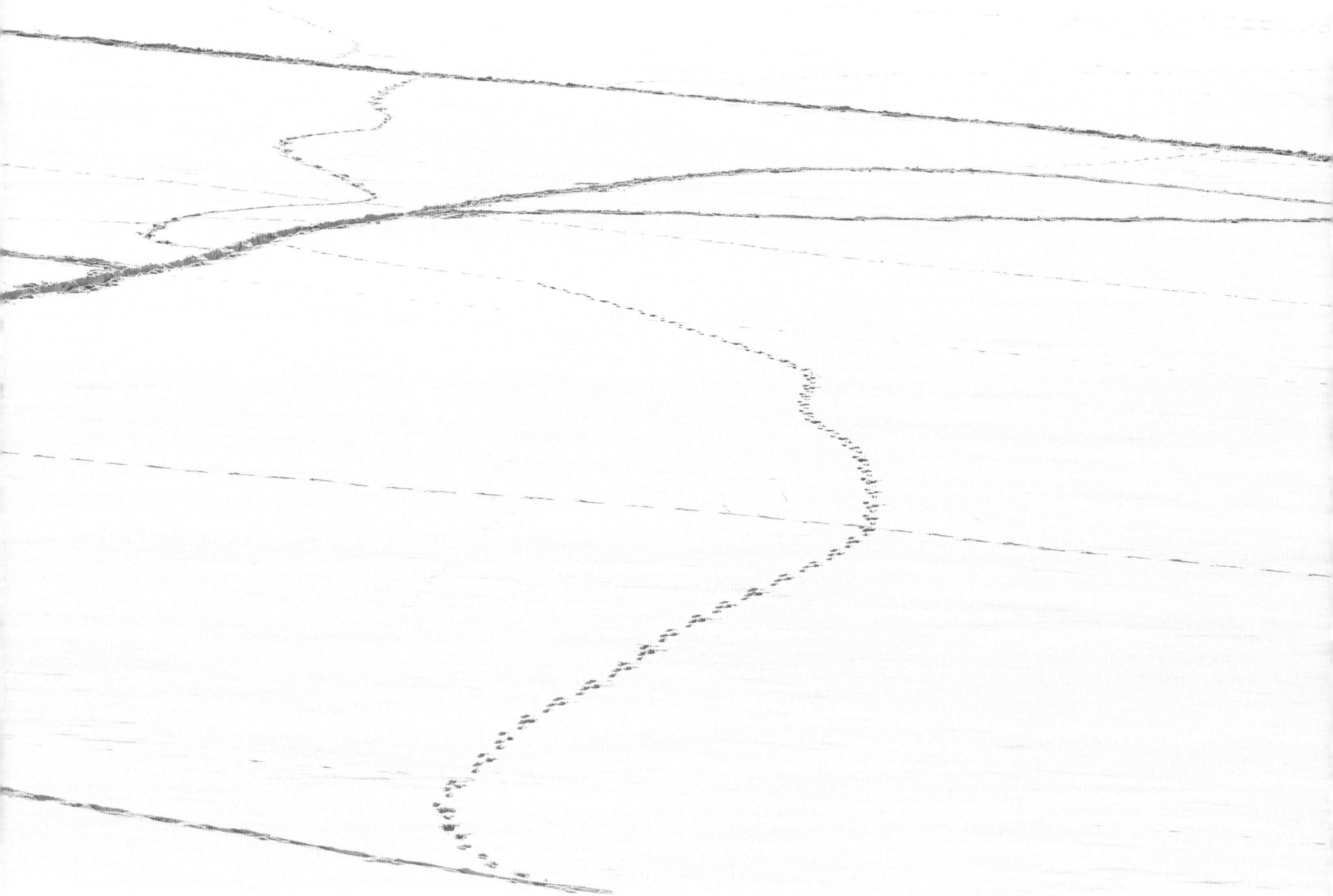

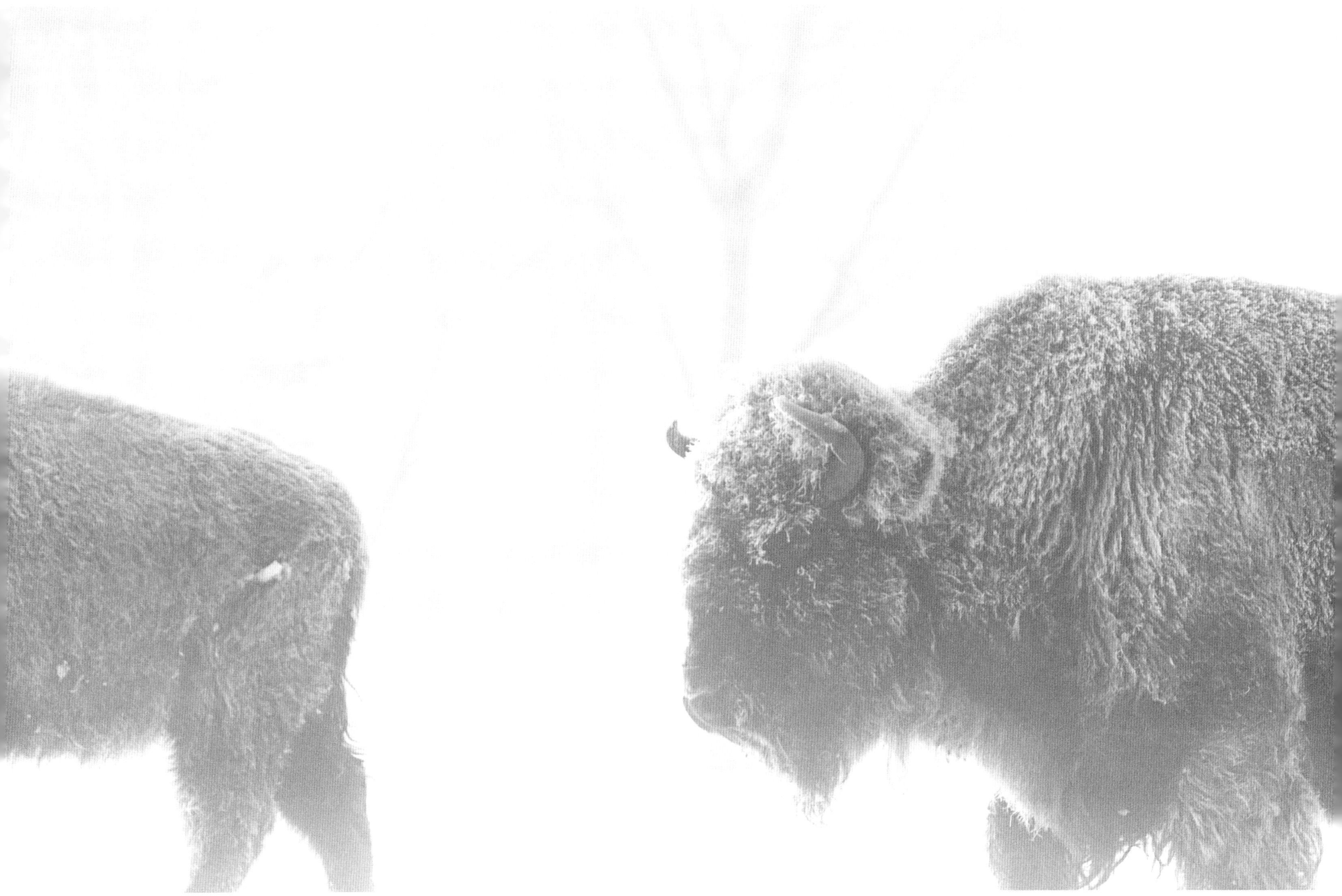

The annihilation of the bison
followed on the heels of
the Lewis and Clark Expedition,
escalating in a relentless rising tide
that was dramatically hastened
by the westward
advance of the railroads.

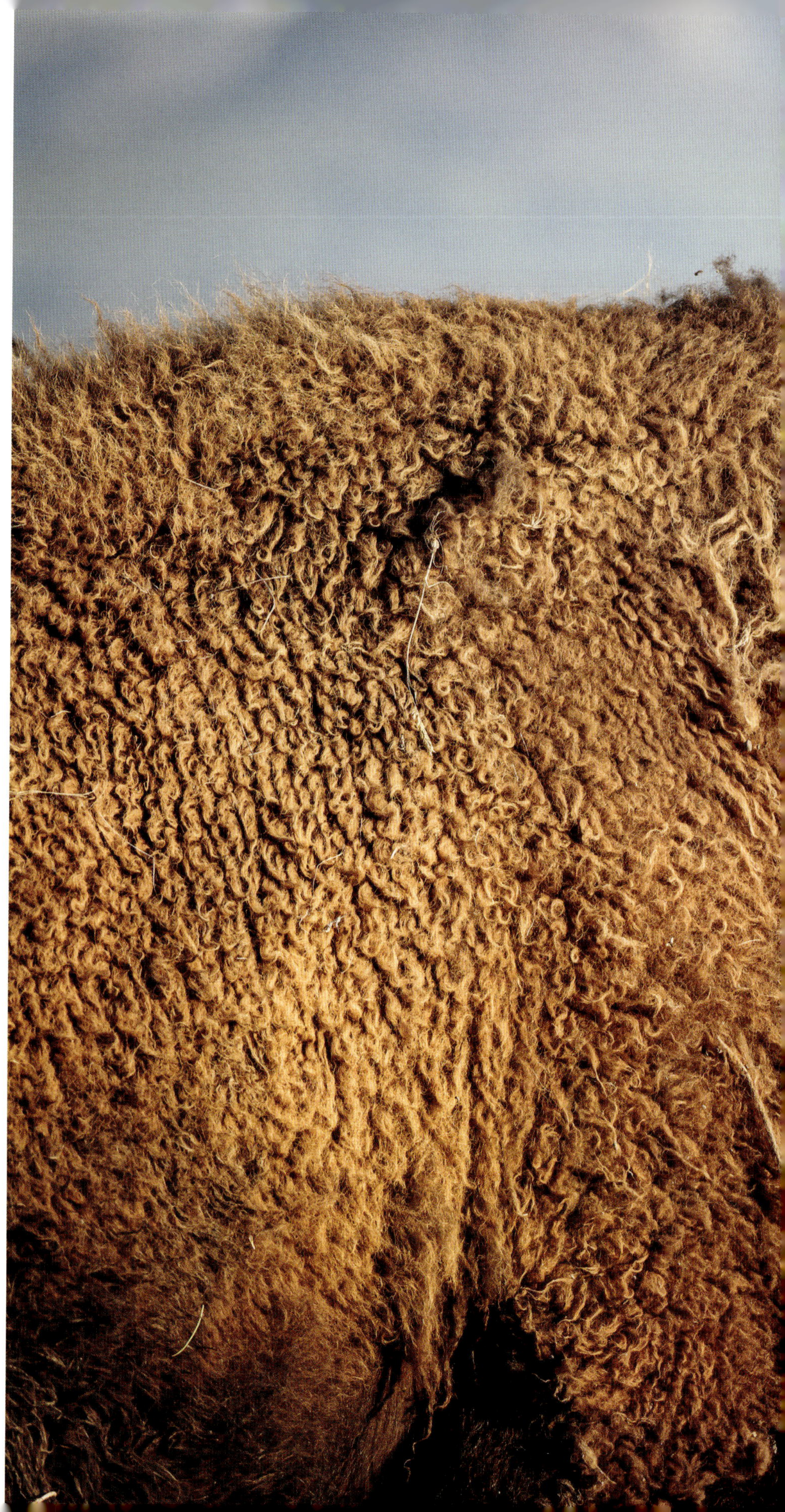

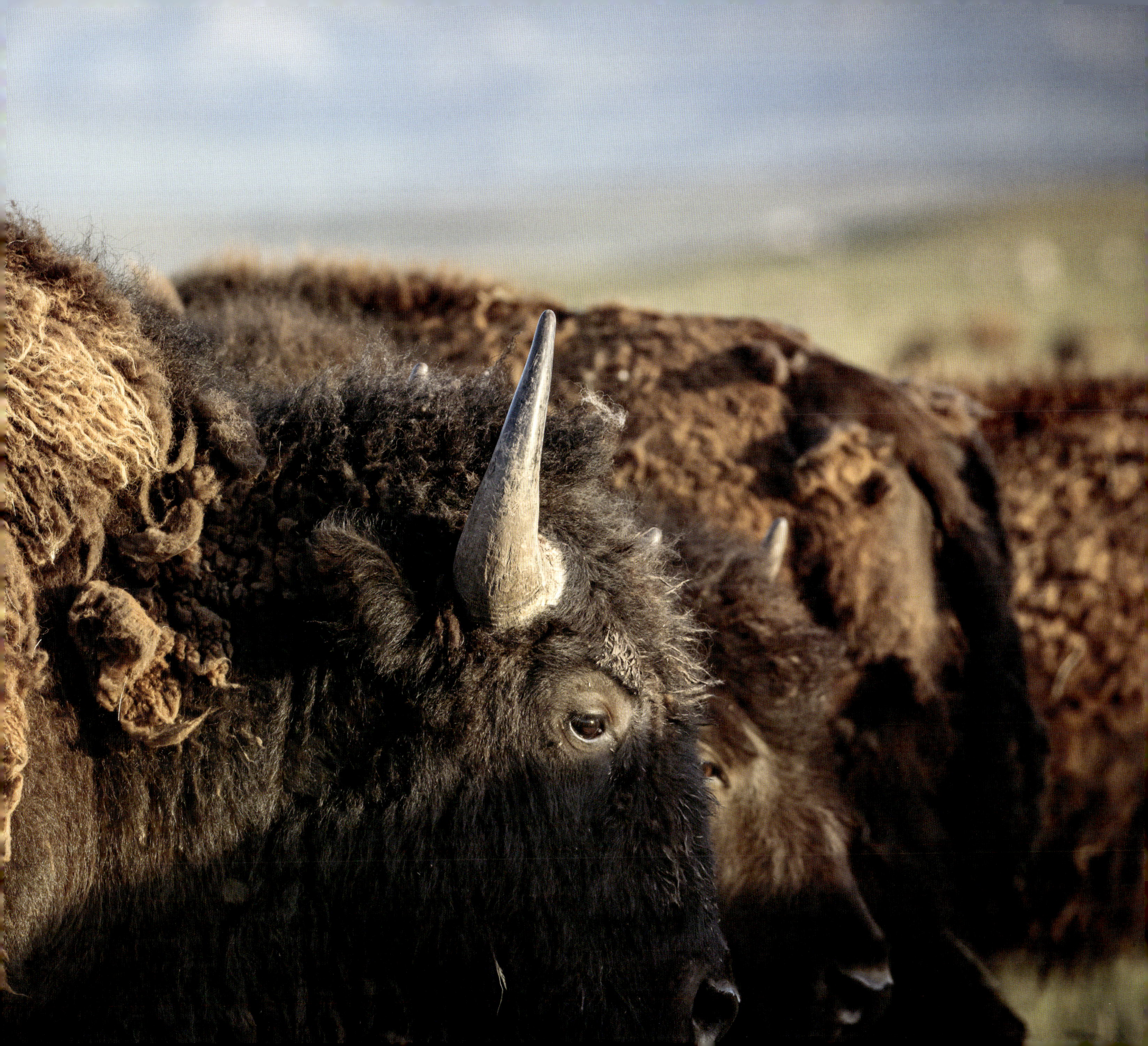

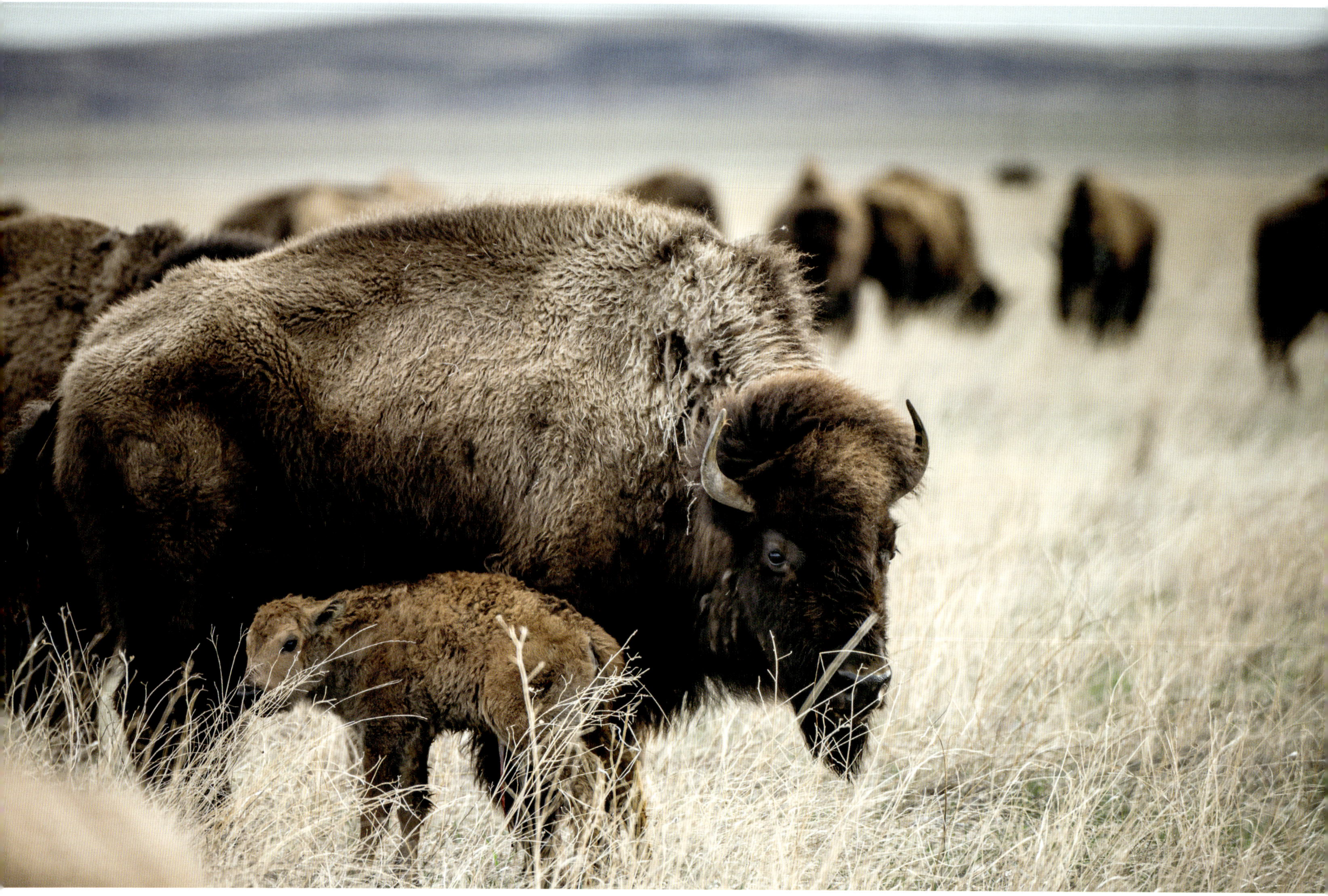

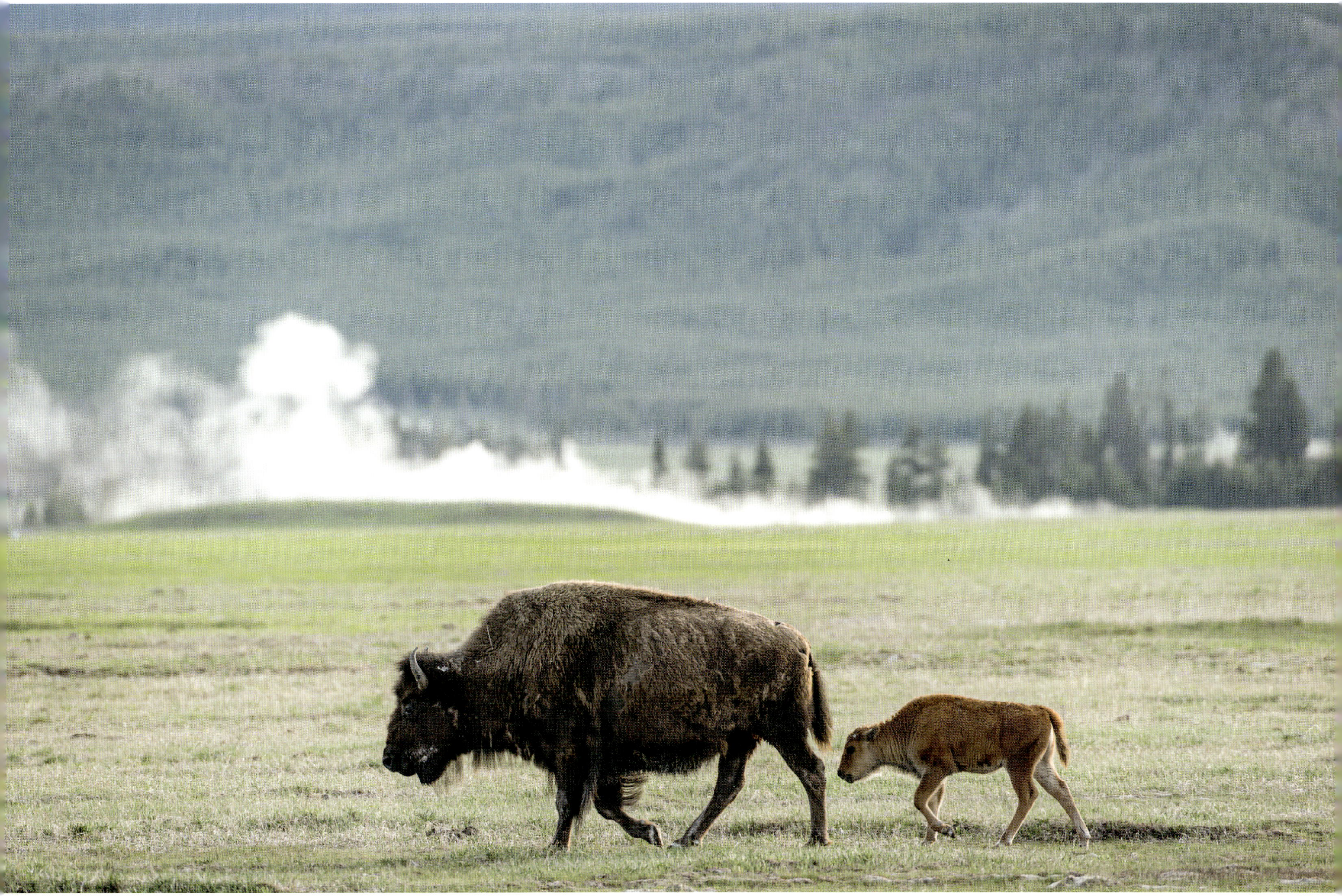

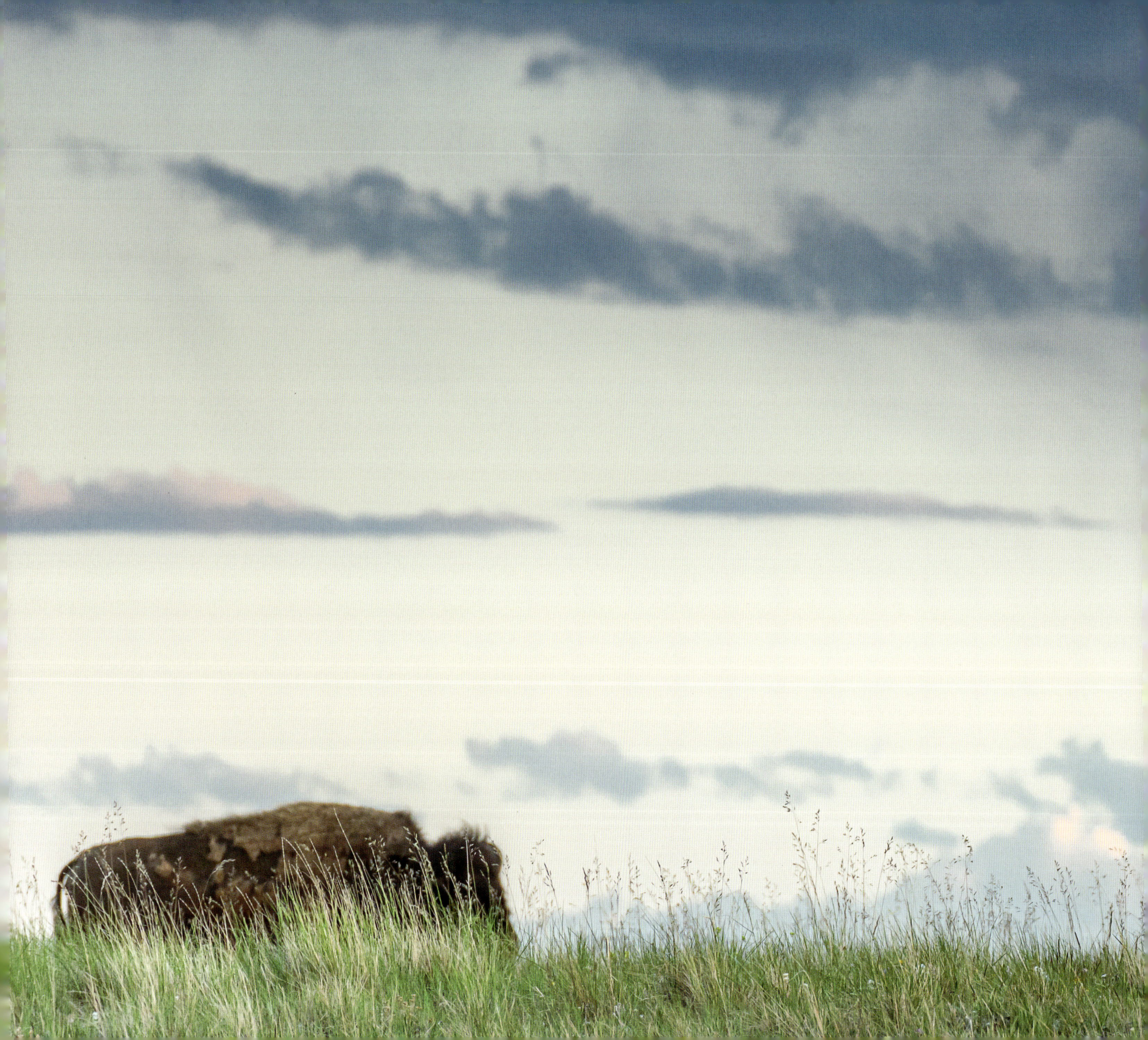

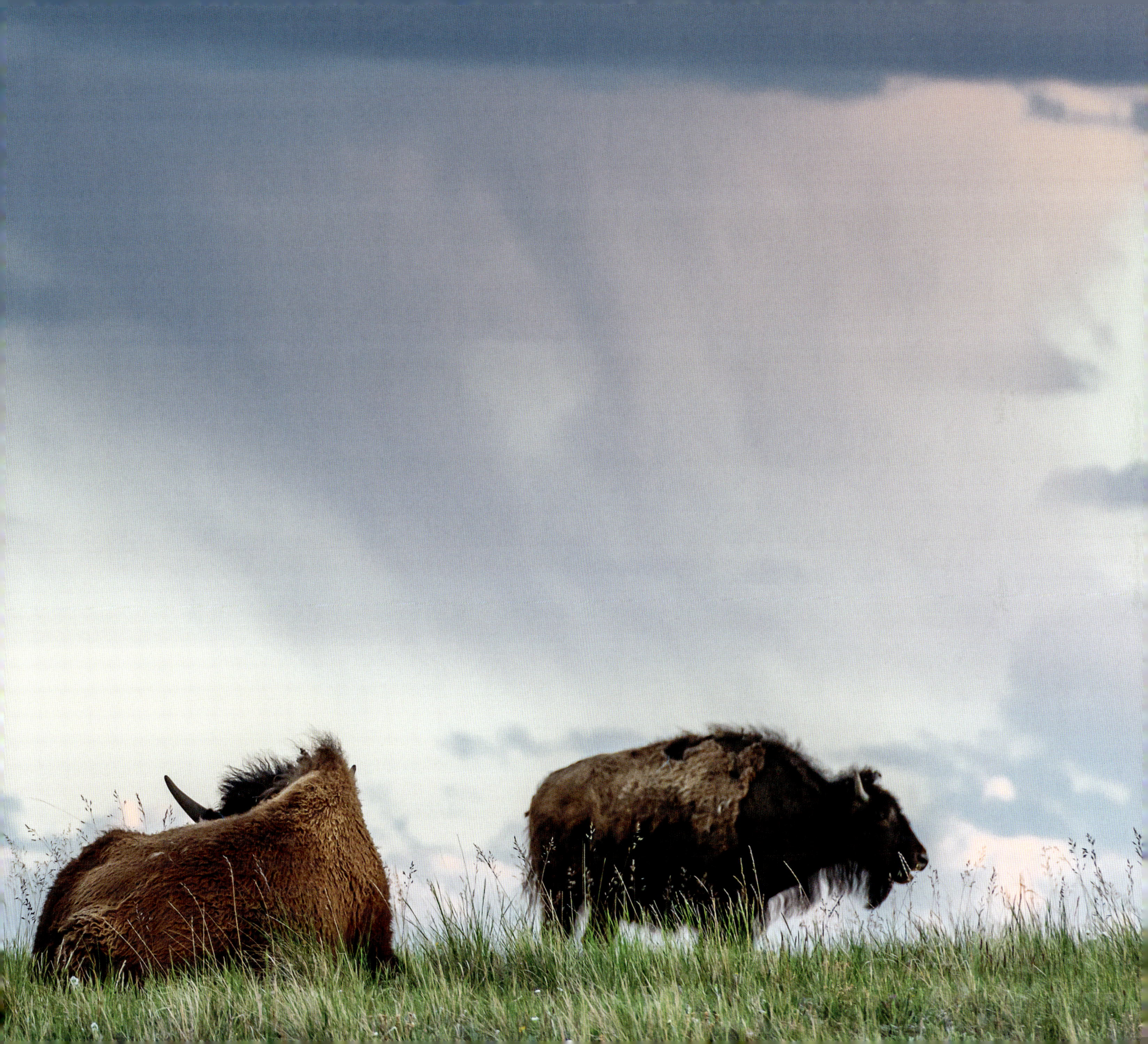

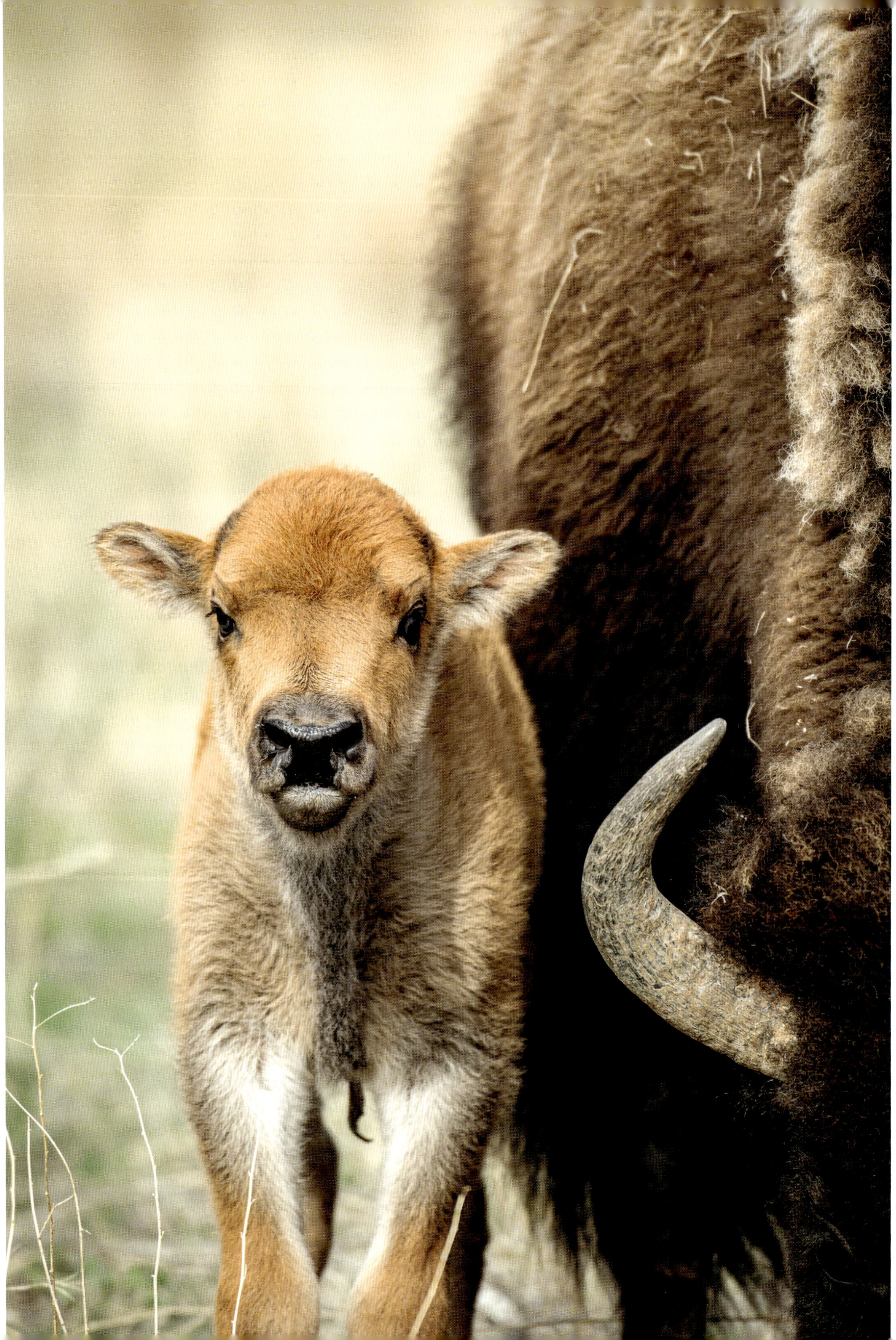

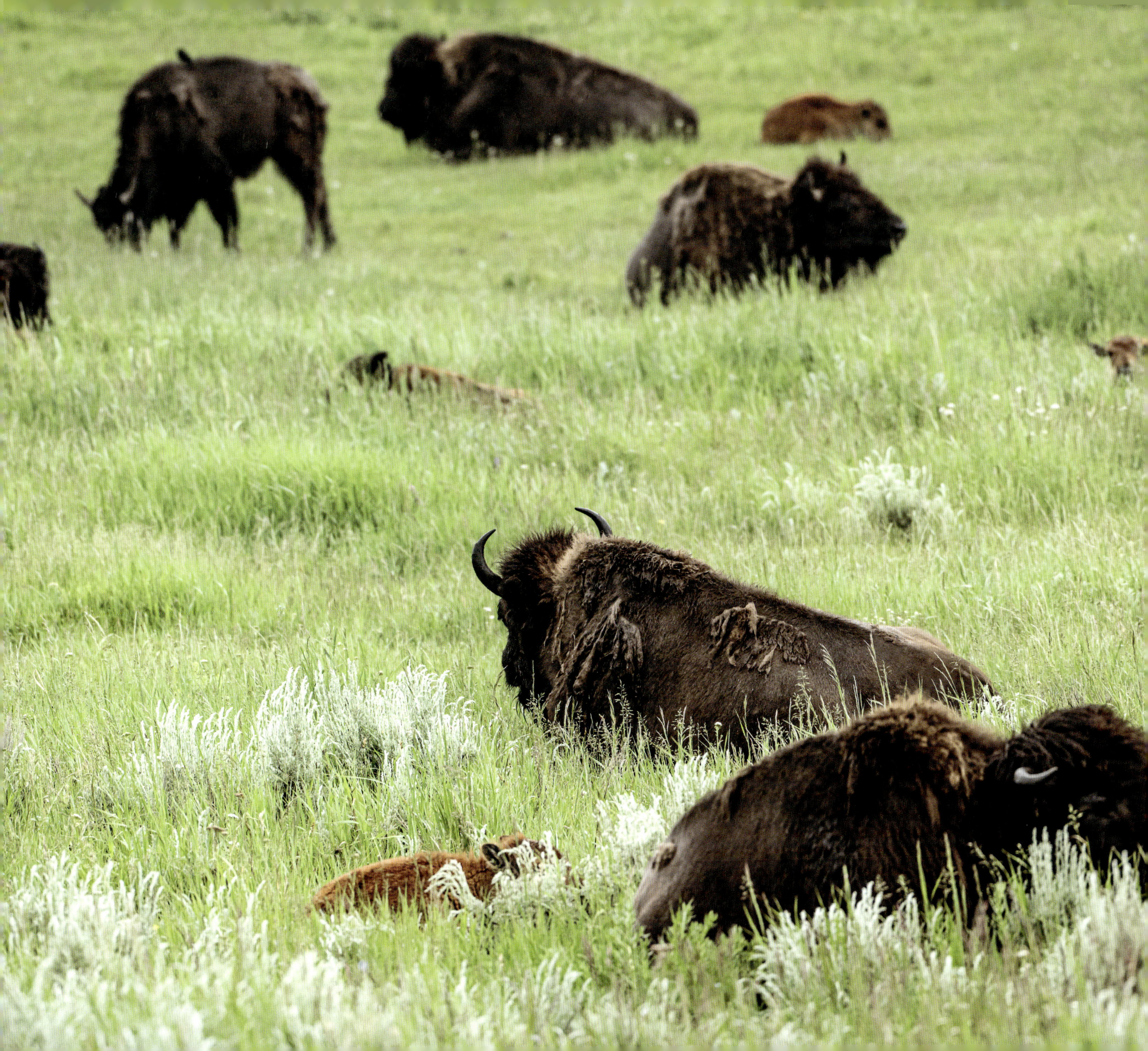

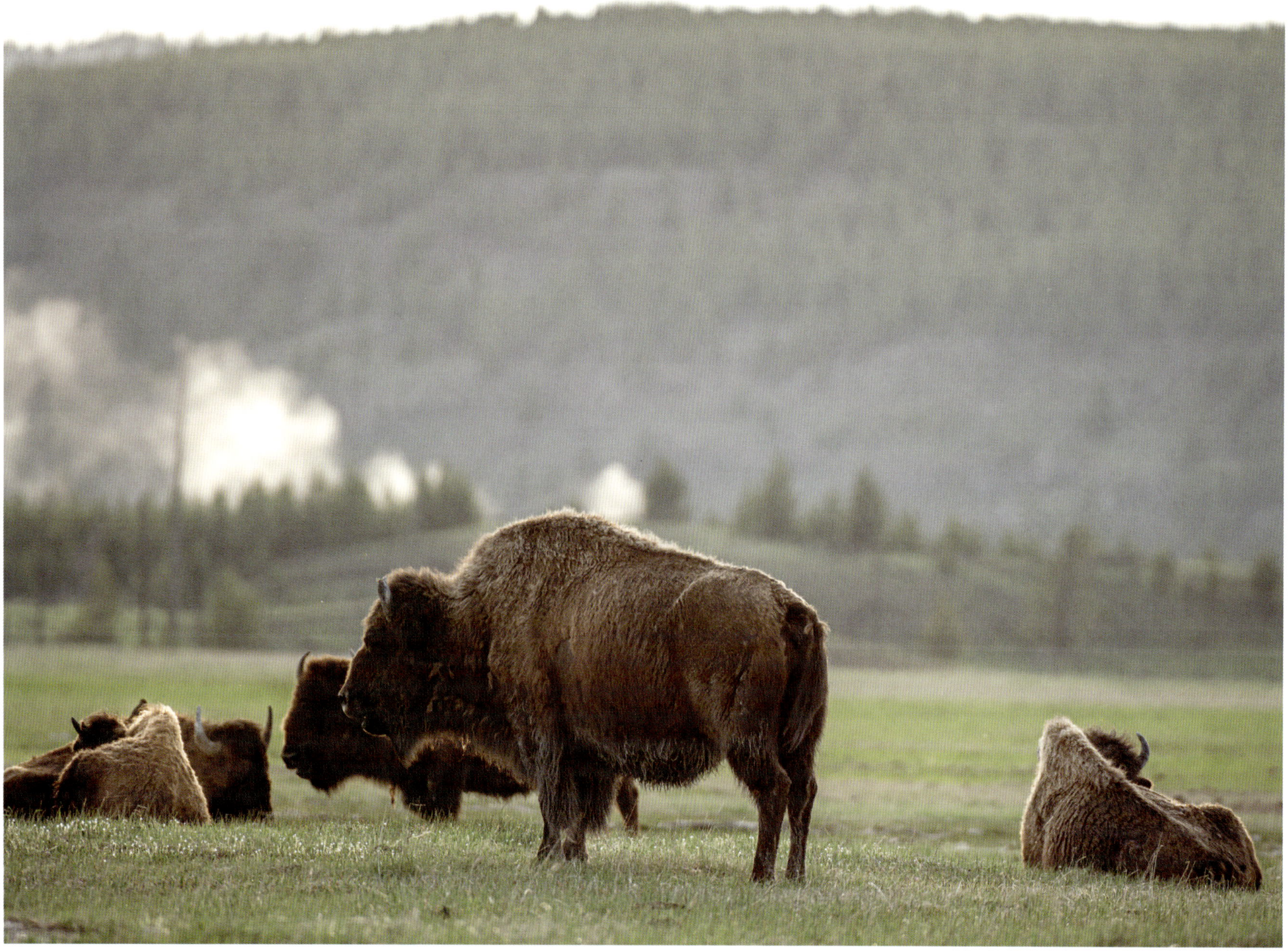

Return of a Native

San Francisco's Golden Gate Park runs in one narrow three-mile rectangle from the heart of the city near the Civic Corner all the way to the Pacific Ocean, where it terminates at a Dutch windmill facing the waves. At more than a thousand acres, the park offers something for everyone, including a botanical garden, museums, a Japanese tea garden—and bison.

A spring visit to the bison paddock involves no jouncing over dirt roads, no question as to whether you'll be rewarded with the sight of bison. They live in a vegetated but overgrazed pasture on undulating ground punctuated with a handful of conifers. Situated alongside John F. Kennedy Drive between a municipal golf course and Spreckles Lake, it's a pleasant spot, with lots of birdsong and passersby walking, biking, and driving, some of whom stop to read the informational plaque.

Bison were brought to the park at the time of its founding in 1890, the placard explains, in homage to the rapidly disappearing Wild West. Bison populations were then at their nadir; barely 1,000 had survived on the entire continent. These days the group numbers only ten (including five calves added in 2020 to mark the park's 150th anniversary), and ever since the day a bison bull got aggressive with a policeman on horseback, all the animals are female. But early in the twentieth century the park's breeding program was so successful its bison were soon providing calves for other zoos, parks, and ranches.

The fact is, bison are not delicate or terribly fussy and they have no significant pressure from predators. They tend to reproduce well and therefore will ultimately overflow the capacity of limited spaces, even spaces as vast as Yellowstone National Park's 2.2 million acres. That overcapacity has worked to the advantage of other budding programs, making Yellowstone bison key players in the restoration story. Descended from an original remnant herd, the bison of Yellowstone and those of Canada's Elk Island National Park especially have been of crucial importance in founding and reinvigorating conservation herds around the country.

The InterTribal Buffalo Council, founded in 1992, is a cooperative group of seventy federally recognized tribes. Its members work together and share resources to help restore buffalo to their lands in recognition of the animal's historic, cultural, traditional, and spiritual relationship with their people. They believe

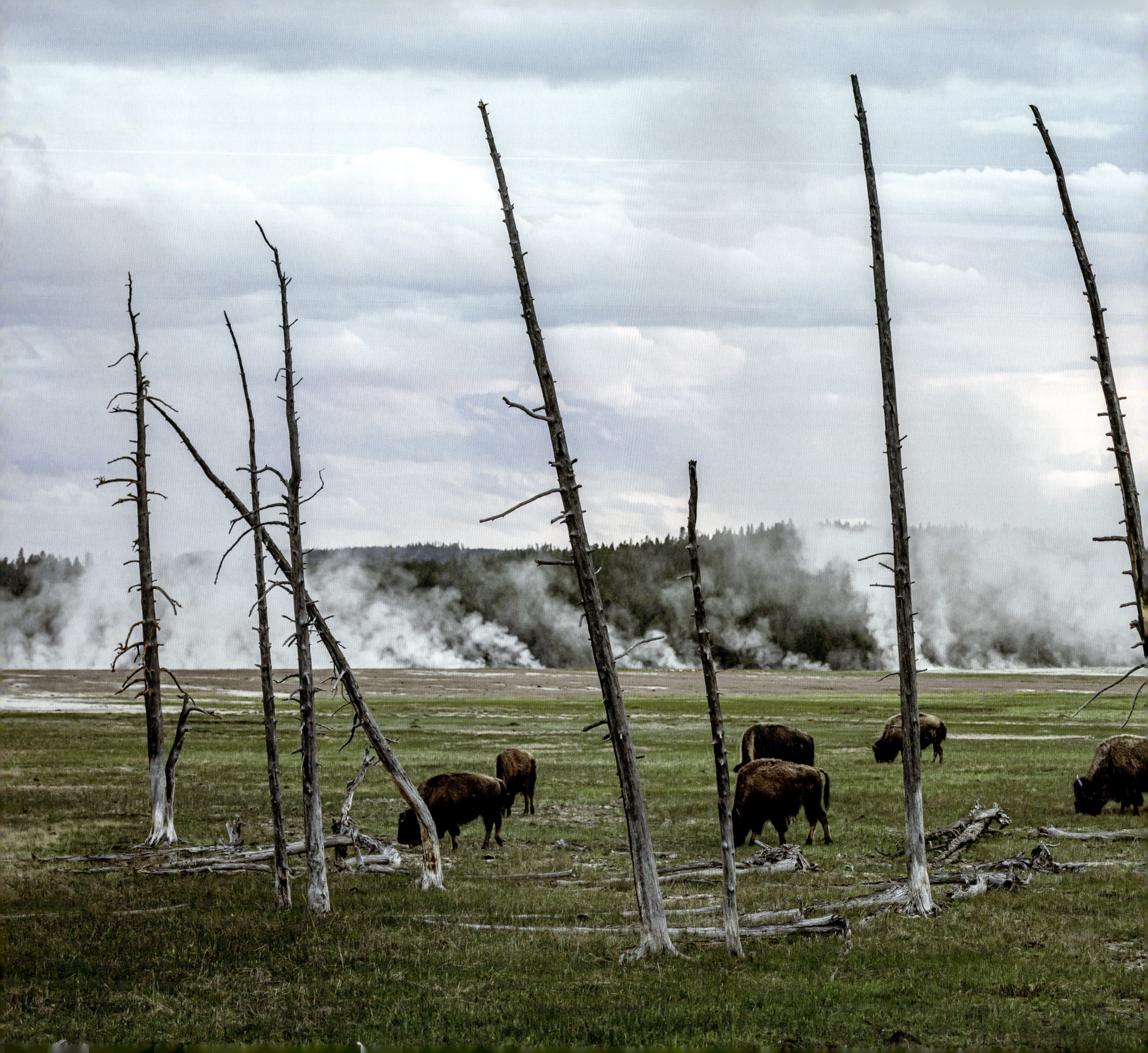

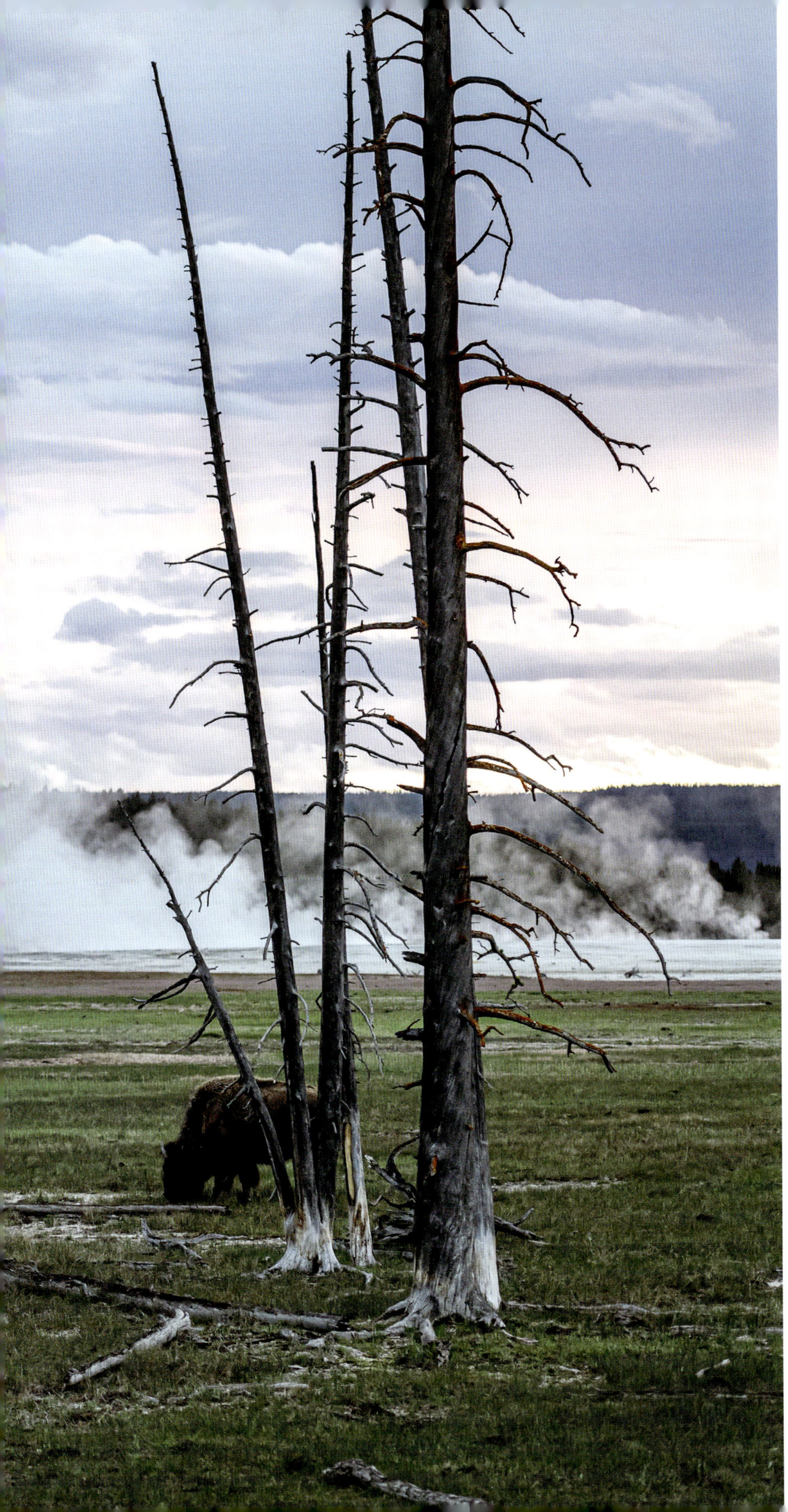

that by reestablishing healthy buffalo populations they're reestablishing hope for the tribes.

Ervin Carlson has served as President of the ITBC for three-fourths of its existence. He also manages his tribe's buffalo herd, numbering about 800 head. In the late 1990s, he was serving as Agricultural Director for the Blackfeet Nation when the tribe's buffalo were put under his direction. At the time, he explains, his people didn't necessarily feel a strong spiritual connection to the buffalo. "We brought buffalo back here in the mid-'70s and it was kind of different back then. We hadn't had the animals here for so long, they just weren't part of our lives anymore. But people are slowly becoming more connected and realizing the significance of the buffalo being gone, and how important they were to us and our culture."

The Blackfeet have been able to rebuild that connection over time, through field trips, school programs, and by providing meat to elders and to tribal members suffering health issues. ("It's the healthiest meat you can eat," says Carlson.) They provide bison parts—hides, head, feet, meat—for traditional ceremonies. There's also immense value in simply having the buffalo visible to people moving through the stunningly beautiful

Blackfeet reservation, a dramatic mountain landscape bordering Glacier National Park.

"In the beginning, it was all about bringing them back because of cultural significance and for health reasons," says Carlson. "But it was also because the animals took care of us. They were our food, clothing, tools; they provided all of that. In this day and age we also want to look at how the animal can help take care of us by creating revenue for our people."

Further east in Montana, on the Fort Belknap Indian Reservation, home to the Assiniboine/Nakoda and Gros Ventre tribes, buffalo range in two herds on 650,000 acres of plains and grasslands. George Horse Capture, Jr., is Vice President of the Fort Belknap Indian Community Council. As Director of Aaniiih Nakoda Tours, he also interprets the area's history and culture for visitors.

The reservation uses its buffalo program to generate revenue, educate schoolchildren, and provide animal parts for local sun dances and other ceremonies. Fort Belknap has its own USDA-inspected meat-packing plant for processing cattle and bison. In hard times, such as those caused by drought or pandemic, the tribe will distribute meat to its own people and, in the interest of being a good neighbor, to off-reservation food banks. Horse Capture describes the buffalo's benefits to the tribes as both tangible and intangible.

"Buffalo, you might say, are environmental engineers. They're not as hard on the land [as cattle]. Buffalo have a real nice habit of moving around, and not eating down to the ground, and they're really great for prairie potholes. When those fill up in the spring they don't last that long but a lot of waterfowl will mate and breed there."

Beyond the practical, Horse Capture adds, "They fit into our culture, our heritage. Some of us just like to go out and look at them. They have a lot to teach, if you watch them."

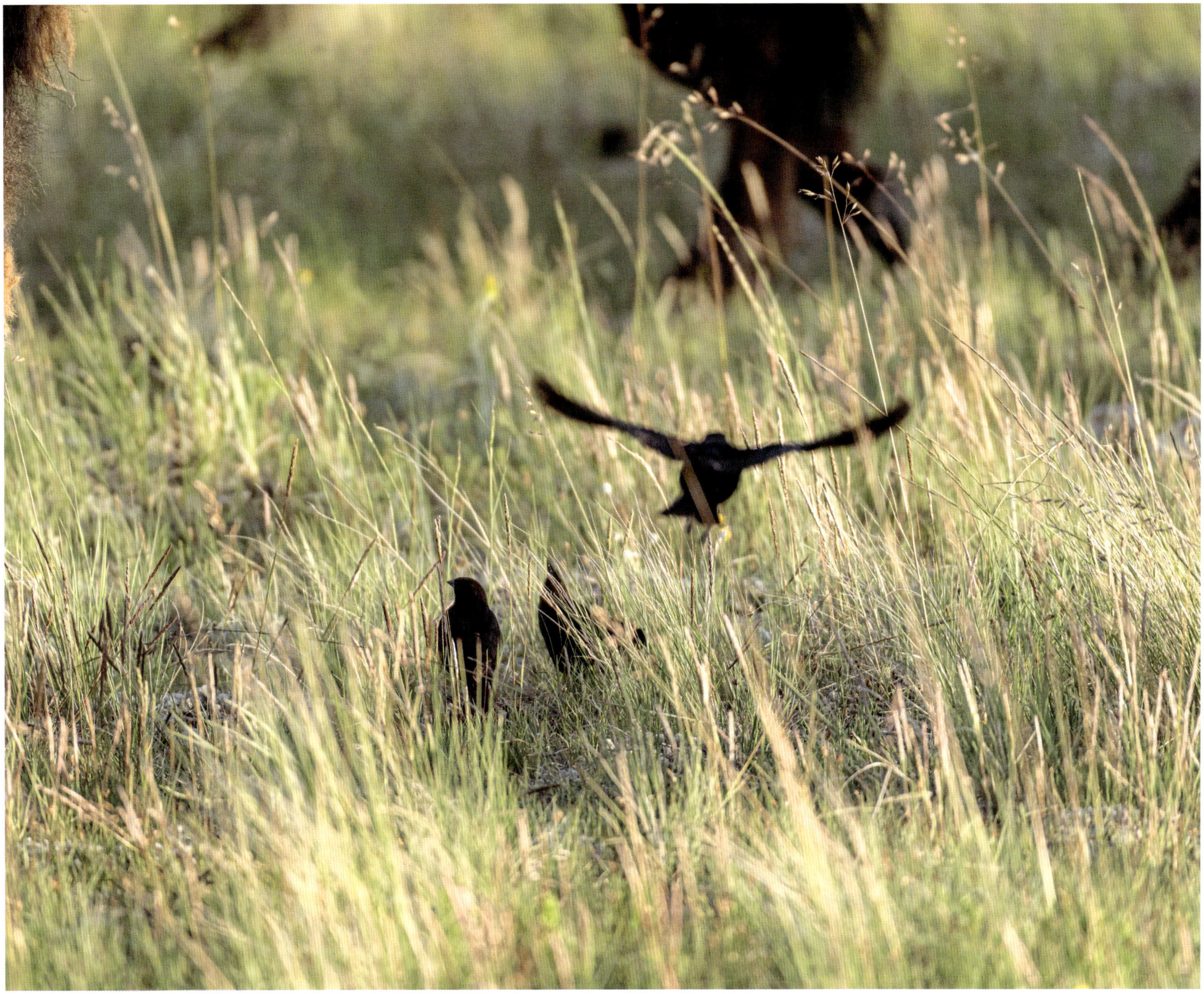

For millennia, the American bison shaped the landscapes of North America and the cultures of many American Indian tribal nations. With settlement, this majestic animal—the largest land mammal native to the U.S., once numbering 40 million—was all but wiped out, slaughtered at the rate of one creature every thirty seconds for forty years. Thanks to the efforts of many tribal nations, conservation organizations, and members of the 114th Congress, in 2016 President Obama signed a law naming the bison America's National Mammal as a symbol of our nation's unity, resilience and health, taking its rightful place alongside the bald eagle, and recognizing the bison's critical role in supporting a balanced ecosystem that nurtures all species, including humans. In shaping our landscapes, this amazing creature, brought back from the brink of extinction, can once again lead us on a path toward a more sustainable future.

—SALLY JEWELL, U.S. SECRETARY OF THE INTERIOR 2013–17

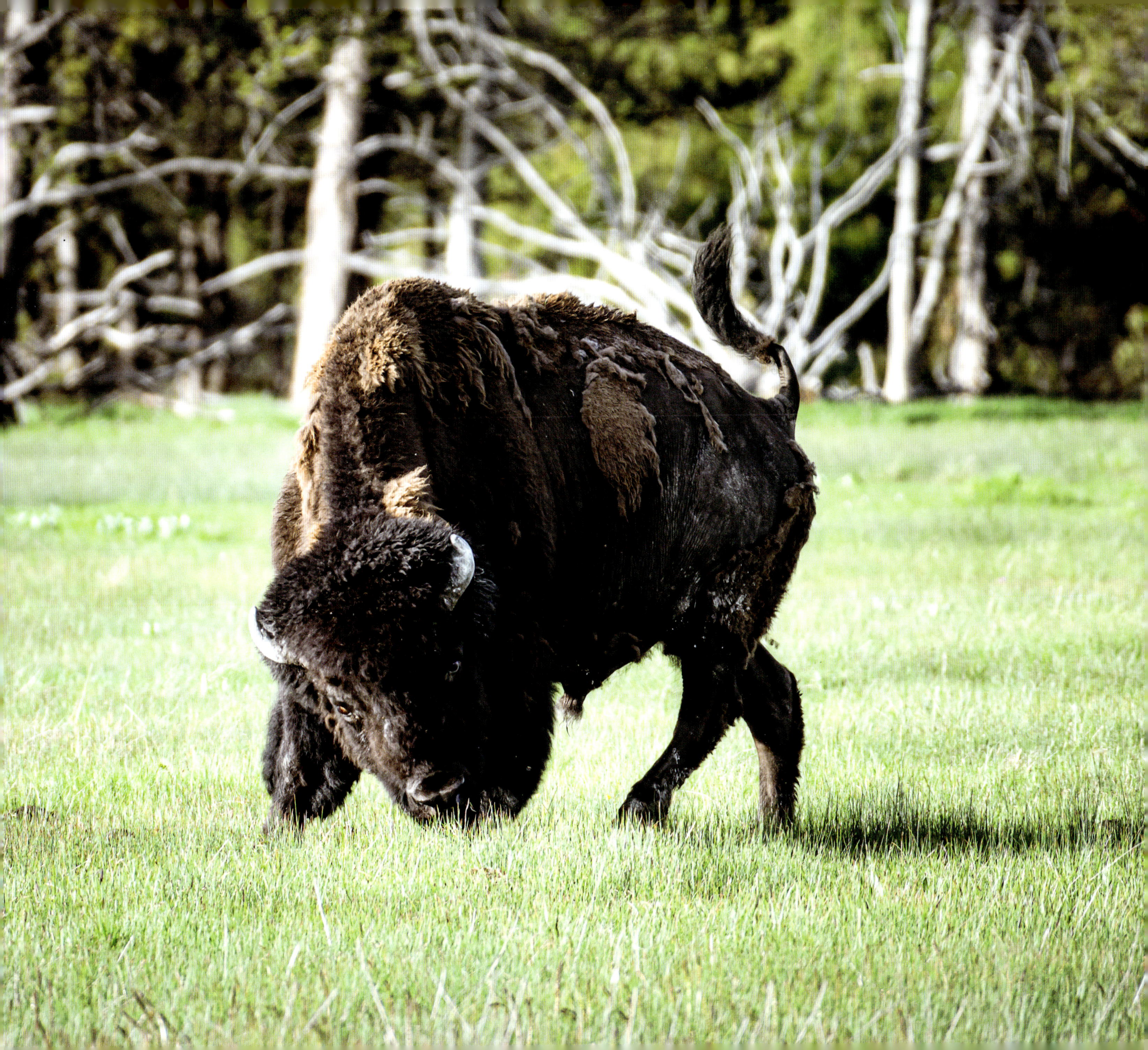

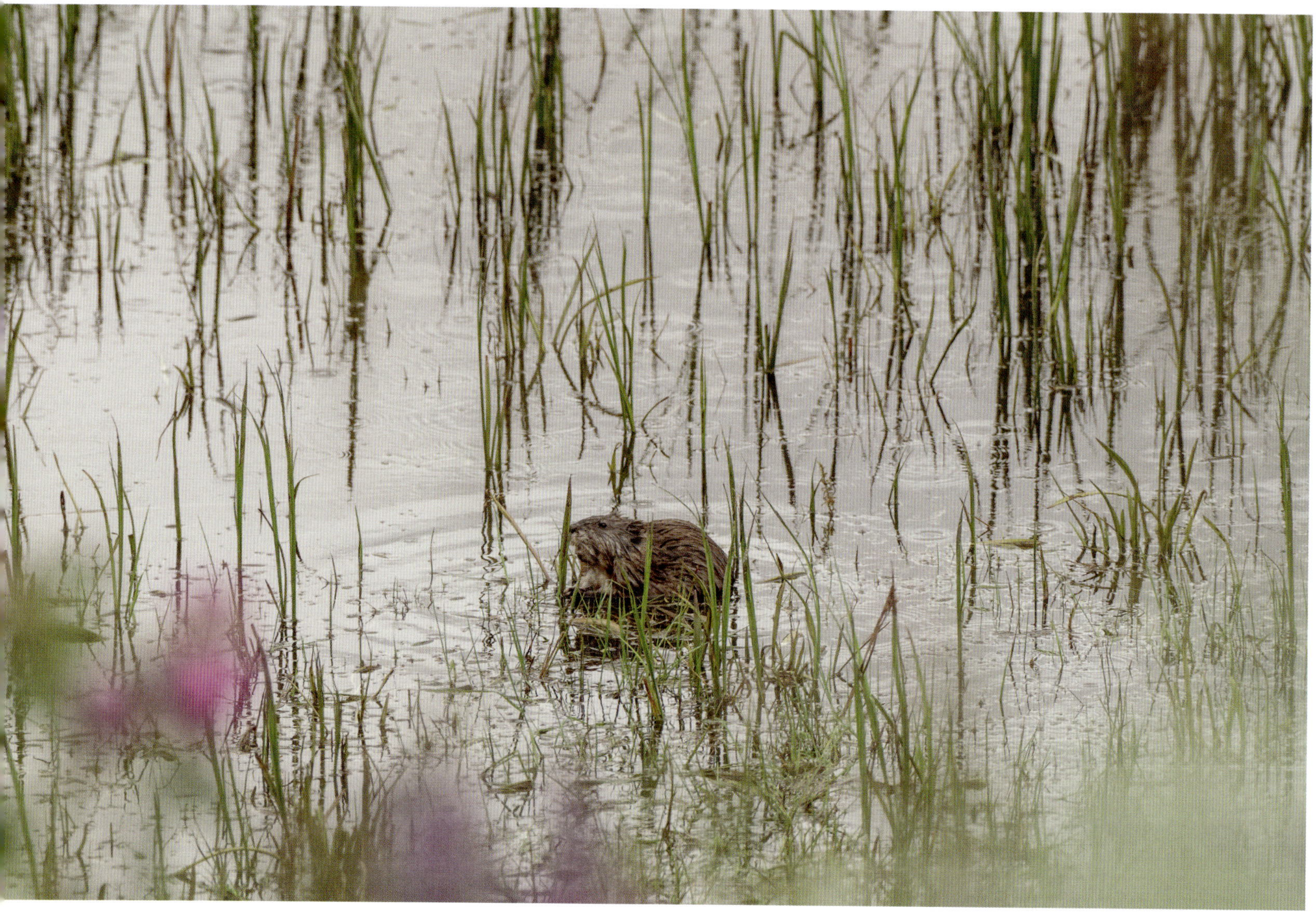

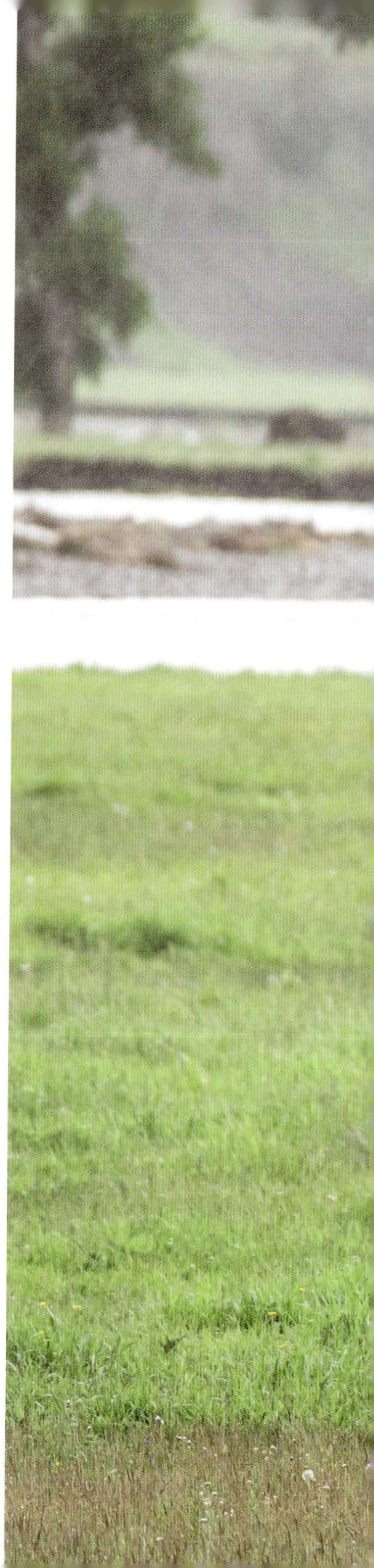

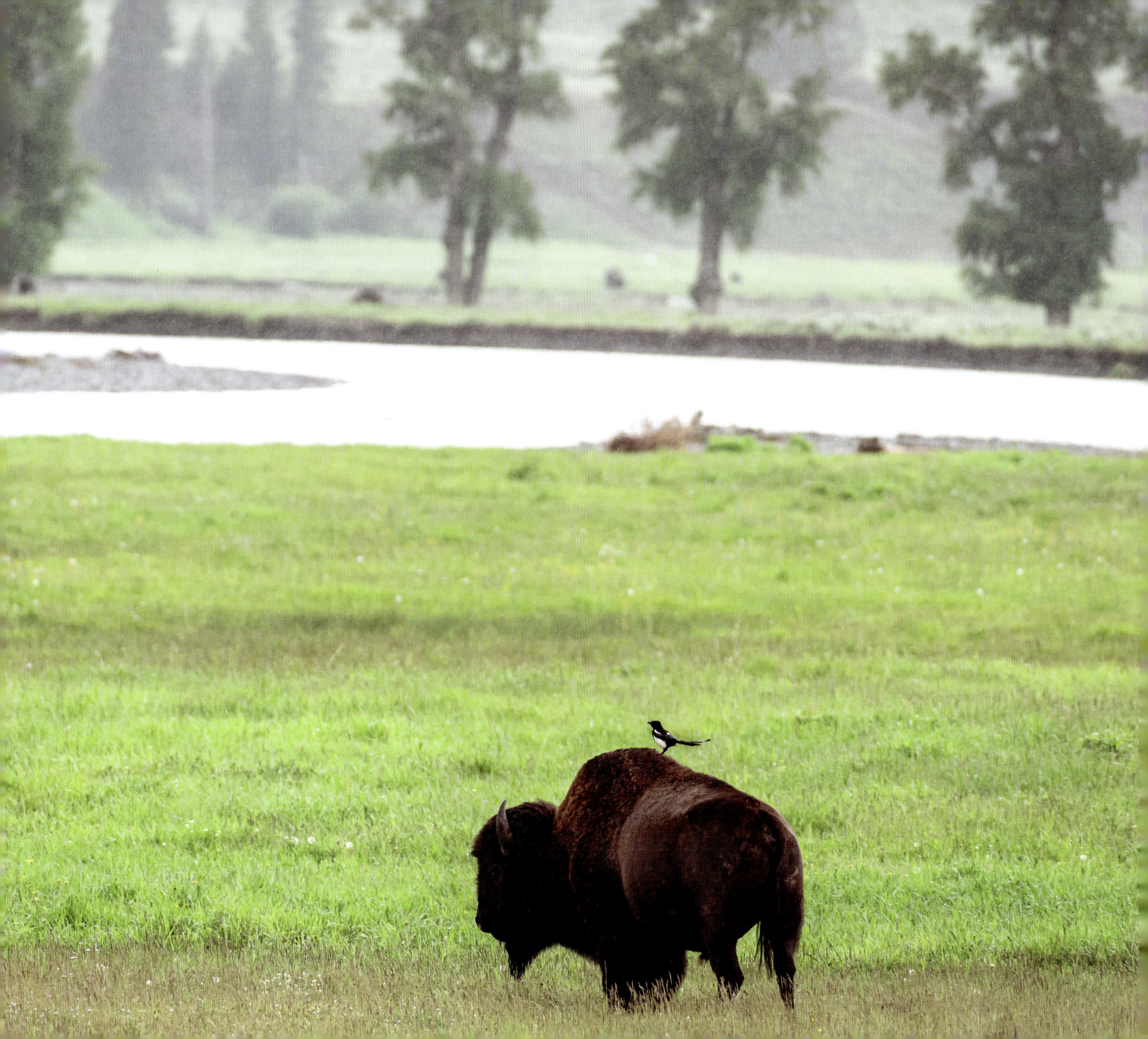

The fact is, bison are not delicate or terribly fussy and they have no significant pressure from predators. They tend to reproduce well and therefore will ultimately overflow the capacity of limited spaces.

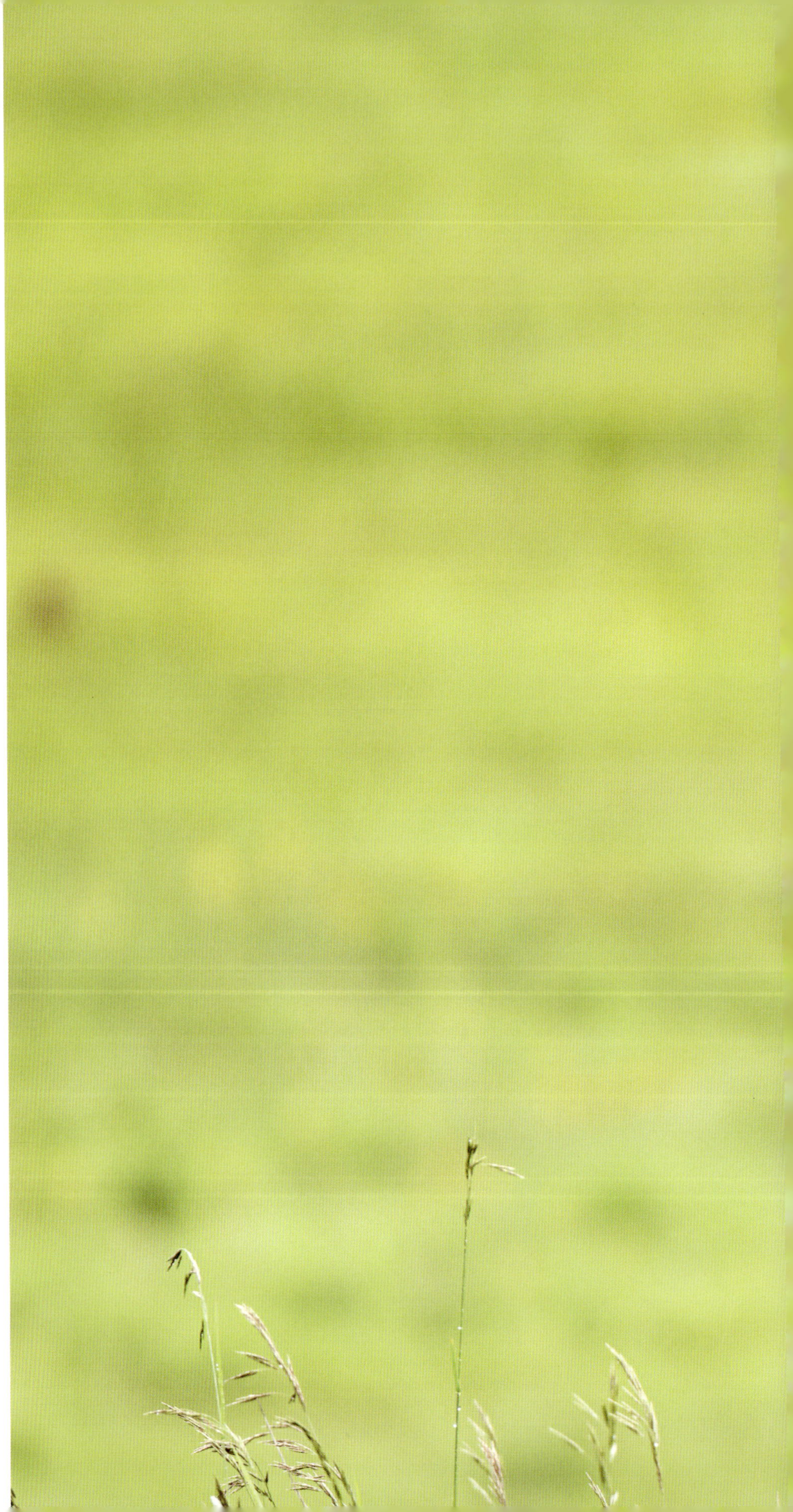

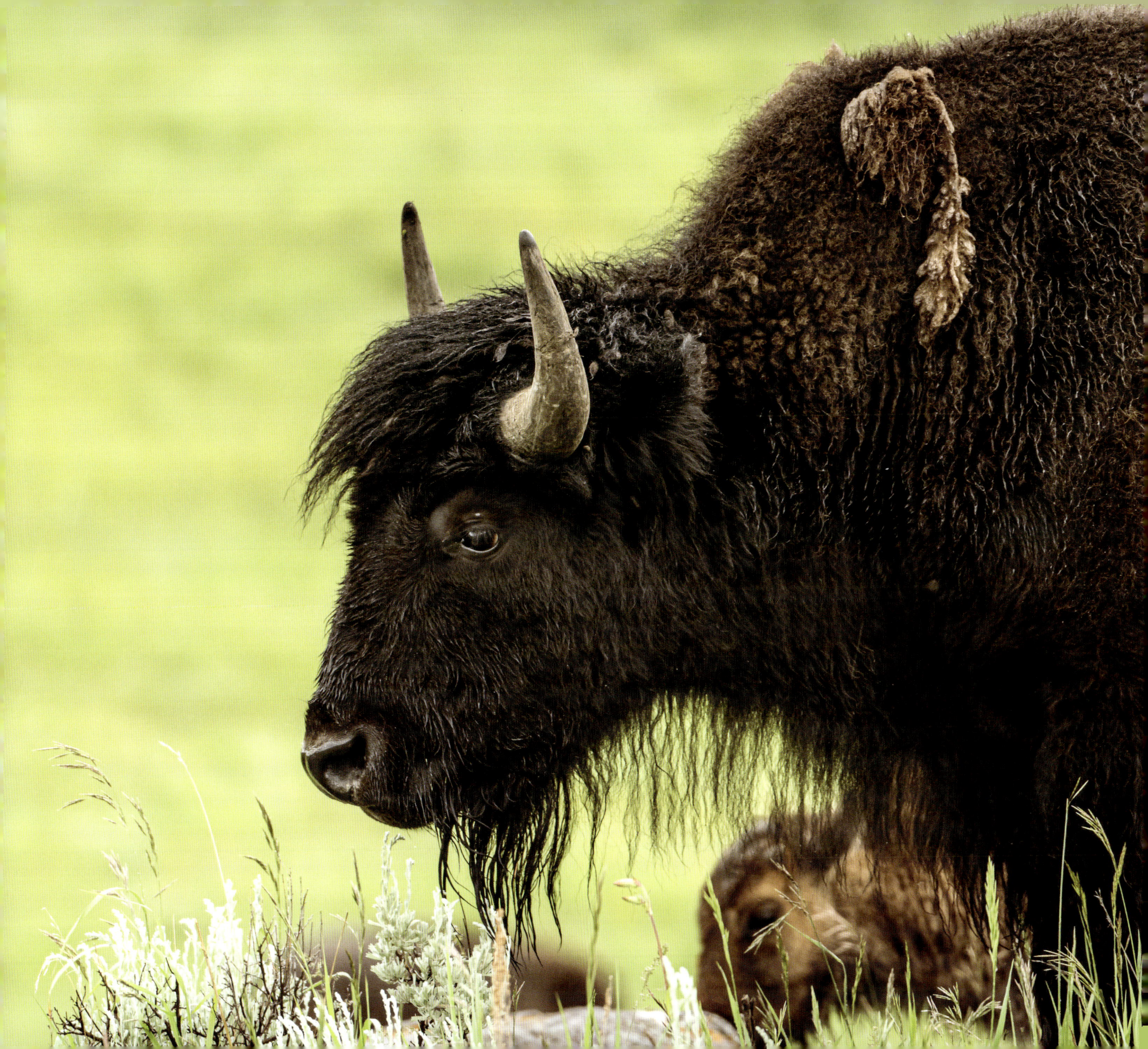

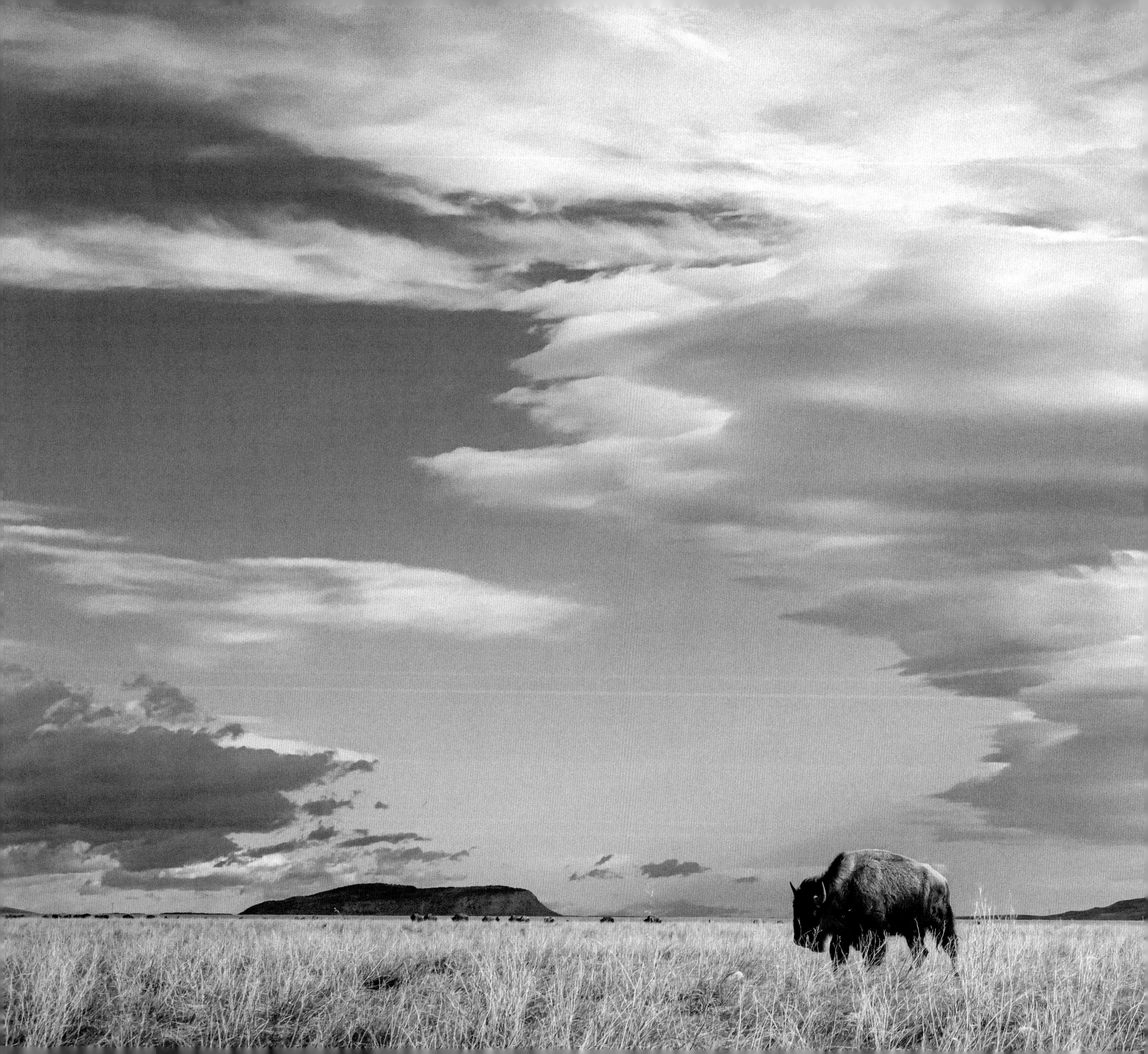

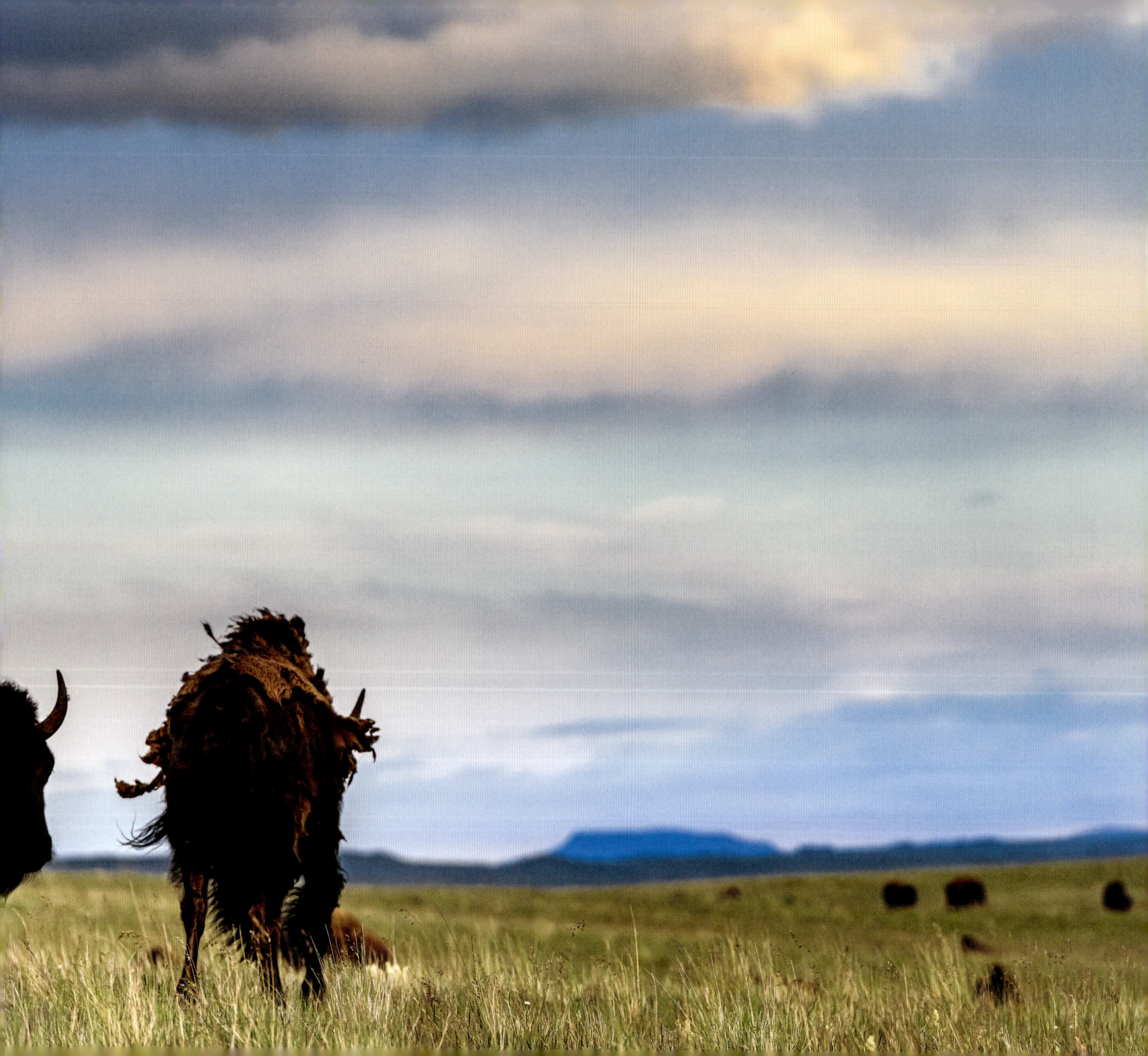

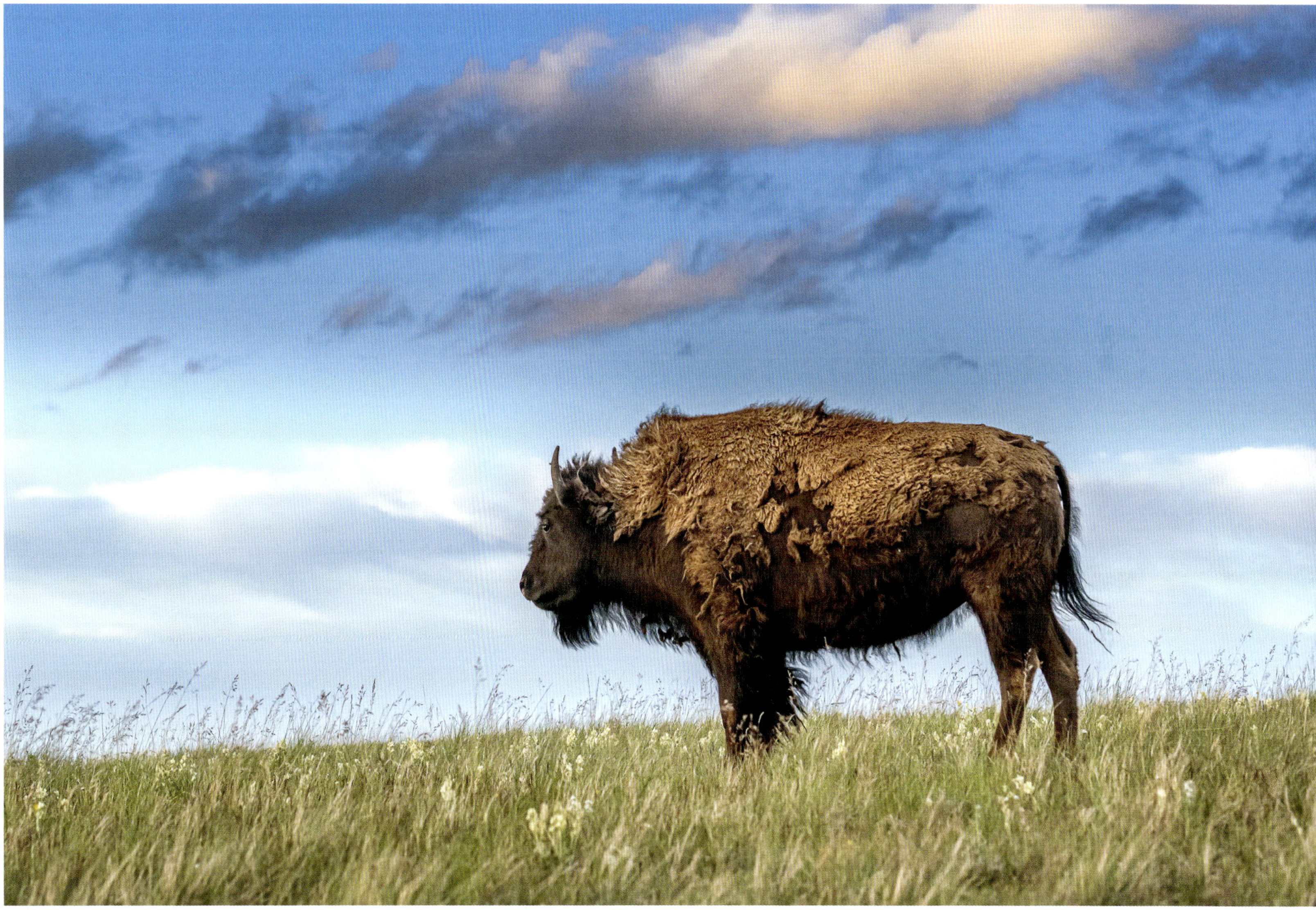

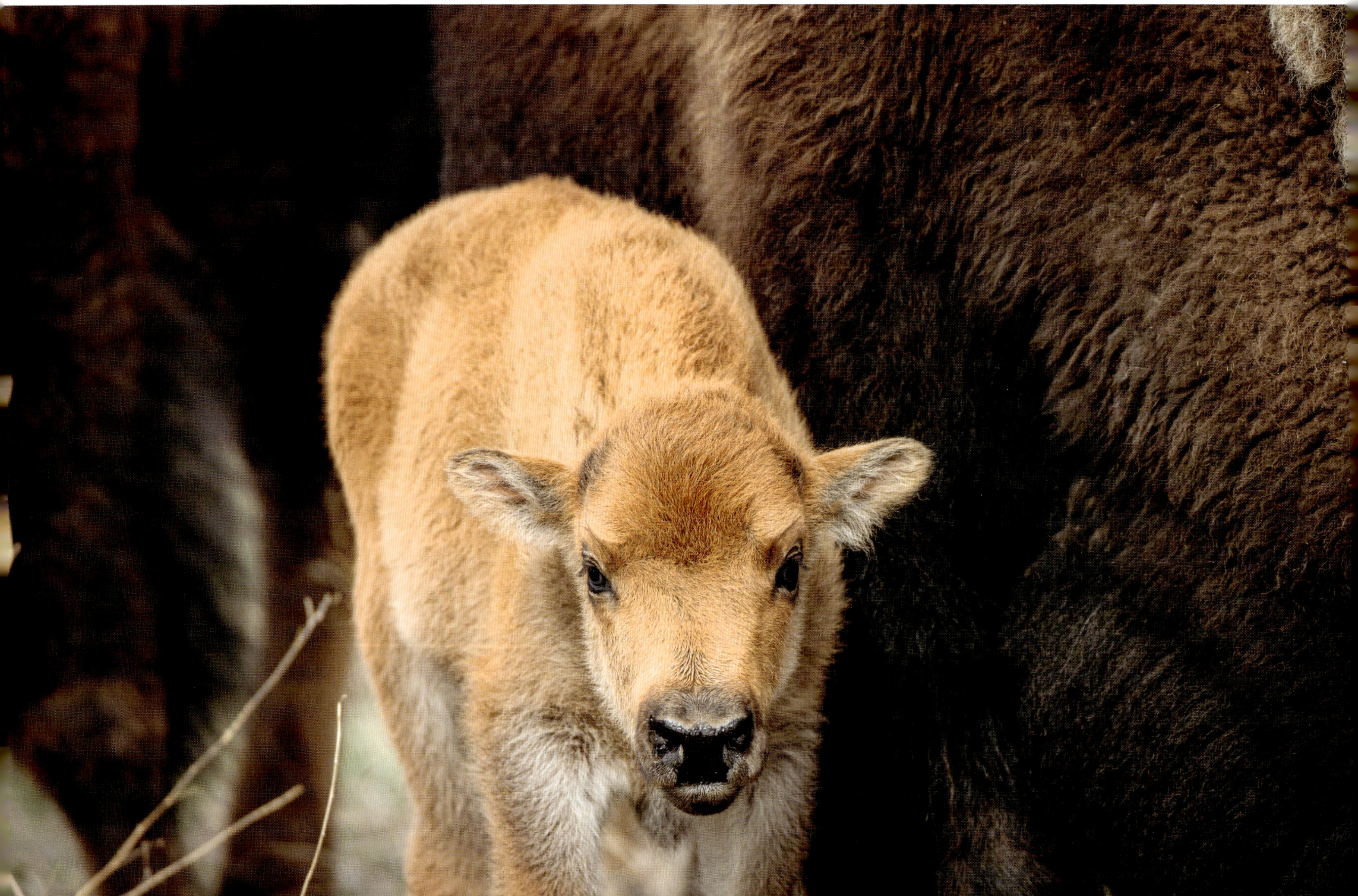

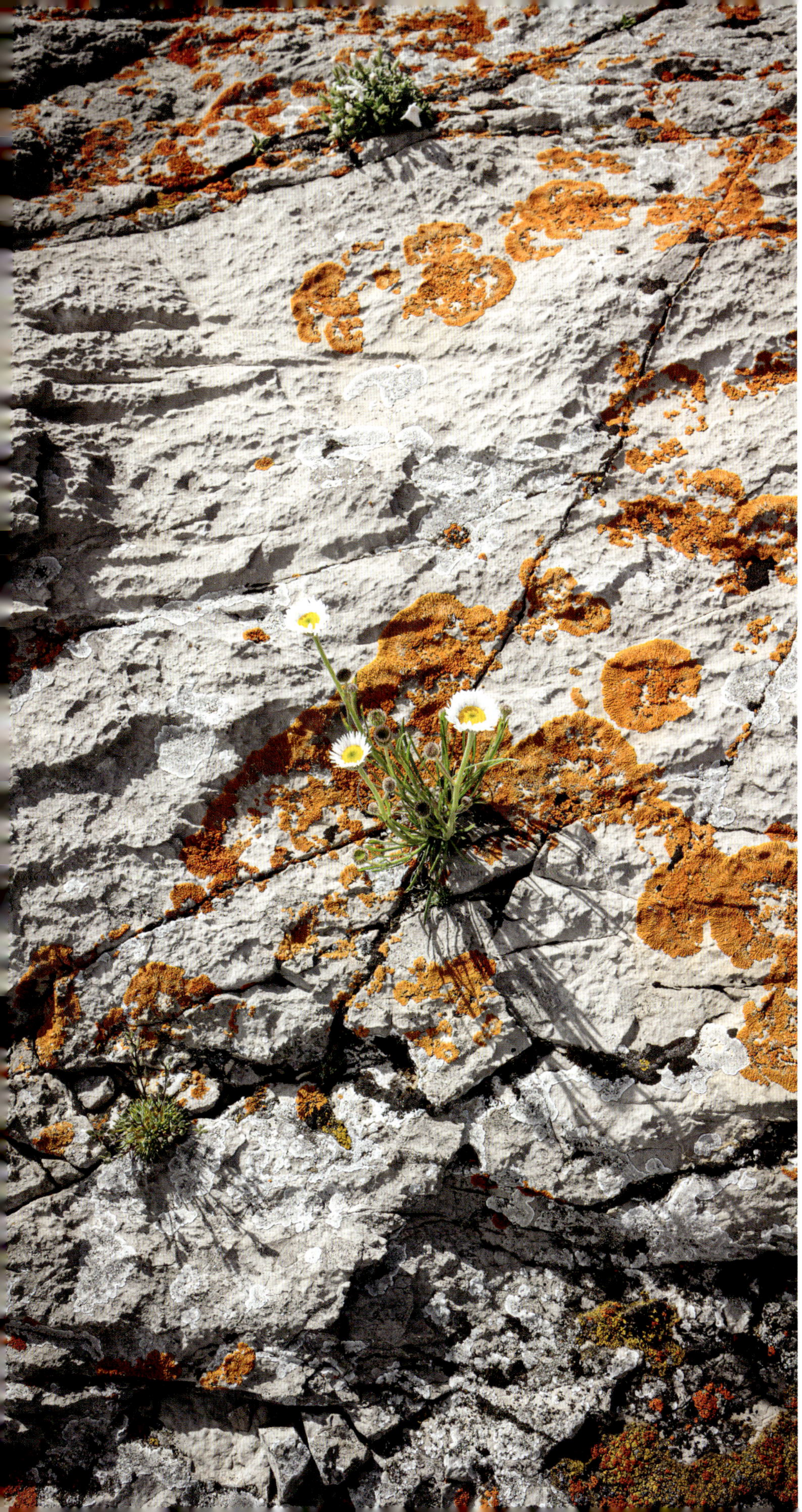

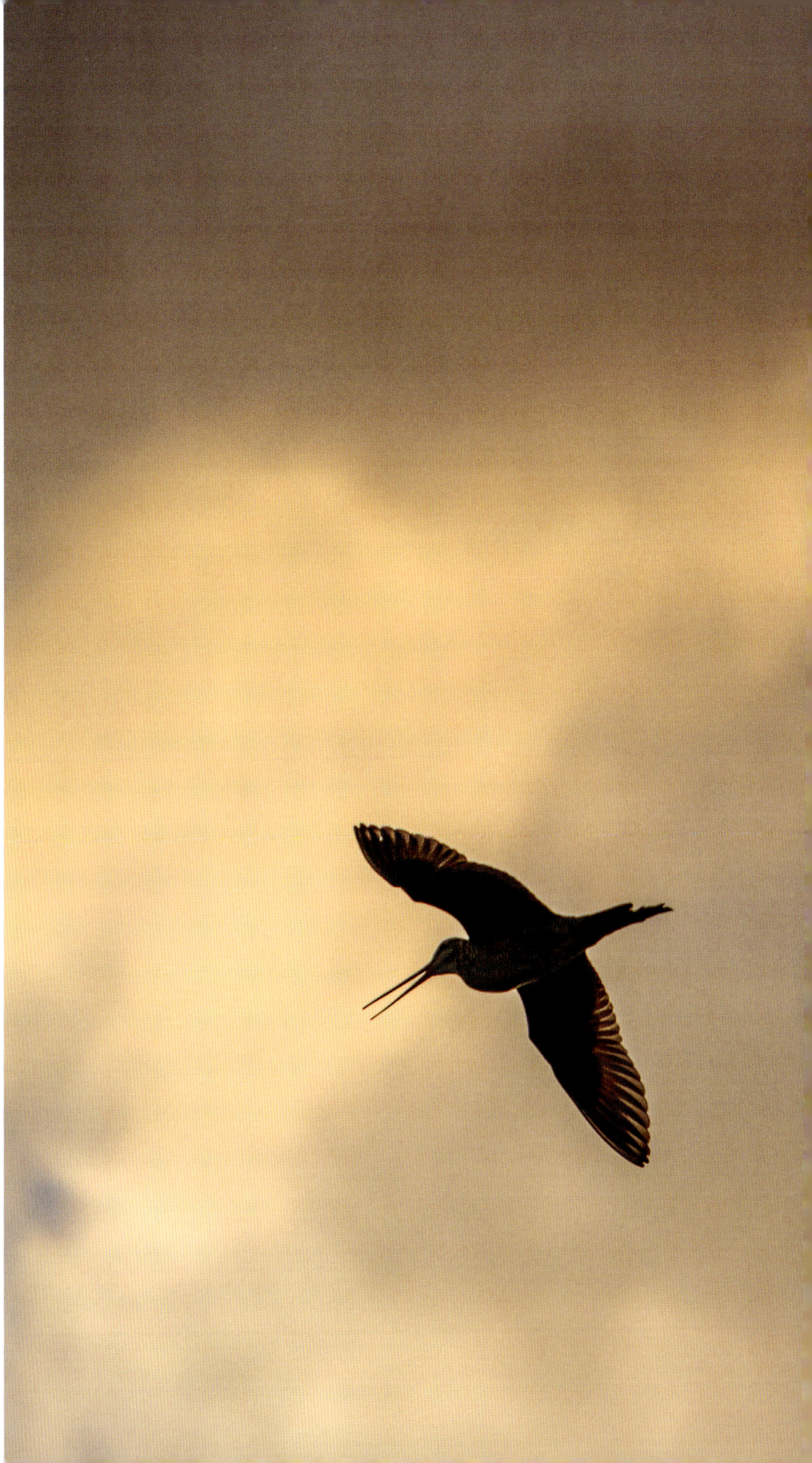

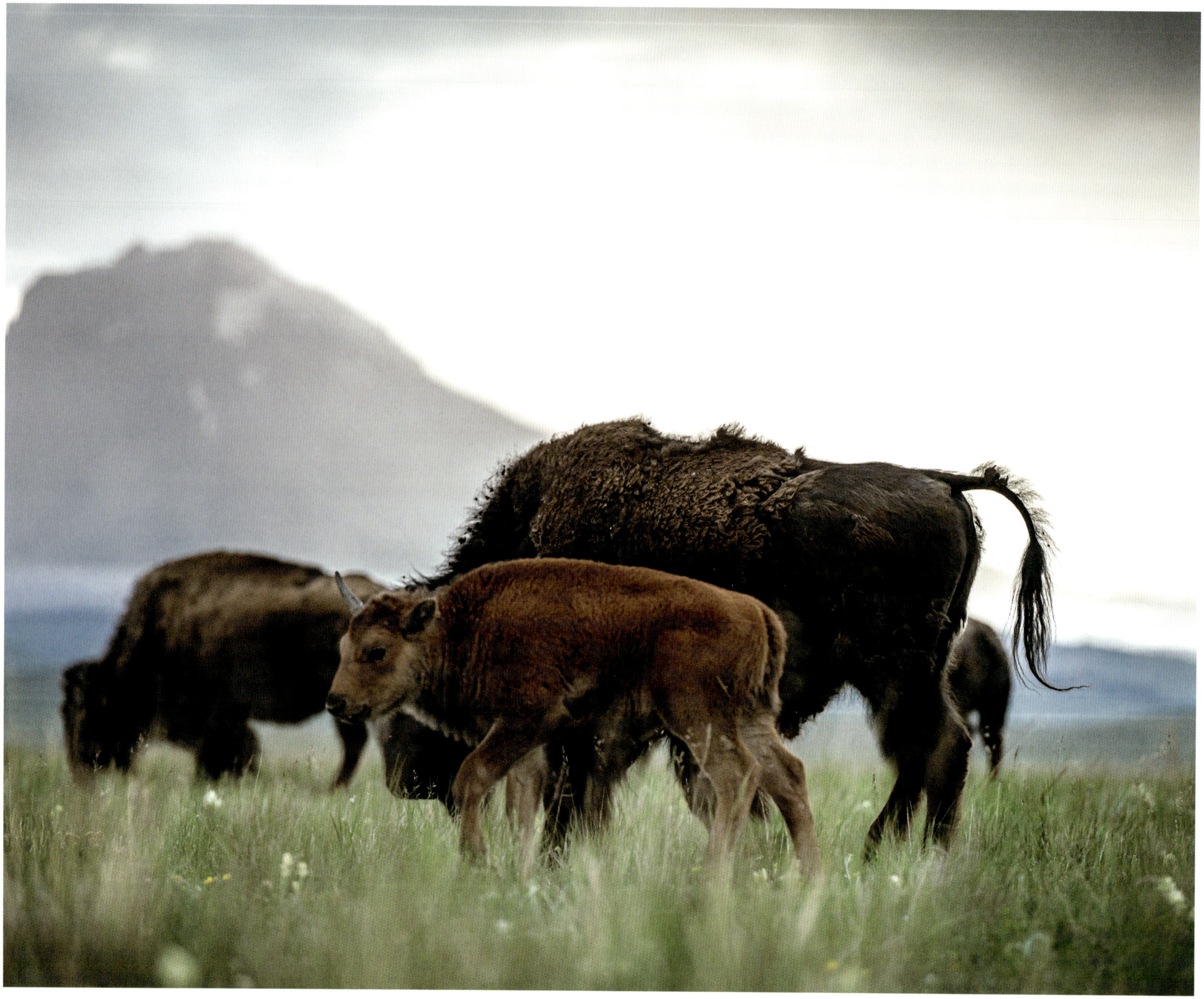

Ecosystem Engineers

Of the roughly 500,000 bison on the continent today, only about four percent of those are in conservation herds. The rest are raised for meat, breeding, and controlled hunting (with management methods ranging from near-wild to cattle-like). But whether the animals are there to roam and breed or are being raised for meat, they do their part for the land, and the people.

Dennis Jorgensen is the World Wildlife Fund's Bison Initiative Coordinator and Program Officer for the northern Great Plains; he works with tribal partners, national parks, and other entities toward a goal of establishing at least five herds of 1,000 animals each on tribal lands by 2025. "My hope in working with tribes to restore bison is that it helps restore the economy for them, particularly in the isolated areas where there are not a lot of economic opportunities," Jorgensen says. "They're bringing back bison in a way that can provide connection and identity. Creating opportunity for sovereignty and [for them to be] self-sustaining is a happy product of helping the tribes restore large herds. I think a lot of rural and more isolated reservations don't have as many ventures lifting their communities. That's where we see bison creating opportunities."

The Bison Initiative's first goal is to establish viable conservation herds. Geneticists suggest 1,000 animals are needed to provide enough genetic diversity for a healthy, viable population, but most conservation herds today number fewer than 400 animals; many have fewer than fifty. The Initiative starts with a survey of community members to ask what they value, what they need, and to what they aspire. "They often say they want access to buffalo just to experience and be around them," says Jorgensen. "They might want access to meat; they occasionally want them for cultural items such as hides."

Ultimately, he adds, "Bison were gone for 130 years but are still central to their ceremony. For people who are traditional, it's extremely meaningful. I don't just want them to know buffalo are back, I want them to feel buffalo are back and part of their lives."

The Quapaw tribe runs a herd of 250 bison on 1,000 acres of its land in the northeast corner of Oklahoma. Shawn Henderson has been the tribe's herd manager since 2017. For the Quapaw people, he says, the purpose of having a bison herd, "is first and foremost cultural." Buffalo meat and parts are made available for dances and powwows, while meat is distributed to elder centers and daycare facilities.

The Quapaw take a holistic approach to their business endeavors. While they do feed corn to their bison, as they do to their cattle, the corn they feed was produced by the tribe on its own land. The tribe invested in a state-of the-art processing plant in order to control the process and ensure quality, and it hired Temple Grandin, a prominent proponent for the humane treatment of livestock, to design a process that's as trauma- and stress-free as possible. The Quapaw tribe has created its own brand of bison jerky, selling it through its own network of convenience stores and casinos, as well as the Quapaw Mercantile, its downtown store that also has an online presence.

Although the bison are managed as livestock, Henderson says, "They're wild animals so we try to treat them as wild animals. We handle them as little as possible, but we've confined them so we have to manage them. Our duty is to manage them as best we can and still keep them wild."

When they harvest bison, every part of the animal is used. The tribe doesn't allow private harvest of bison but makes the hides, horns, and bones available to the people. Slowly things are coming full circle, Henderson says. "Back in the old days the people would make dice out of bison bones and play games with them. Some people have started to do that again. "

The benefits of raising bison go well beyond the economic and cultural, says WWF's Dennis Jorgensen. "By restoring bison, you're creating an umbrella for all the other occupants of the land."

Dr. Ian Thompson has witnessed this firsthand. The bison has been important to the Choctaw people throughout their history. Dr. Thompson serves as the Tribal Historic Preservation Officer for the Choctaw Nation of Oklahoma, helping preserve Choctaw cemeteries and historic sites in nine states, repatriating sacred objects, conducting research into tribal history, overseeing the nation's museums, and providing classes and presentations on Choctaw history and culture. He's also President of the Oklahoma Bison Association. "I grew up brain-tanning hides and making bowstrings from bison tendons," says Dr. Ian Thompson. "Raising buffalo has been a dream of mine since I was a teenager."

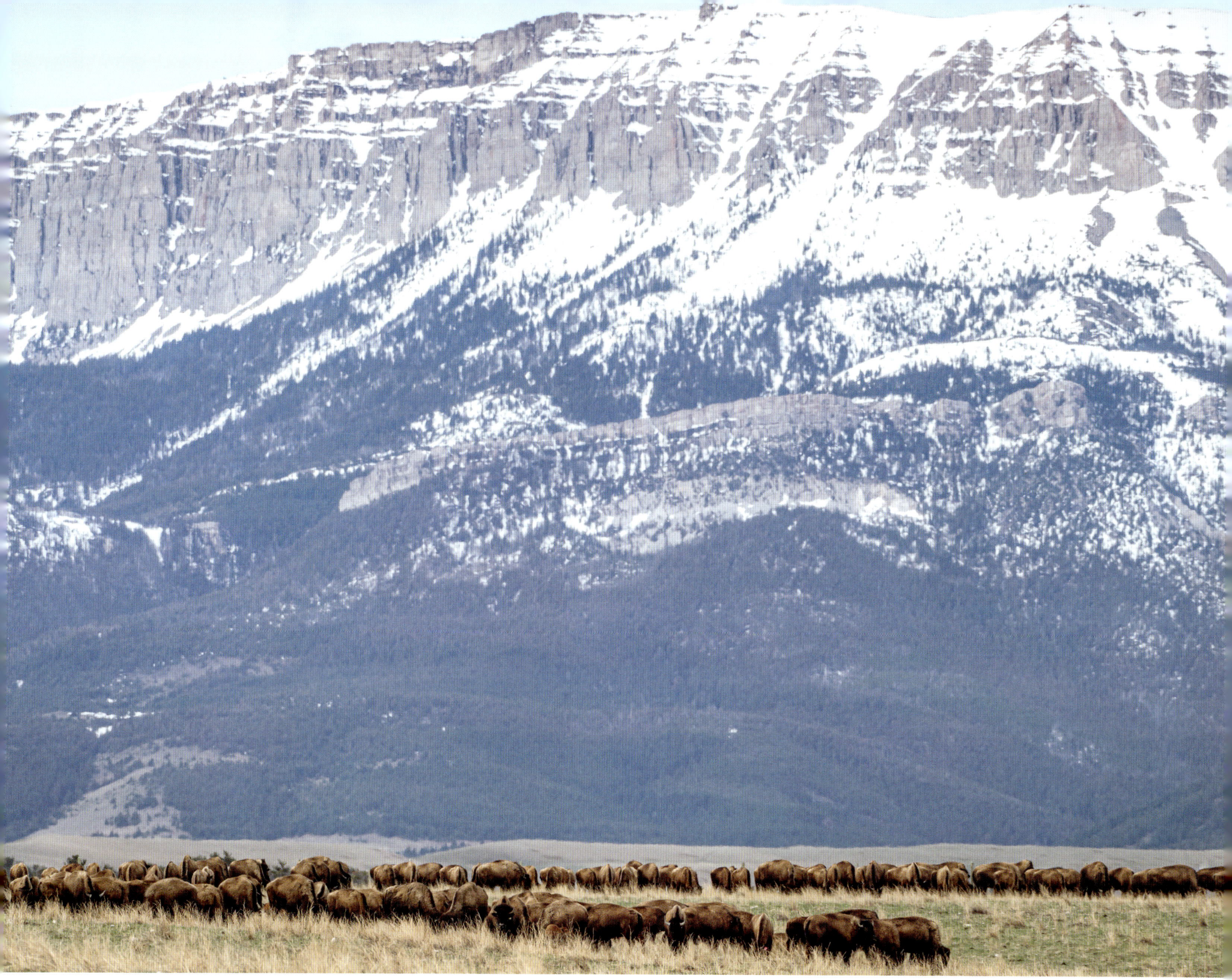

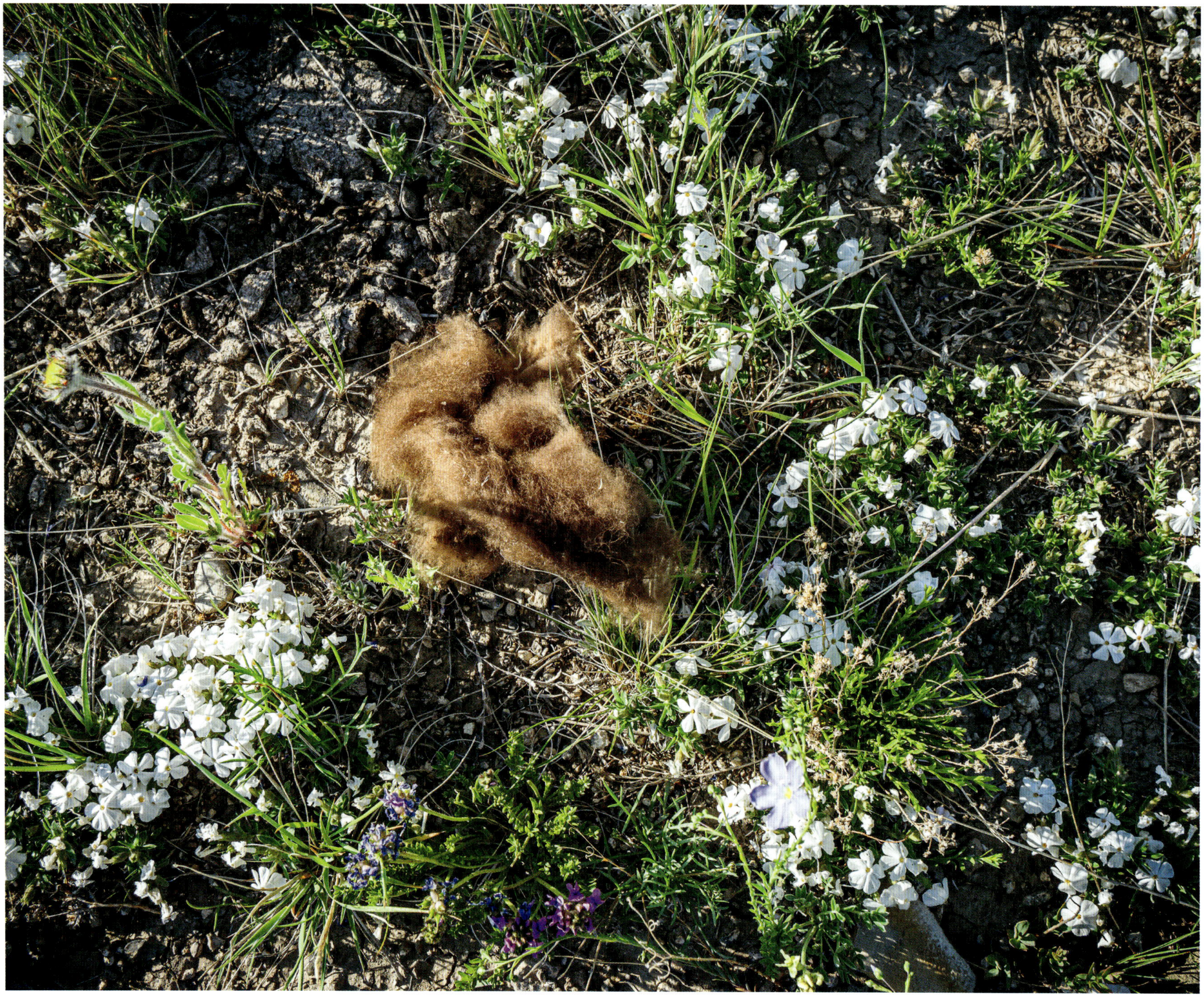

Six years ago, Thompson and his wife, Amy, founded their family farm, Nan Awaya Heritage Farmstead. The name refers to the tribe's creation story, in which the people were given rules for how to live in balance with the land and with other people. When the Thompsons bought their land, he recalls, "It was almost nothing but ragweed and goatweed. The land was so heavily grazed that certain keystone plant species were gone. It was out of balance." The farm's goal, he adds, is "to heal our land, support our community, and bring to light the traditional knowledge of our Choctaw ancestors. One tool we have for doing that is bison."

To encourage their community members' involvement and interest, the Farmstead hosts activities ranging from campouts and cultural plant walks to classes on Choctaw traditional arts. The Thompsons have also produced a book on traditional Choctaw foods and landscape management. Recently a group of traditional skills practitioners collected buffalo wool on the farmstead and elsewhere then made a traditional Choctaw skirt from the wool. That particular traditional expression, Thompson says, "hadn't been made in our community for 200 years."

Meanwhile, in just five years of careful management, the land is starting to bounce back. After letting it lie fallow for a year then putting bison on the land and managing their movements, the Thompsons now count 120 species of native prairie grasses and forbs. "When you think about indigenous cultures," says Thompson:

> *they're based on 15,000 years of living on the land. They contain timeless insights that can help modern people have a higher quality of life. Food is the most direct connection between people and the land. Societies managed to live on the land without degrading it for thousands of years. In 200 years of industrial agriculture we've obliterated it in huge chunks in the world. Our goal is to connect our farm with the broader cultural revitalization and hopefully get people interested in indigenous foods and in doing things that bring balance and sustainability to the community.*

The bison's symbiotic relationship to the prairie is key to its recovery, explains South Dakota writer and rancher Dan O'Brien in his book *Buffalo for the Broken Heart*. "From the beginning," the former wildlife biologist wrote, "I saw that the welfare of all the Great Plains' wildlife species depended ultimately on grass. The health of the grass dictated the health of the birds I was studying and, ultimately, all the animals on the prairie food chain."

"Great Plains plants," he continued, "evolved to be healthiest when herds trample and graze them intensively and then move on. The large native herbivore, the buffalo, evolved to stick close together, initially as a form of protection from predators, and so give the grasses of the Great Plains exactly what they need to thrive."

Grasslands that thrive carry an invisible benefit for all: carbon sequestration. The green vegetation in a healthy buffalo pasture represents only the visible portion of the plant life while the roots beneath the ground are perennial and live through the winter. That plant matter draws carbon dioxide from the air, breaks the carbon atom from the oxygen atoms, sends the oxygen back into the atmosphere, and holds the carbon in the ground. As long as the carbon remains sequestered there, it can't exacerbate global warming.

When ranching cattle, O'Brien's approach to land management was to divide his ranch into smaller pastures and practice rotational grazing in an effort to force the cattle not to graze their favorites down to the ground. But buffalo, he explains, "evolved on the northern plains at the same time that the plant species they depended upon evolved. Their existence is mutually beneficial. Whatever pressure buffalo put on the native plant species, those species have had eons to evolve strategies to survive and flourish."

O'Brien took the leap in 1988 with his own modest ranch and a group of thirteen orphaned bison from a roundup at the 777 Bison Ranch (another key player in the bison game, having been humanely raising grass-fed bison for more than three decades). Despite many sleepless nights spent calculating the cost of bison fencing and worrying about risk, O'Brien's enterprise, Wild Idea Buffalo Company, has thrived. Similarly, his land, previously battered under the effects of cattle grazing and drought, has bounced back. Wild Idea now manages its own bison herds on several ranches on the North Plains and has forged partnerships with another half-dozen bison ranches in four states, helping bison thrive on as many as 200,000 acres of grasslands. Wild Idea bison are kept as close to wild as possible, handled only once a year and, when harvested, taken down in the field by one extremely precise shooter from a long distance

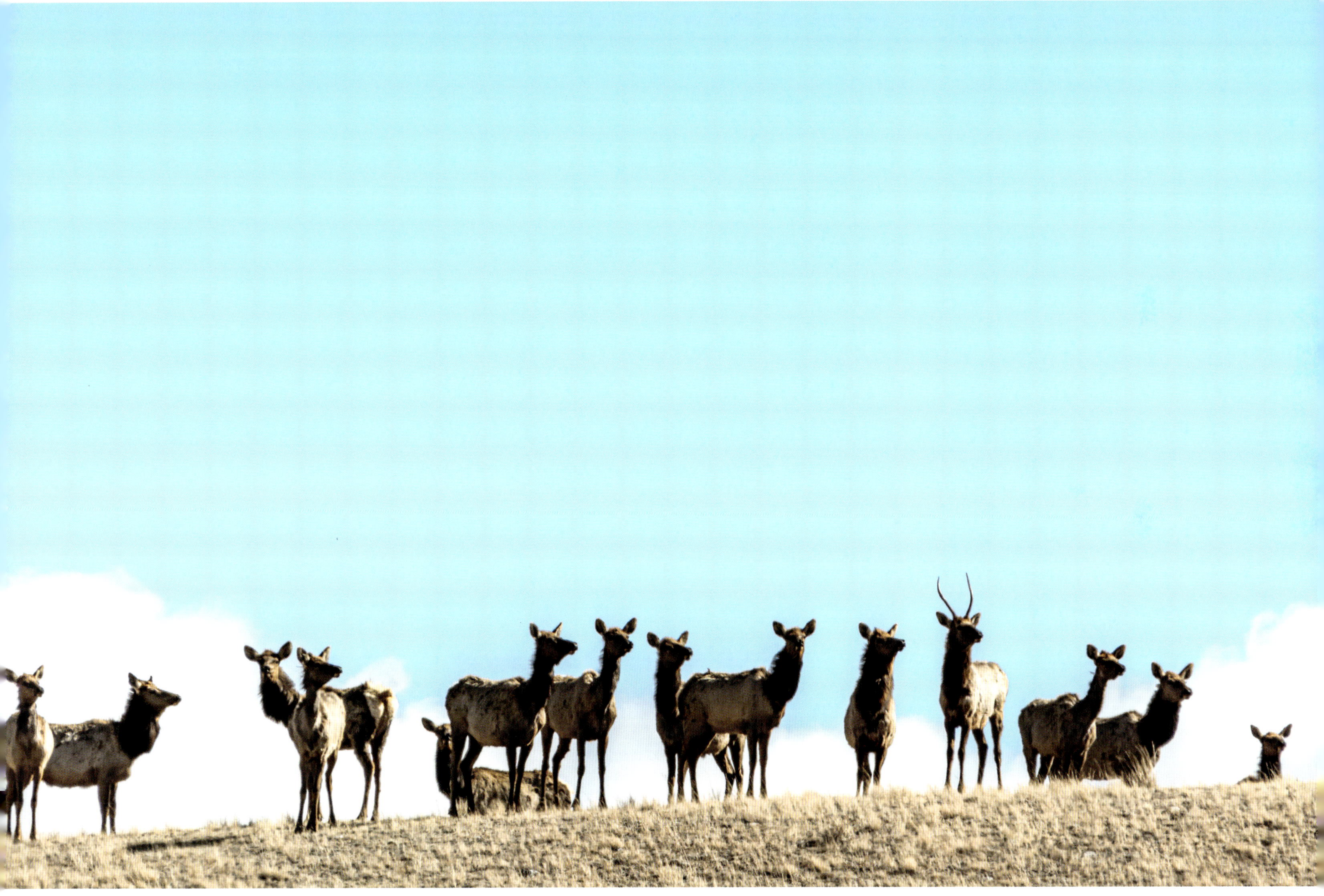

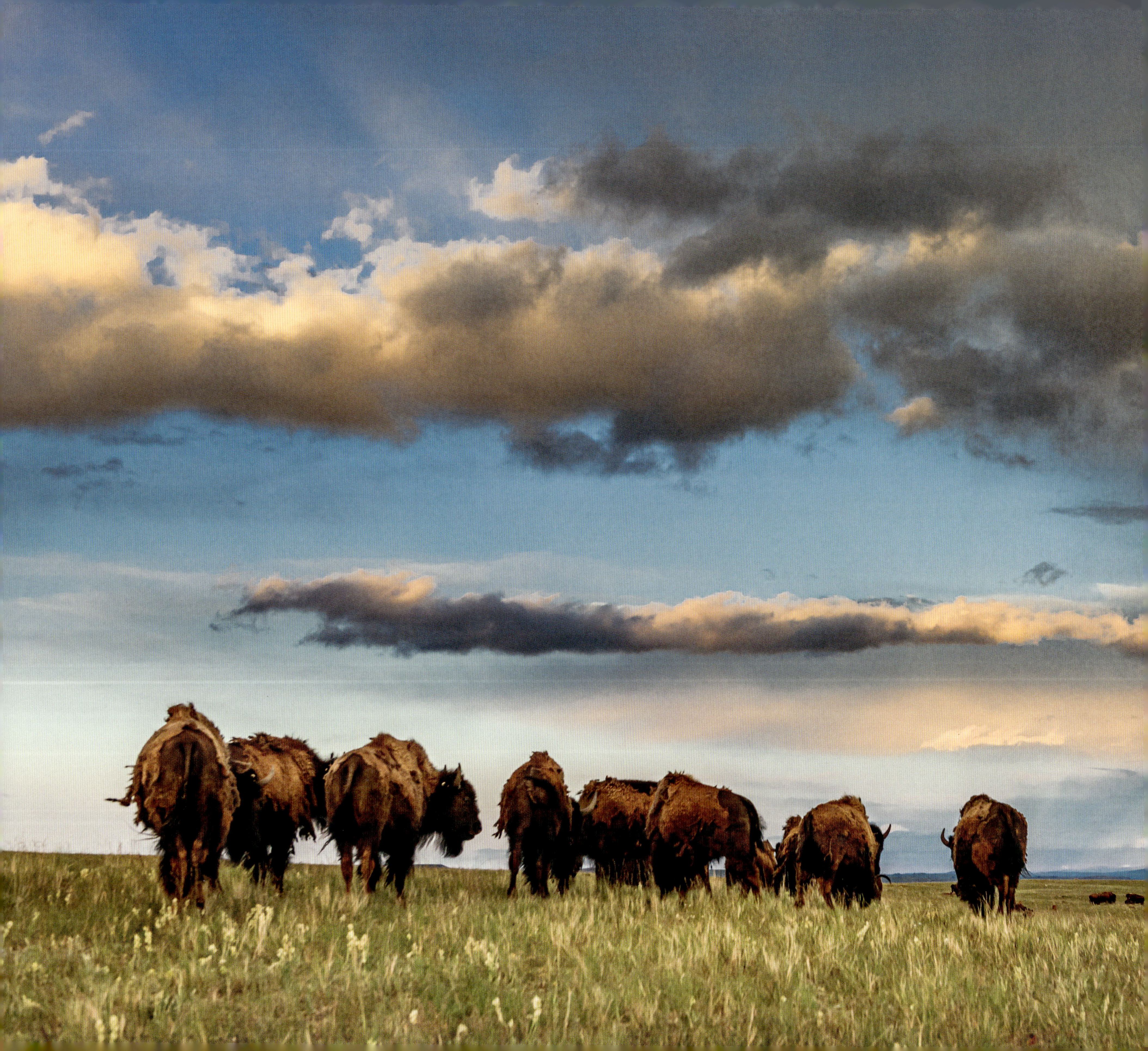

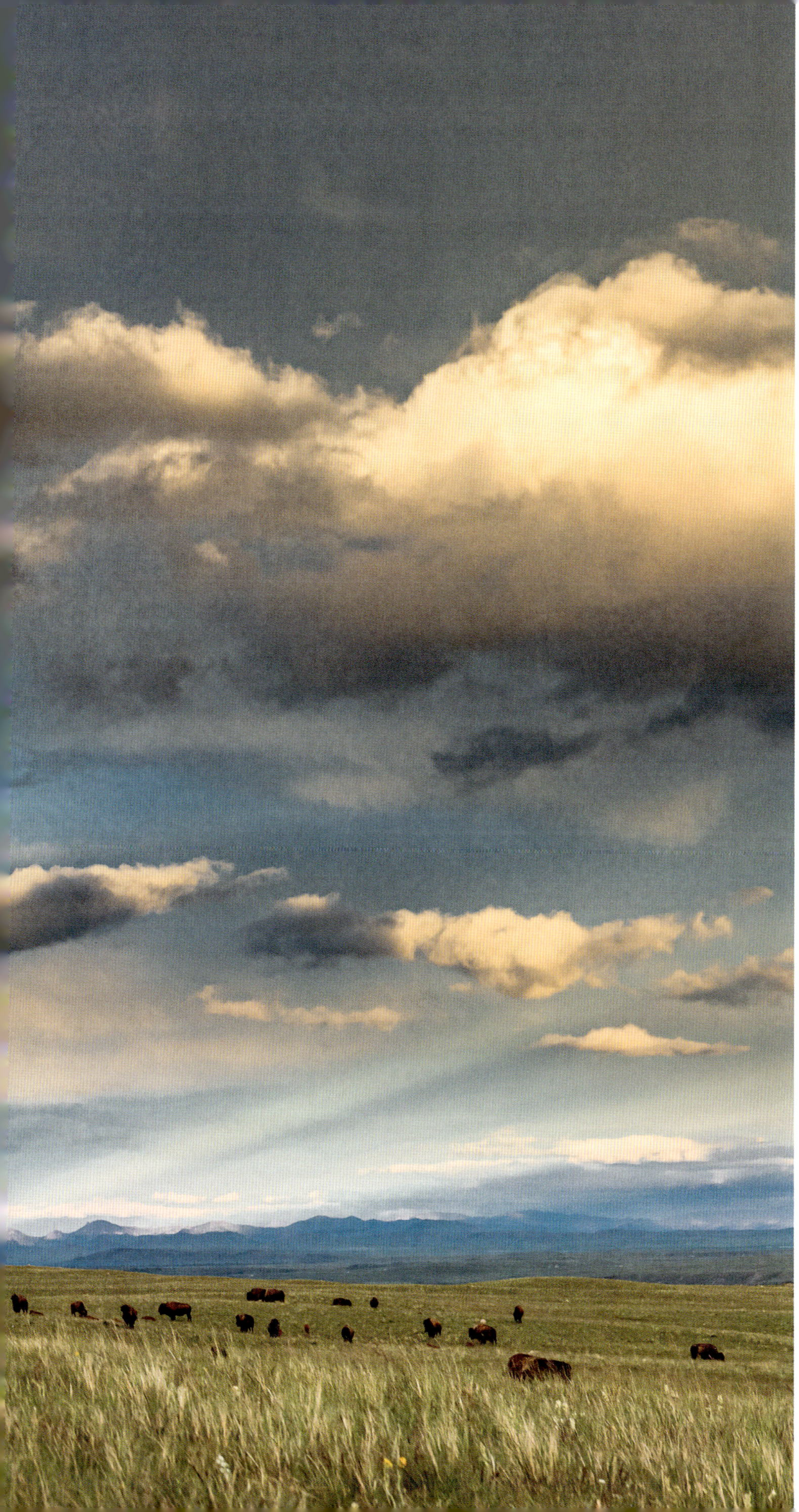

away so as not to upset the herd. (For some reason, bison tend not to react when one of their herd mates suddenly falls down.) Hormone- and antibiotic-free, grass-fed, grass-finished, and humanely harvested, the meat is about as healthy and close to wild as possible. Meanwhile the prairie is responding.

O'Brien's son-in-law, Colton Jones, wears many hats at Wild Idea. He's in charge of procurement and the mobile abattoir; with his wife, Jilian, he runs the Cheyenne River Ranch's 1,000-head herd on 50,000 acres near Badlands National Park. "We want to raise bison in the most favorable condition for bison, in big open country," Jones says. "We use them as a rehabilitation mechanism after the land had been used for industrial agriculture in the '30s and '40s. We introduce grass species and we use bison as a tool."

Jones, who has a degree in biology, says in ten years on the Cheyenne River Ranch he's seen a flourishing of wildlife diversity and density. "It's been fun to see everything evolve and watch the land change beneath your feet," he says.

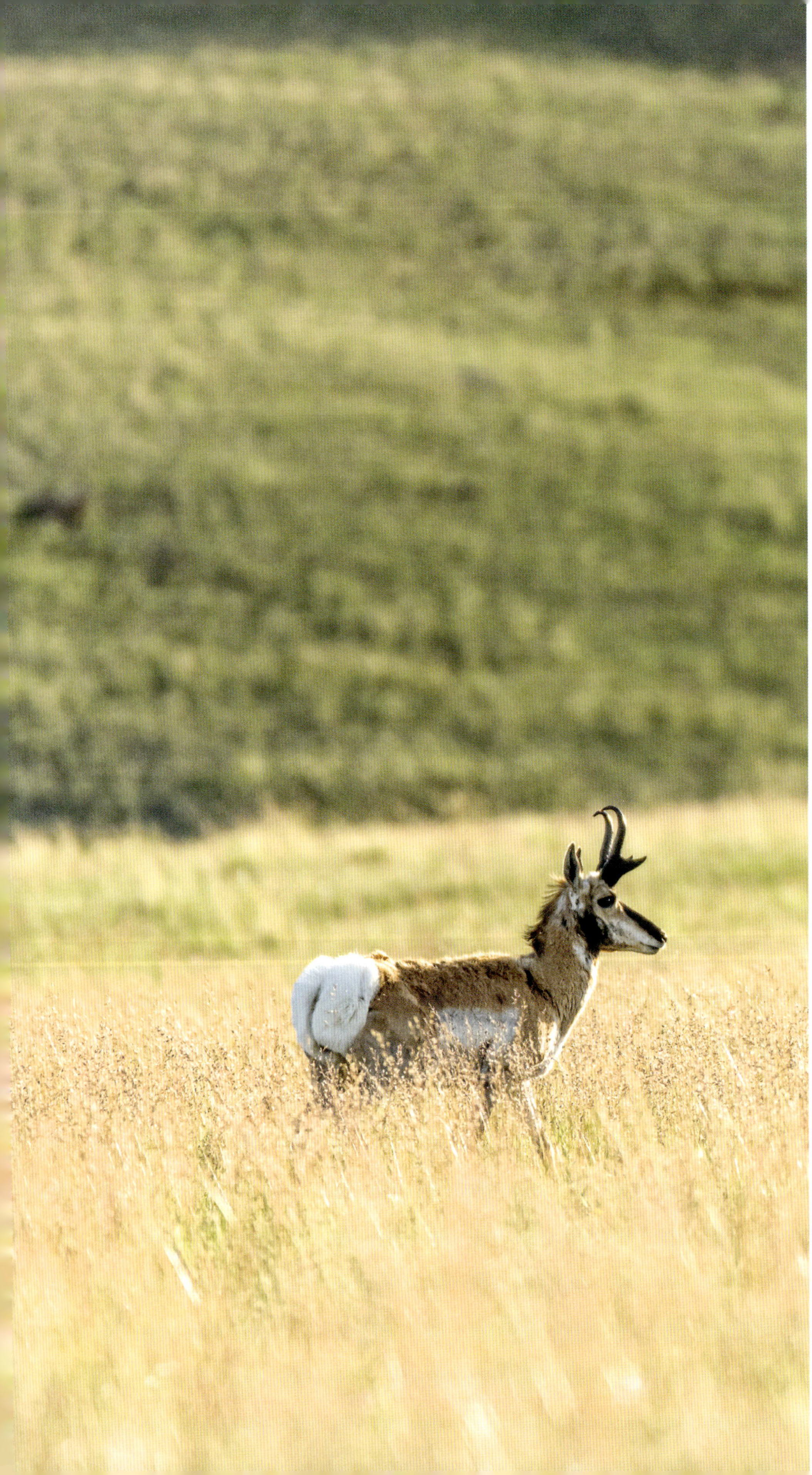

We never used to see magpies; now we see them daily. Short-tail grouse are much more plentiful. Plants species diversity has increased; there are more native grasses and in drought years they do better. And the population density of wildlife has increased {such as with} pronghorns and coyotes and prairie dogs, which keep the grass consistently shooting up with new growth. Wallows {made by the bison} provide a place for water to pool and you see more wildlife around those areas. Even if a puddle is there for just a couple of weeks, it provides a place for a duck to stop on a fly-through. A lot of that is pasture management, but I believe that putting a keystone species back on the land is like putting a missing piece back into something that was broken.

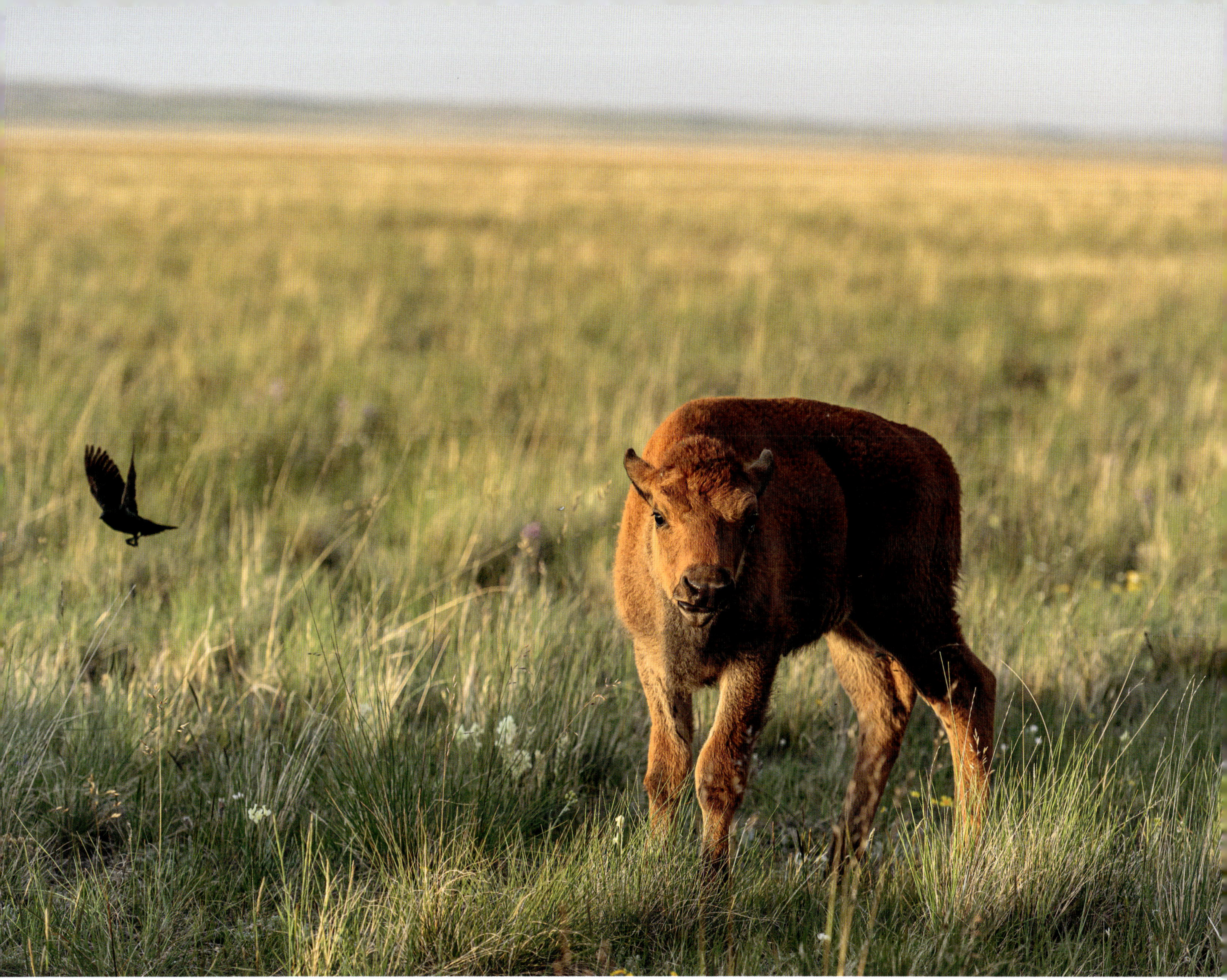

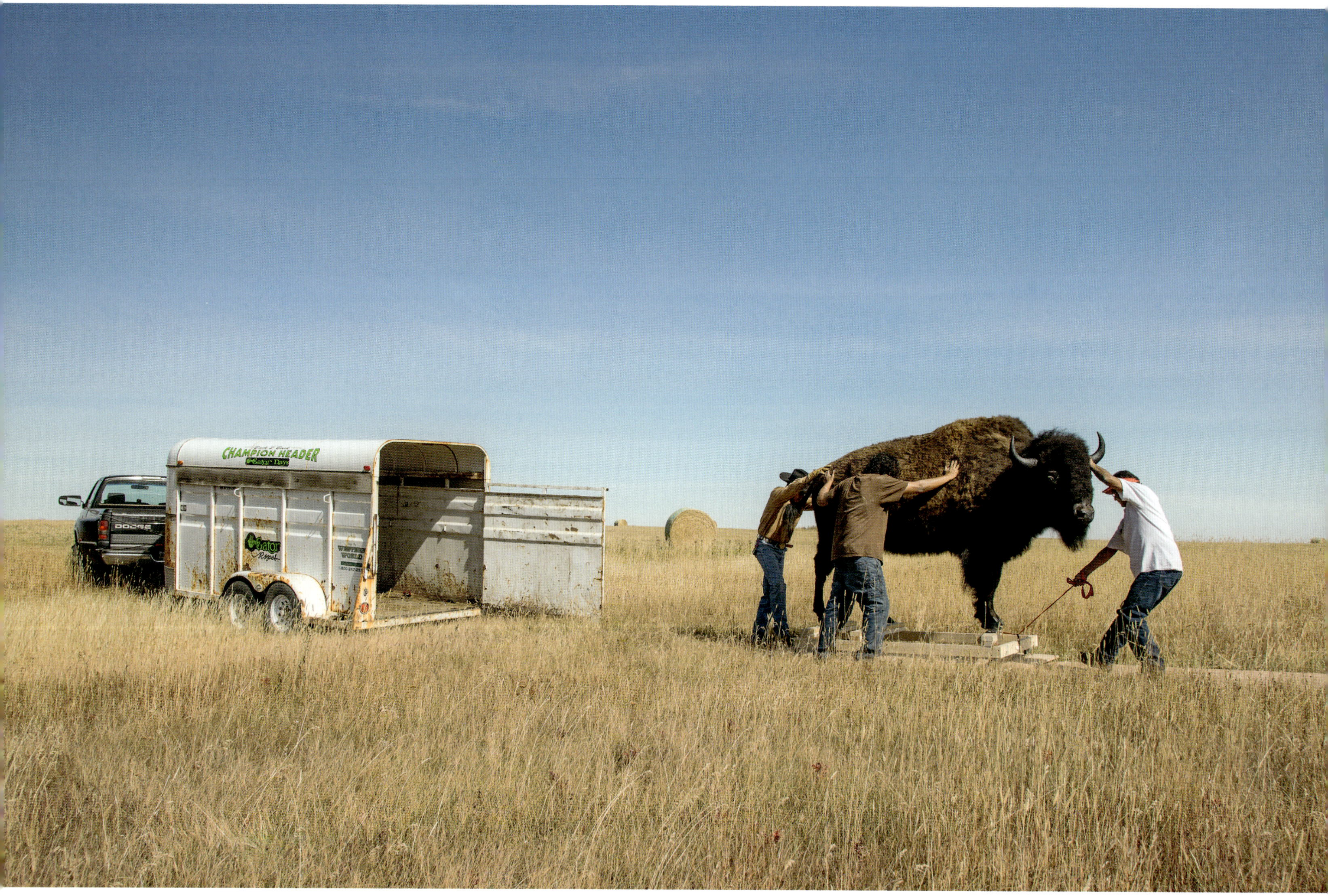
CHAMPION HEADER

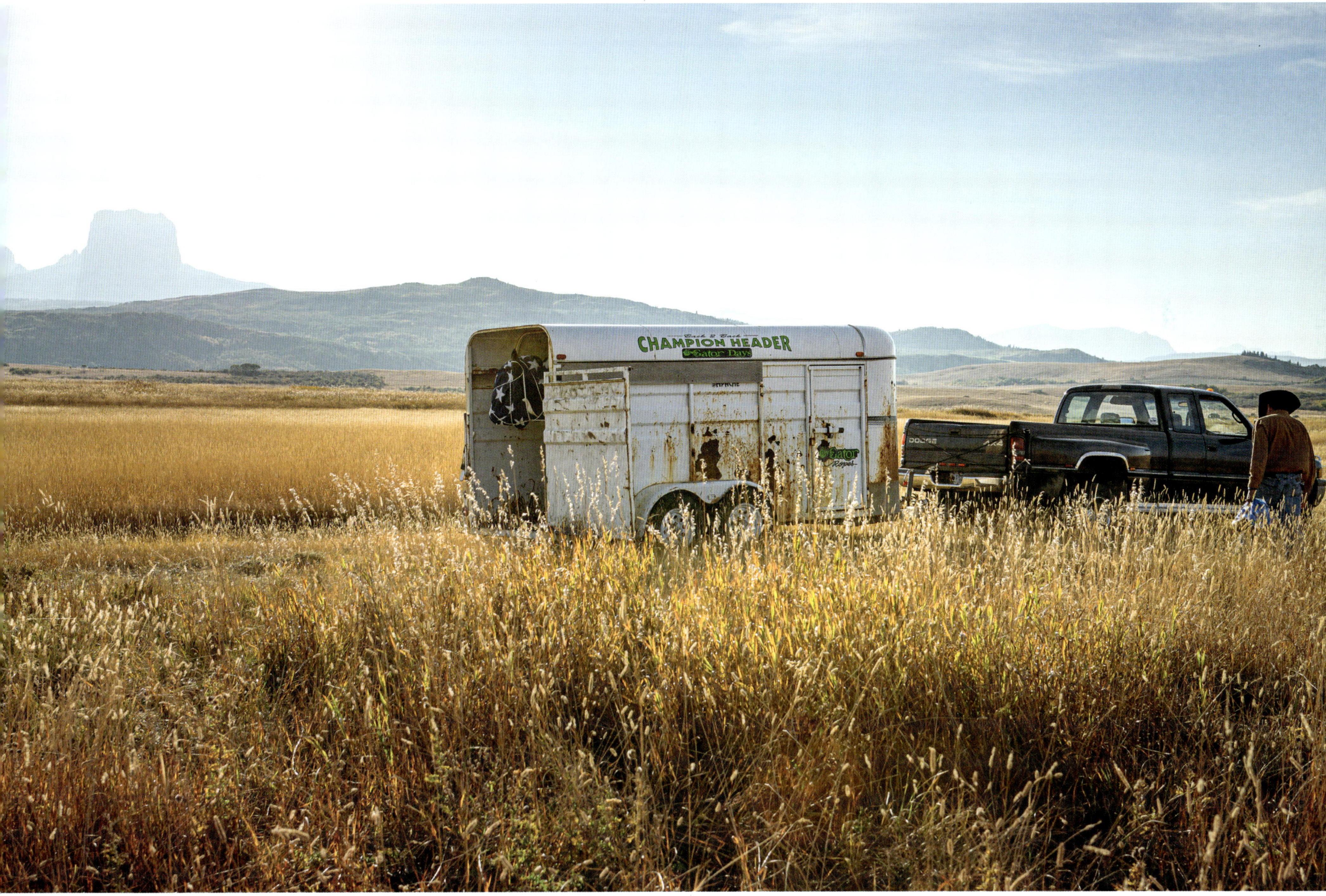
CHAMPION HEADER
Gator Days
DODGE

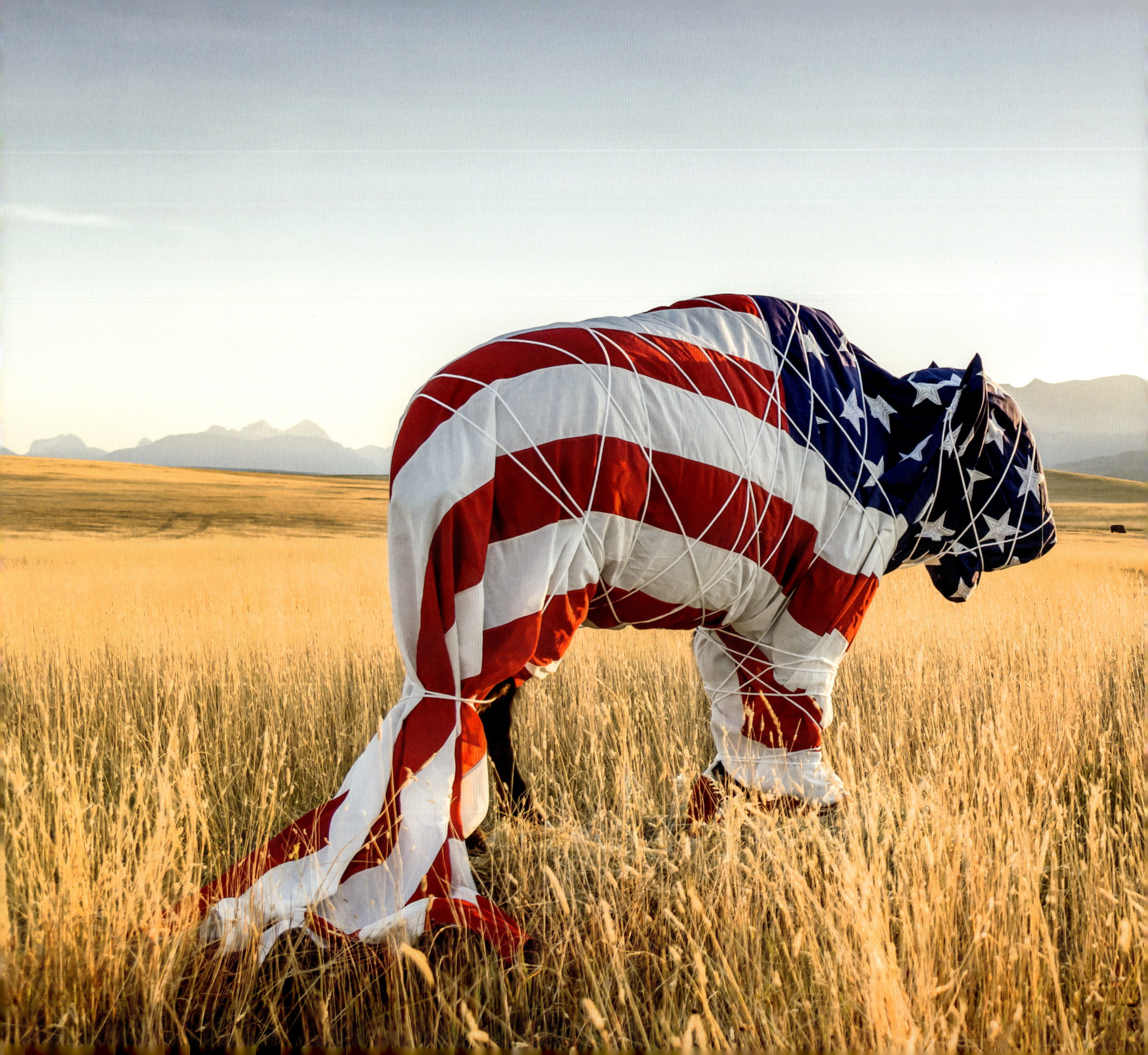

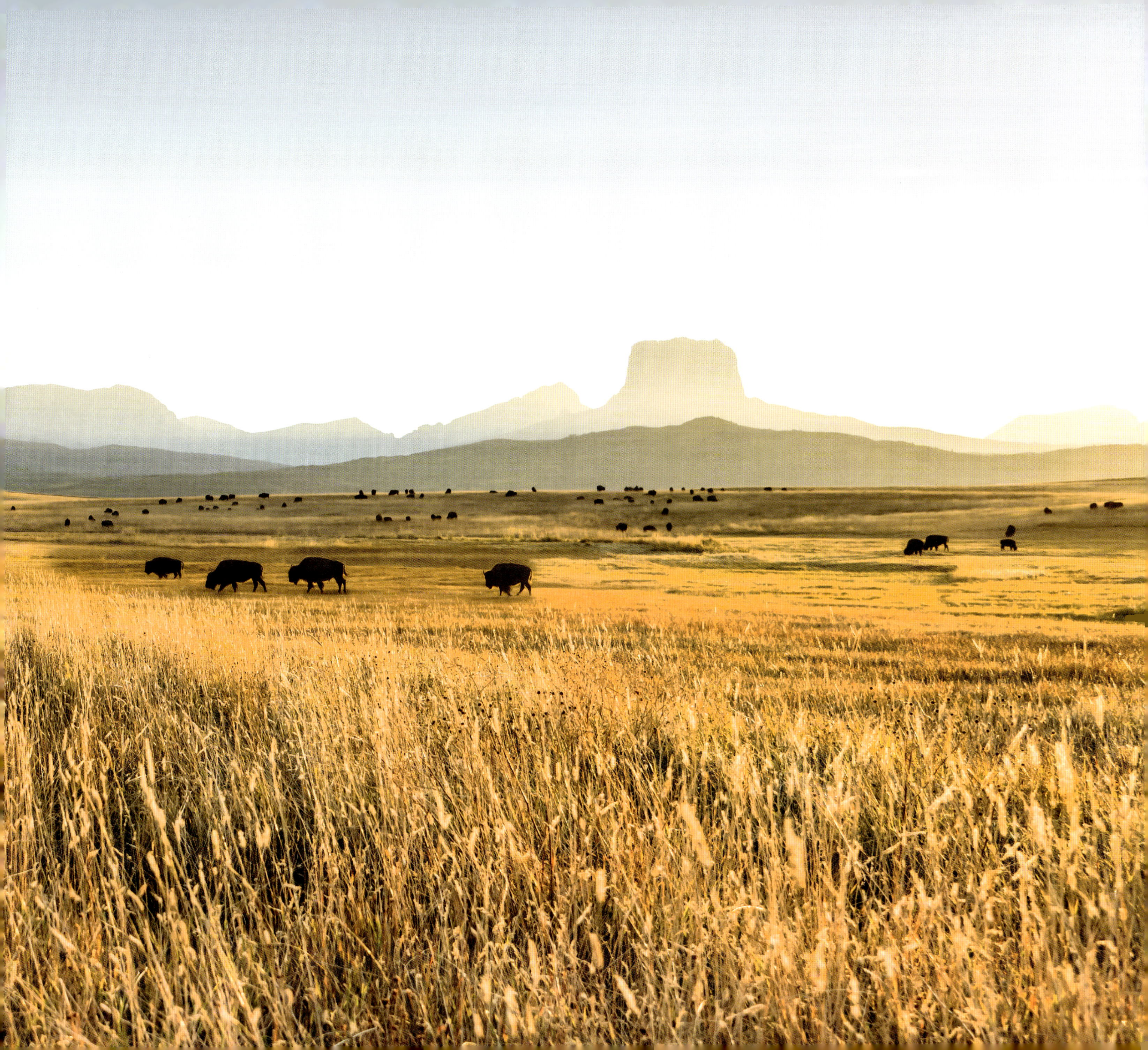

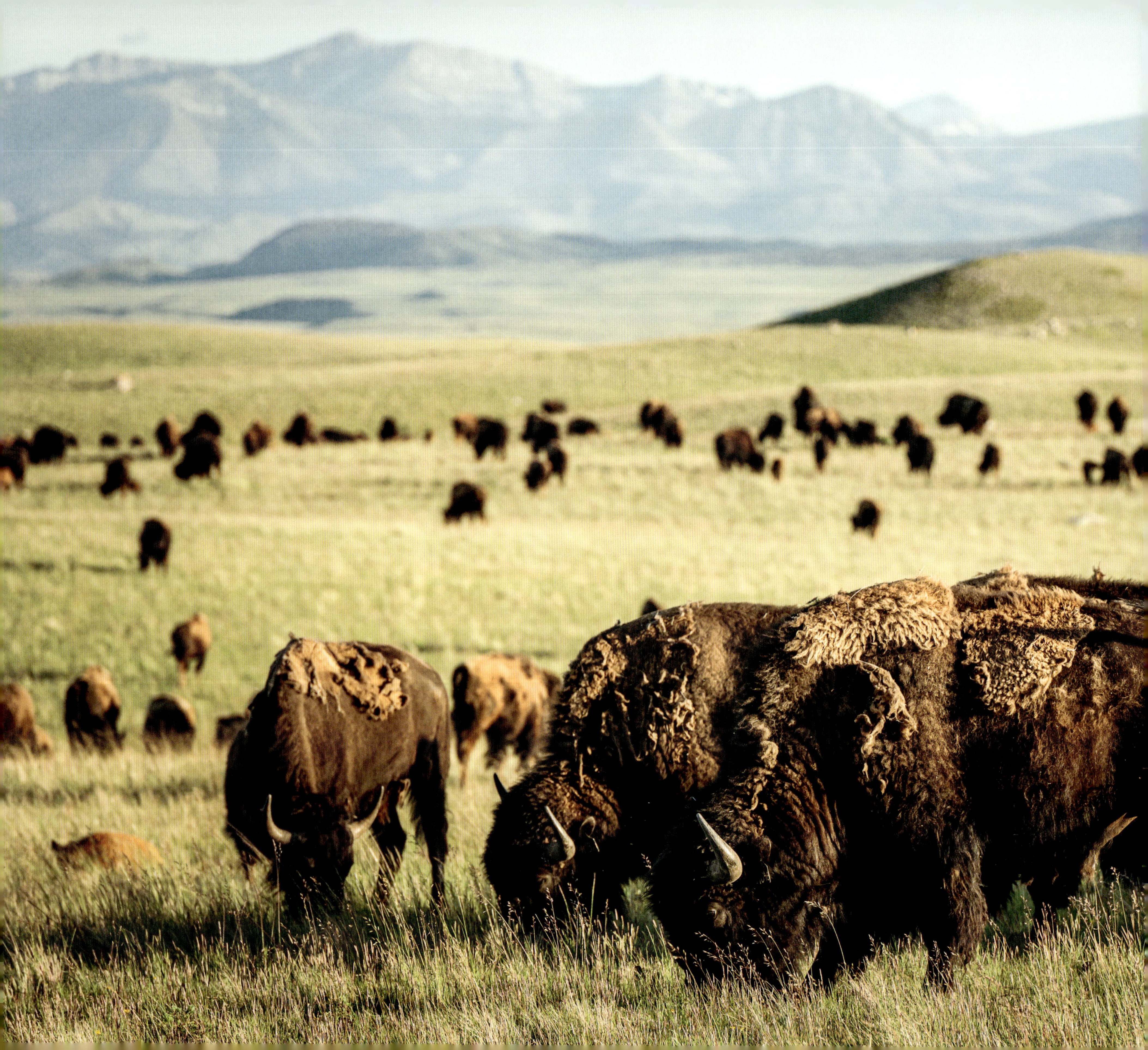

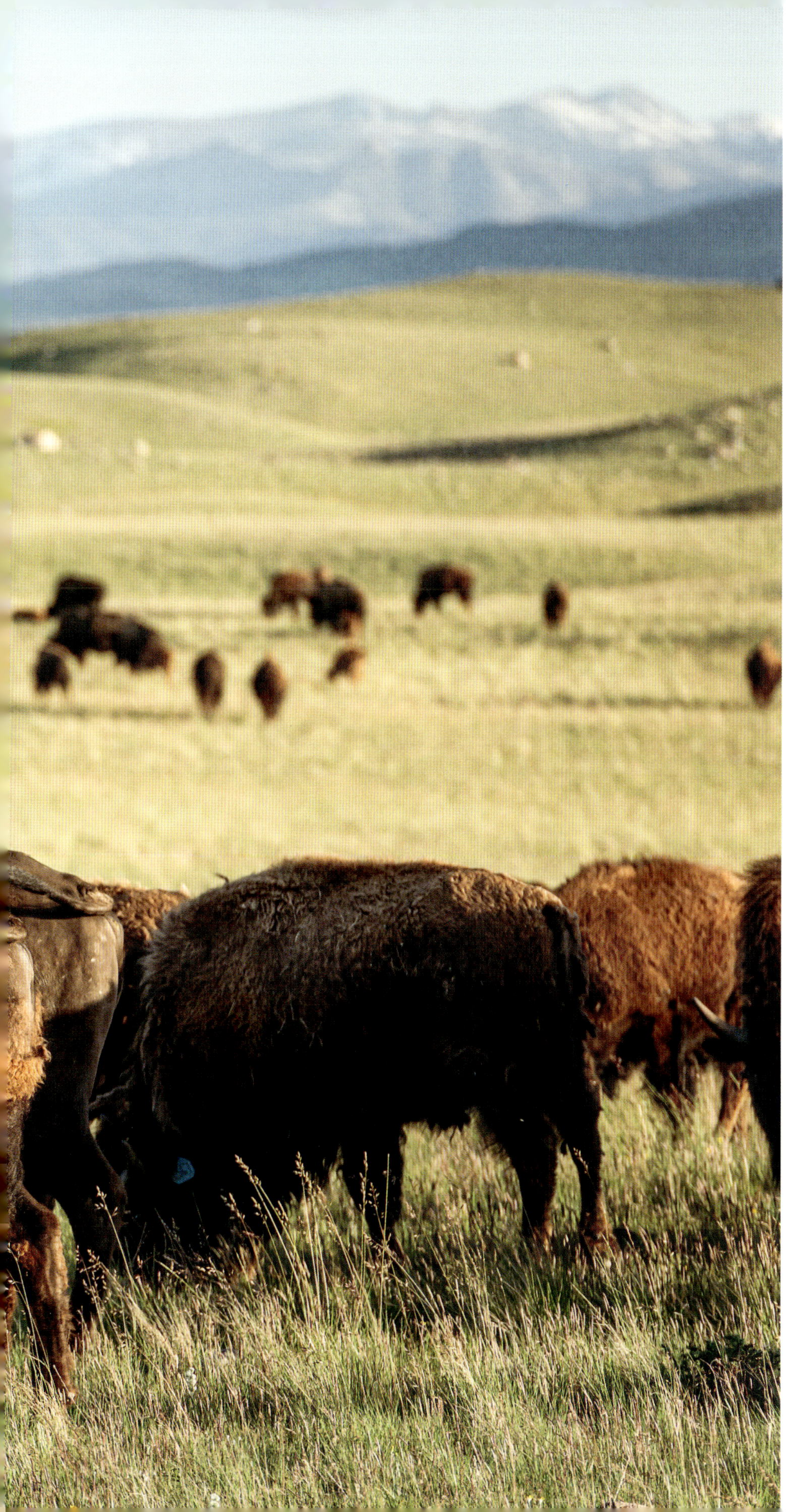

Excerpt from Thought Ways: Buffalo Jump Preparation

The buffalo that had been singled out to be driven over a buffalo jump were from just the edge of a pretty good-sized herd that came over Little Mountain from Teepee Lining Ridge (in present-day Montana). Most buffalo jumps were on the trail of the buffalo leading to the nearest good water, and we were camped and preparing a jump near where the buffalo crossed the Big Horn River to go to the many rivers that leave the Big Horn Mountains (locations in present-day Montana and Wyoming), away from the Mountain With Something Beyond (Cloud Peak, in present-day Wyoming).

Maneuvering these buffalo into place meant coordinating with charcoal and sweet sage being burned on top of the buffalo jump's rock pile wings, where the buffalo would bounce off to ricochet, running out of control from wing to wing down a slope. These wings, up to 40 steps in length and made up of just-below-the-knee-high piles of rock about five to ten steps apart, were slanted inward in a staggered formation and got closer together as they funneled through end piles of rock at the edge of a cliff. The angle of a wing became very important when the buffalo did not come straight at these walls of smoke, but turned in toward the bottoms of them, since the angle would direct the herd to the next wall of smoke.

—HENRY REAL BIRD

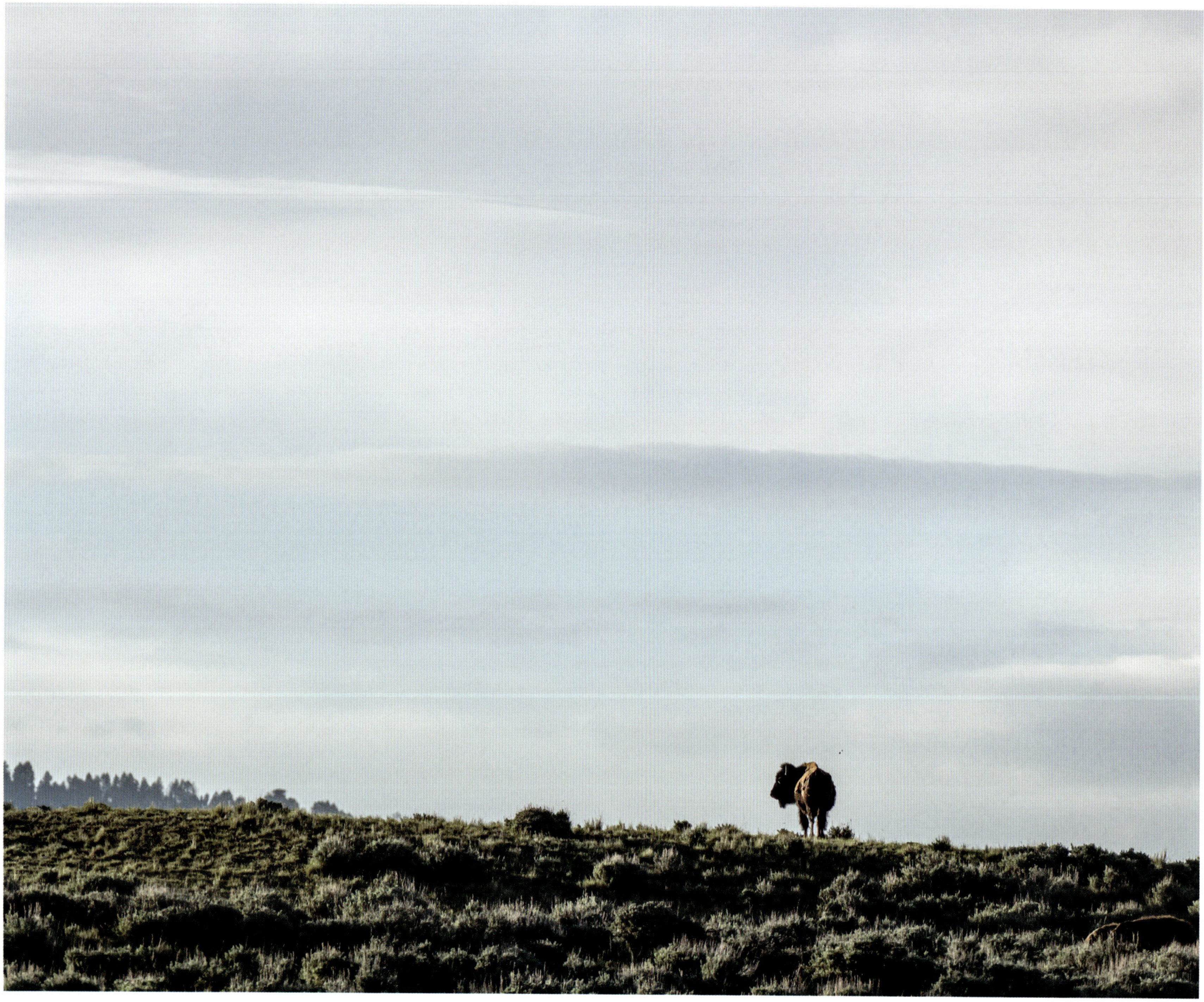

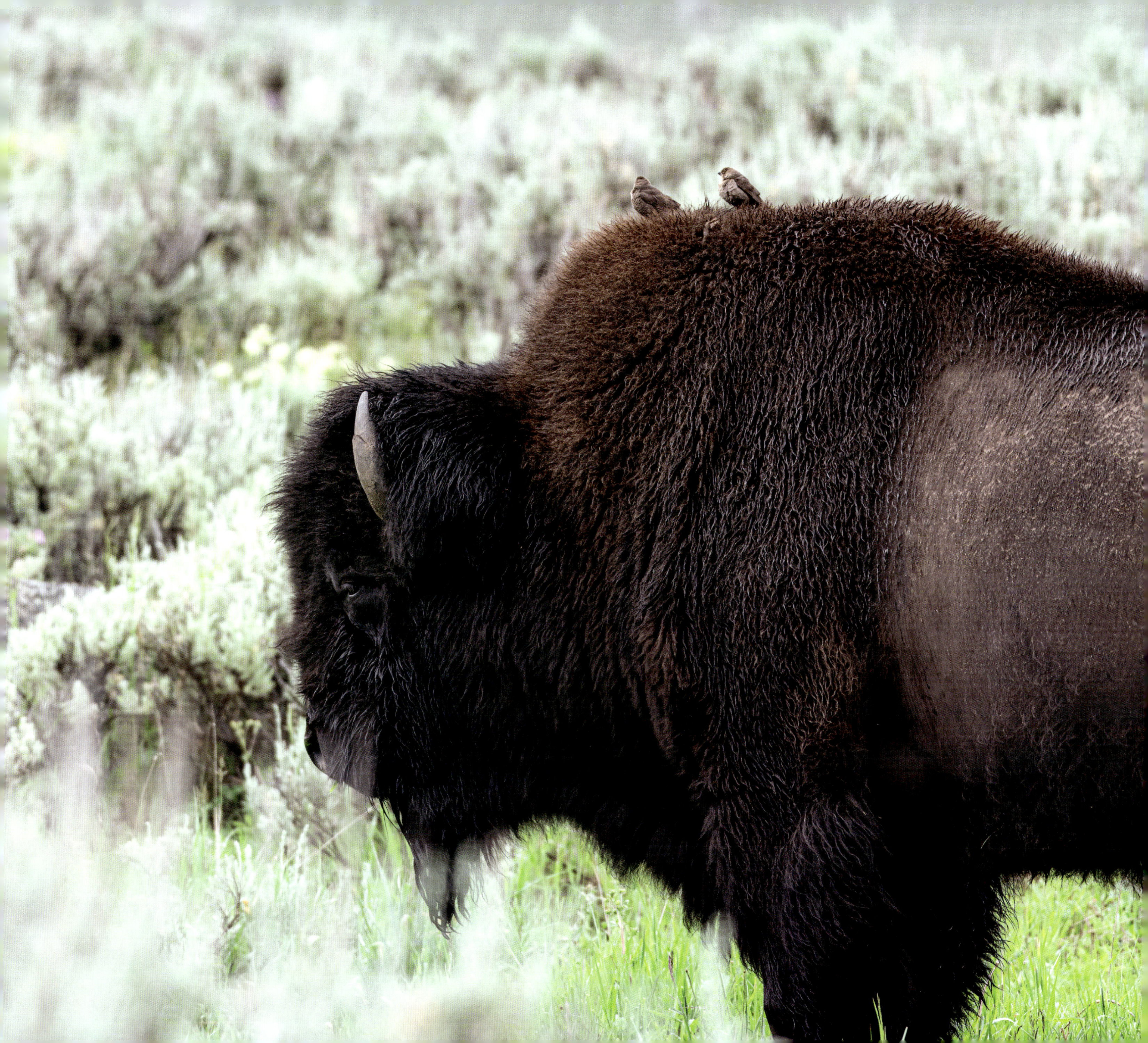

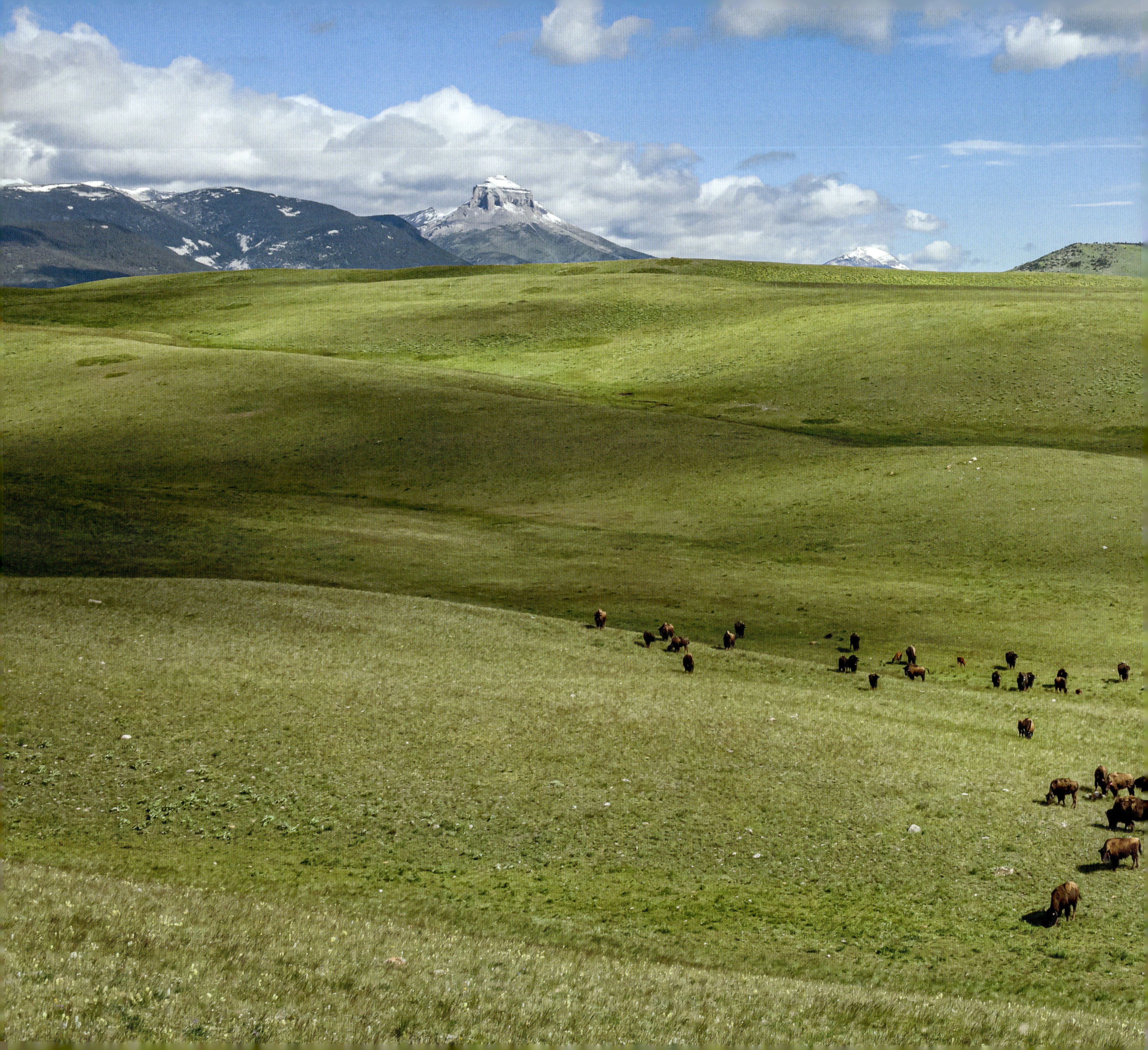

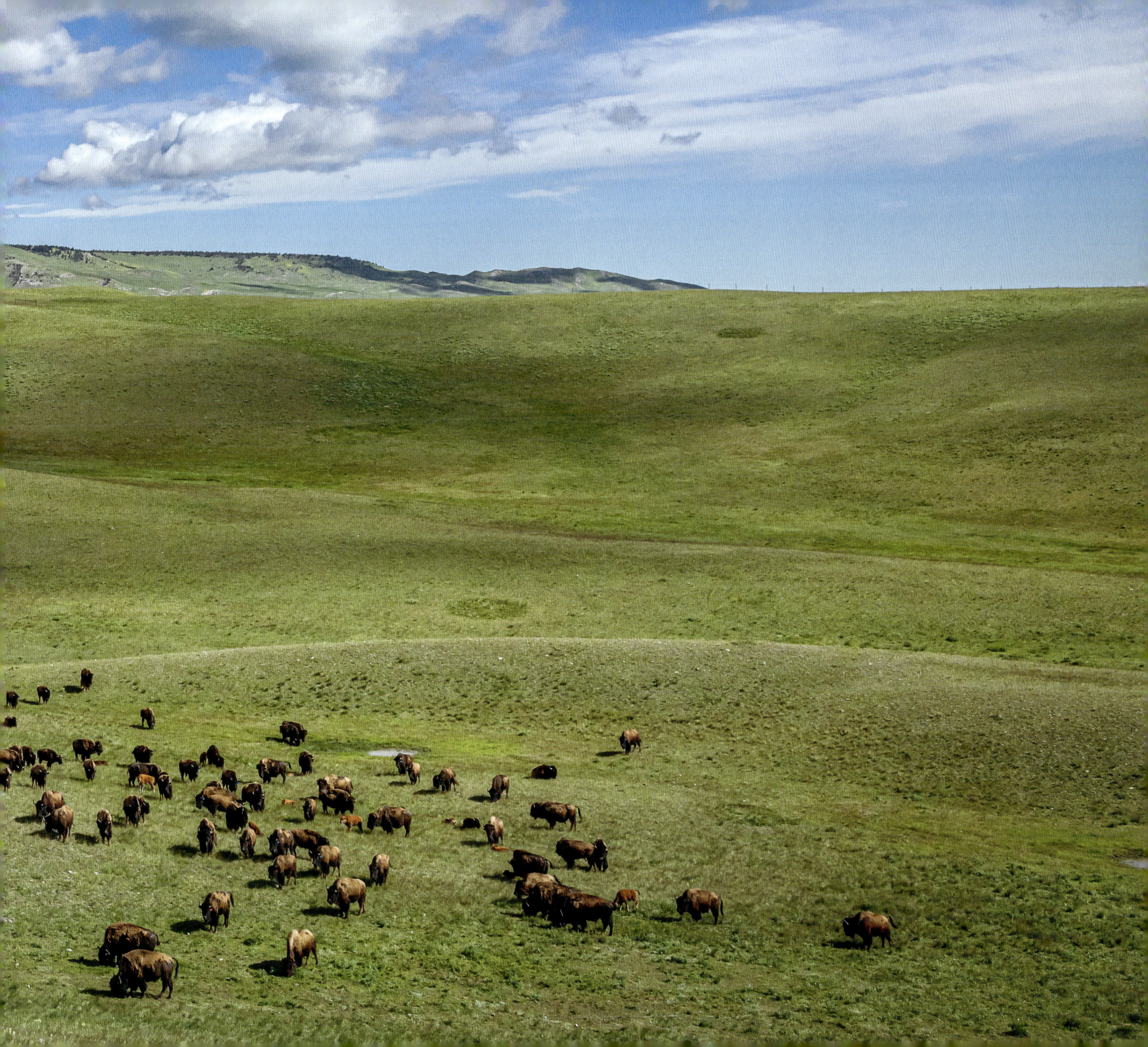

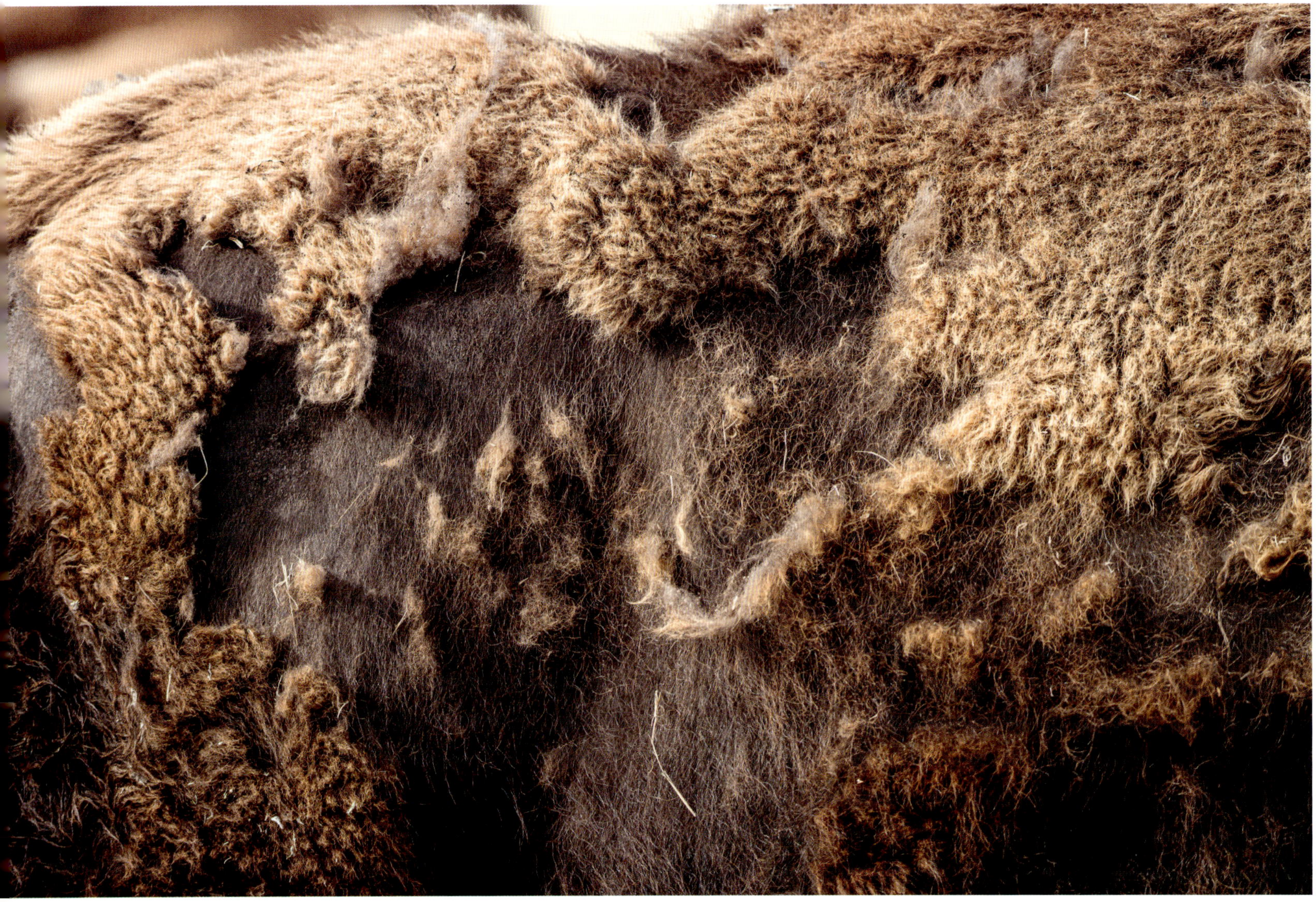

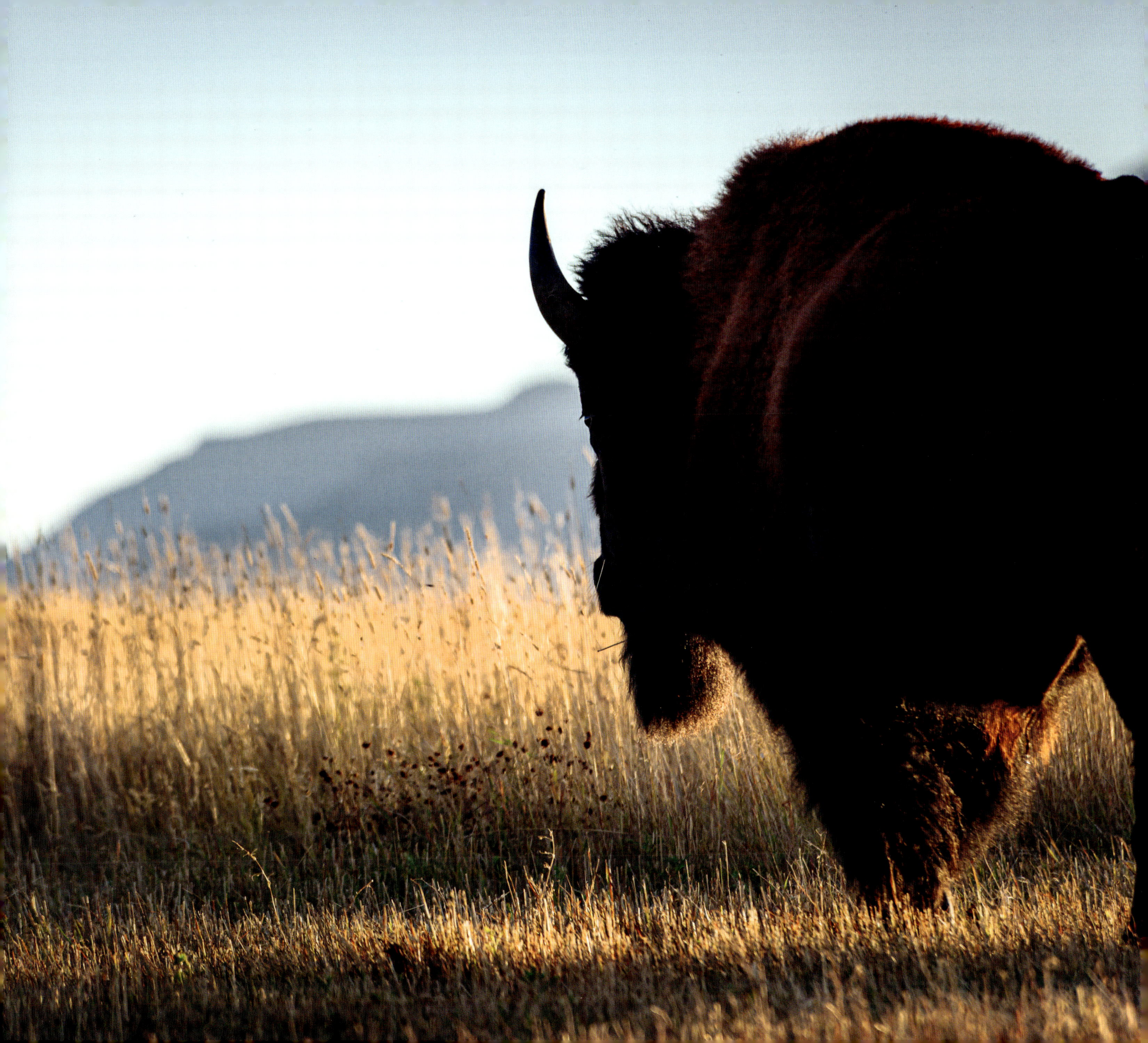

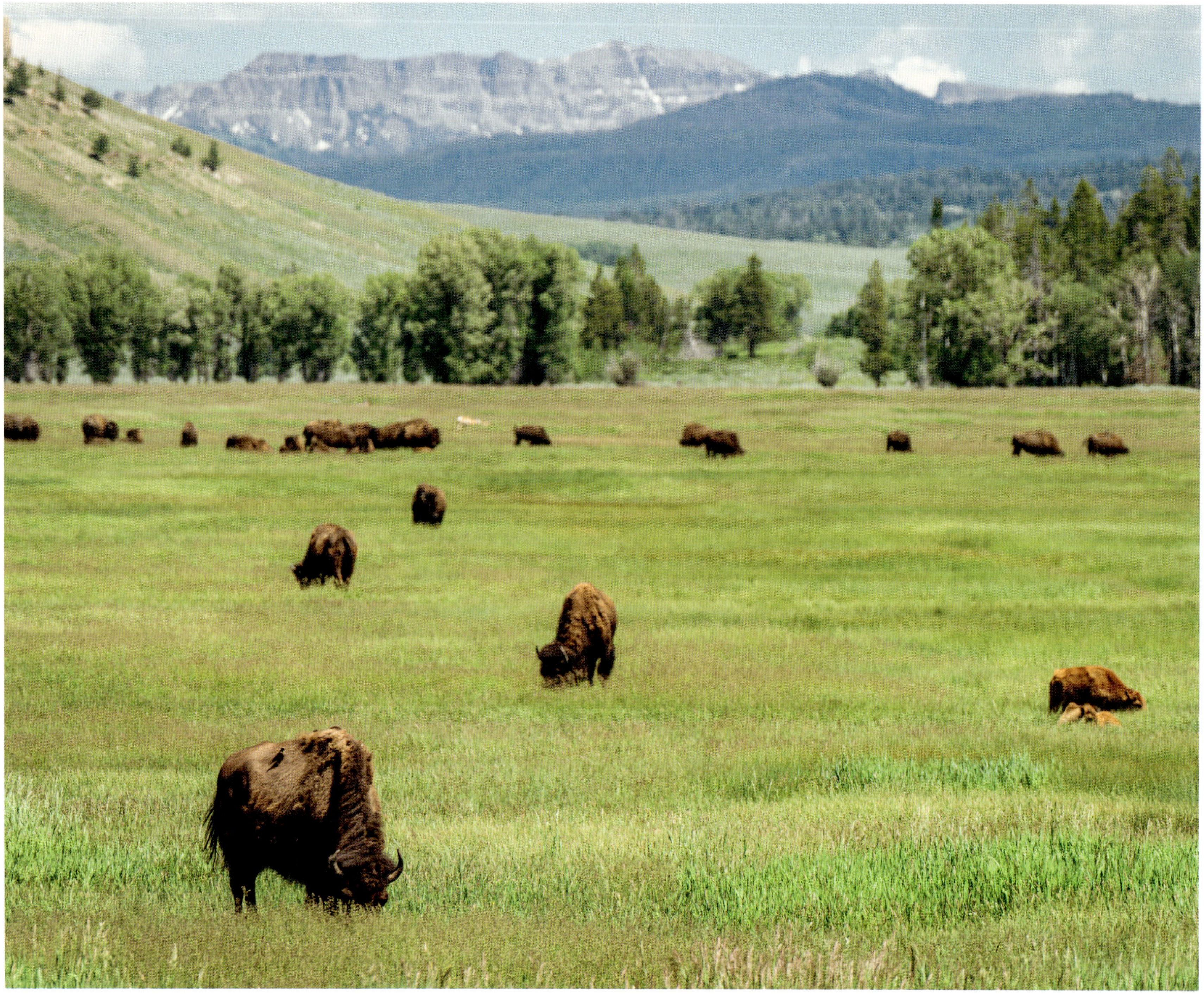

The Missing Piece

Perhaps no single person has had more to do with the recent return of bison on a large scale than Ted Turner. The billionaire CNN founder owns nineteen ranches totaling two million acres in the United States and Argentina. Turner got interested in bison ranching early, buying his first bison in 1976. Turner Enterprises' mission statement—to manage Turner lands in an economically sustainable and ecologically sensitive manner while promoting the conservation of native species—governs the management of his bison herds, which now encompass approximately 50,000 animals in six western states. Turner's organization also helped, along with other organizations, drive the successful campaign to have bison designated as the national mammal.

Montana writer Todd Wilkinson wrote about the impact of Turner's efforts—especially as applied through his chain of forty-plus restaurants in sixteen states—in an article for Mountain Journal subtitled, "How Ted Turner's Bison Restaurants, Inspired by Montana, Have Cast Big Green Ripples Nationwide."

At Ted's Montana Grills, diners are introduced to bison as meat in the form of bison burgers, chili, meatloaf, short ribs, brisket, and chopped steak. For those who read the fine print on the menus, they'll learn not just about the health benefits of the nutrient-dense bison meat. They'll read about bison's impact on the land and how healthy soils, forests, and grasslands store carbon dioxide, helping slow the effects of climate change. The meat tastes good, of course, and that can change shopping habits. As Turner's restaurant partner George McKerrow told writer Todd Wilkinson, "We have statistical proof that wherever we put a Ted's Montana Grill and touted the healthy virtues of bison, we saw demand increase dramatically at the grocery store level."

Helping meet this demand are producers around the West, some small scale, some large scale, some raising grass-fed and field-harvested animals, some allowing bison to live as close to wild as possible before finishing the animals on grain.

Chris Bechtold had a degree in wildlife management and worked with grizzly bears before he signed on to work on cattle ranches and guide hunters in the Bob Marshall Wilderness. He was hired to run the Diamond 4D Ranch, located on the Rocky Mountain Front in northern Montana, a property whose land had historically been used for cattle. As he and the new owners delved into the facts about bison, he recalls, "The more we learned, the more enamored we became. It's a healthier meat, they're easier to take care of, and they're smart. I dove in head first."

The Diamond 4D now runs about 700 head in a place where conditions can be extreme. "Bison are the perfect animal for this environment," Bechtold says. "Yesterday we had fifteen inches of snow and forty-mile-per-hour winds. It's damp, wet, and dipped below zero for a couple of days in a row. That's pretty harsh for mid-April. But I don't worry a bit about them. The first calf hit the ground yesterday and it's healthy. I went out today and they're digging through the snow, perfectly content. If I were still a cattle rancher I'd be worrying. But bison evolved to do what they're doing right now."

Bison manage their own calving, something that keeps most cattle ranchers up in the middle of the night in inclement weather for weeks on end. According to Bechtold, bison have smaller calves than cows do, but those bison calves have a faster growth rate. Also, bison don't rely solely on hay for their winter feed. If enough grass has been left behind for winter grazing, small amounts of supplemental feed can easily get them through

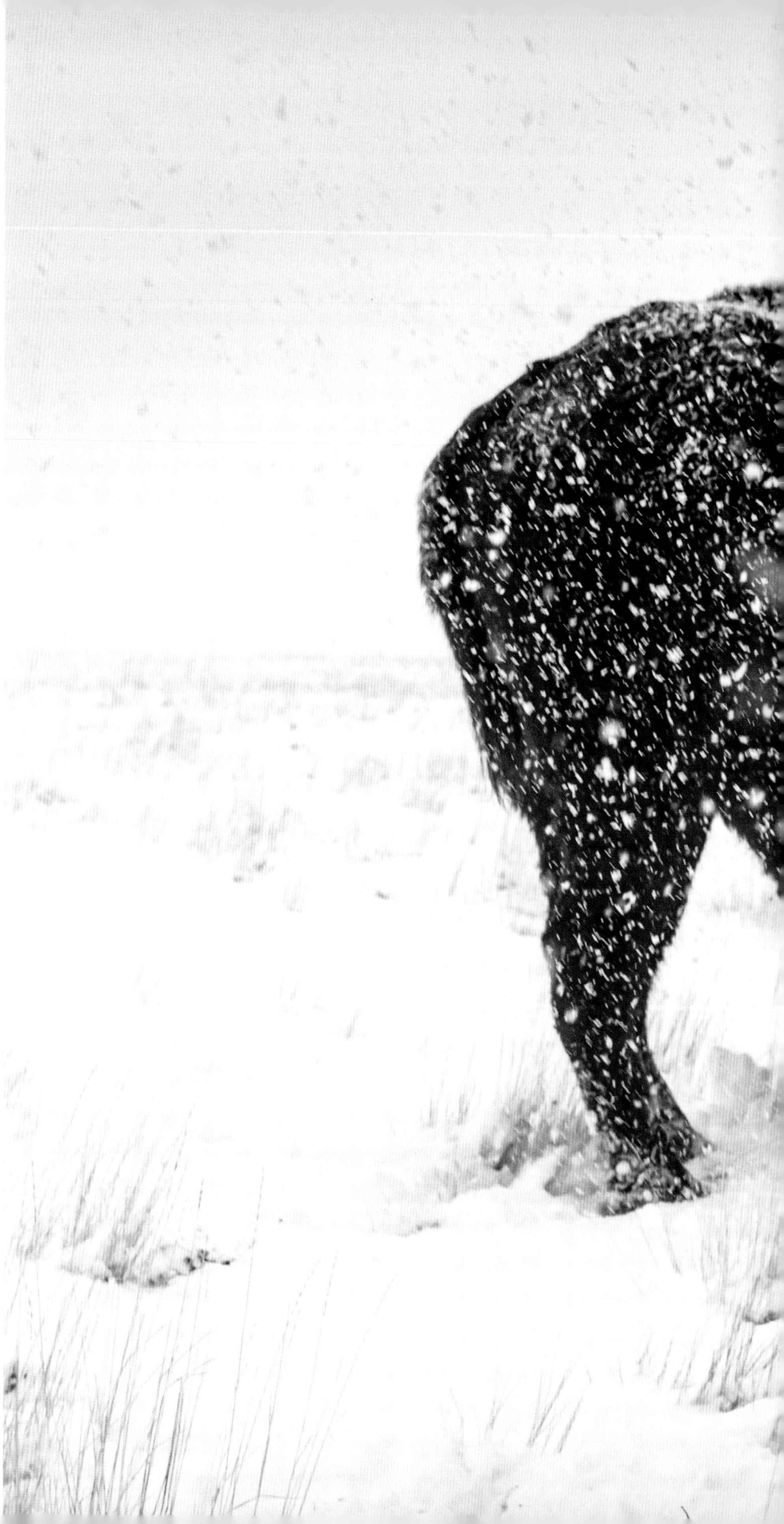

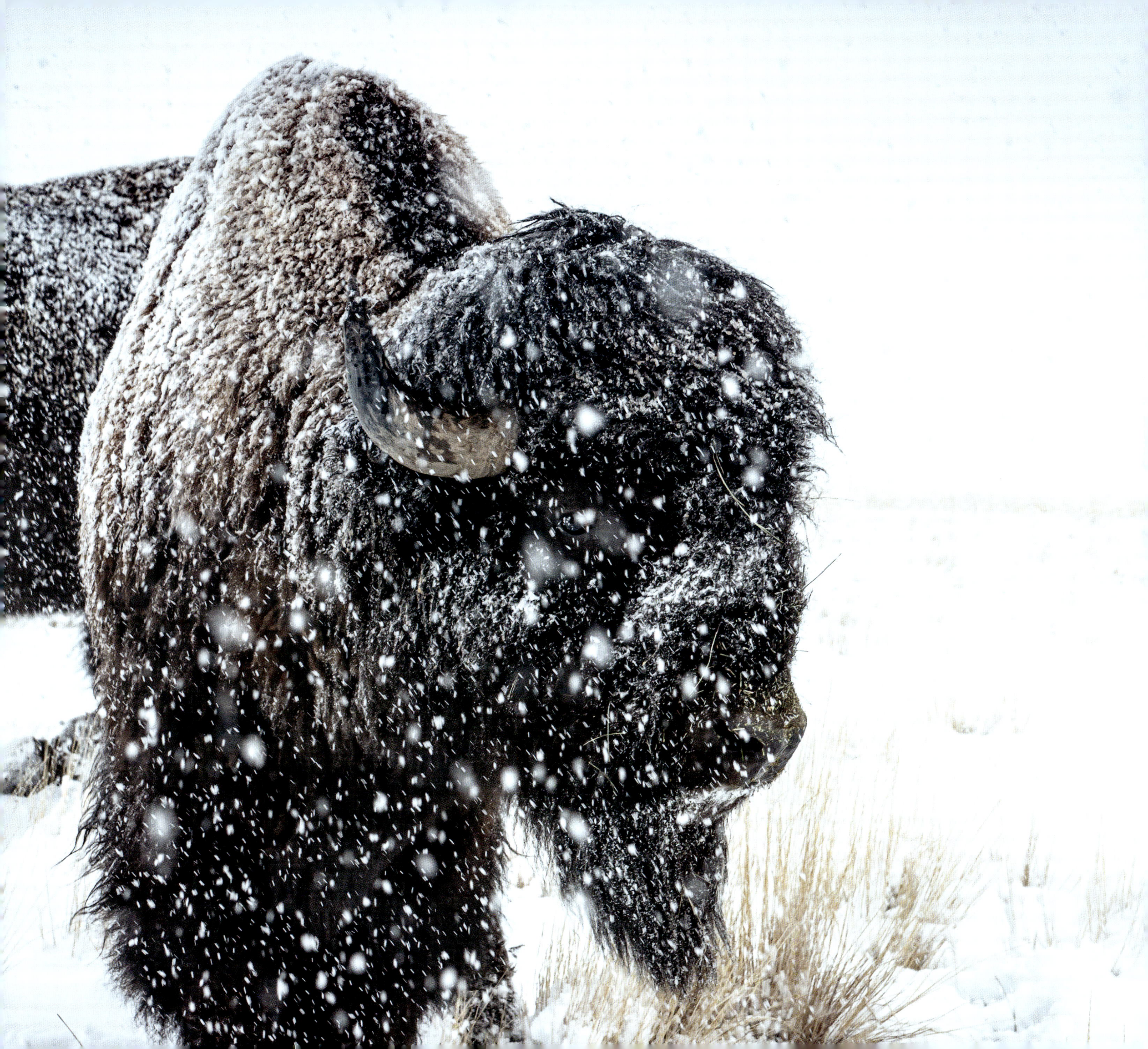

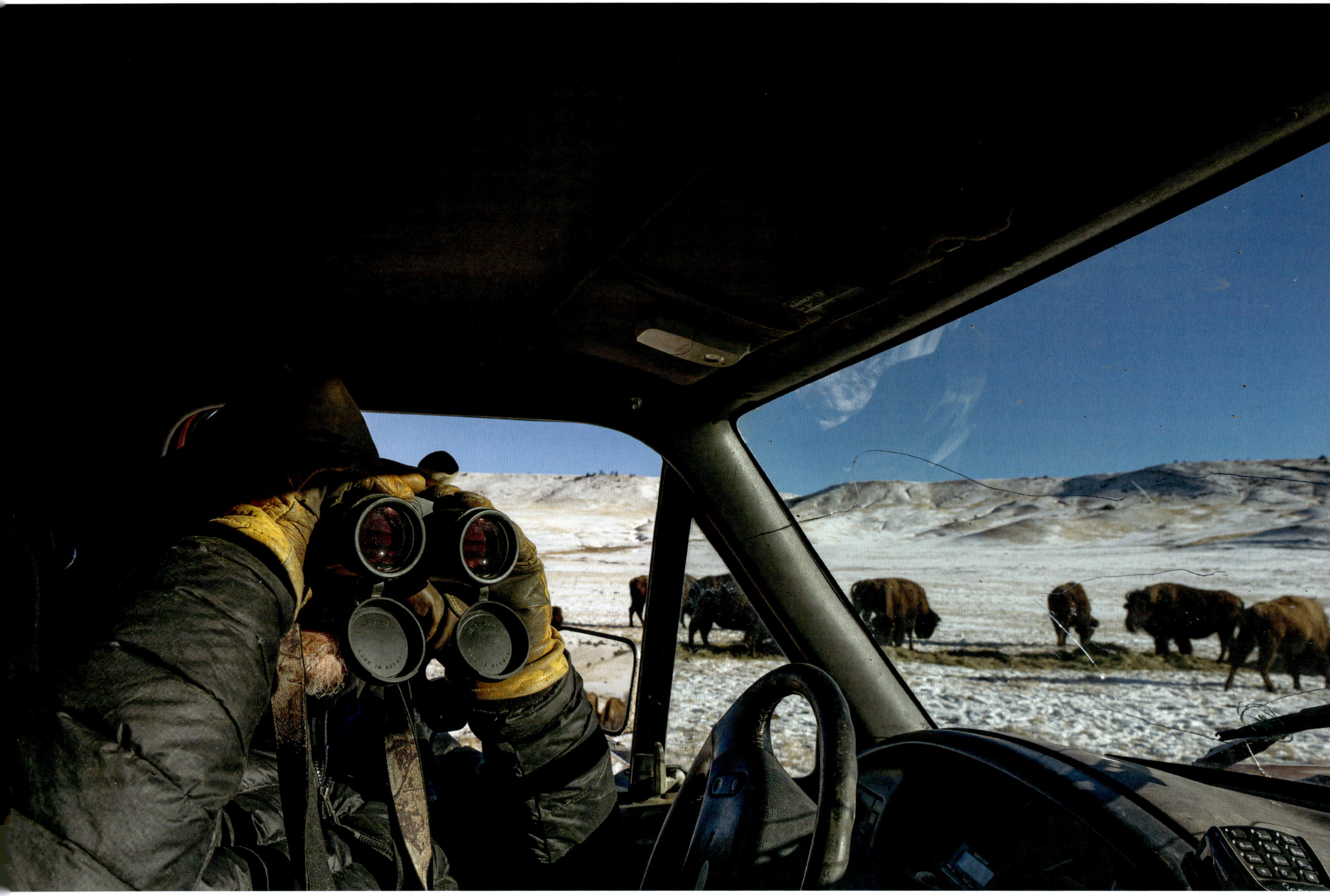

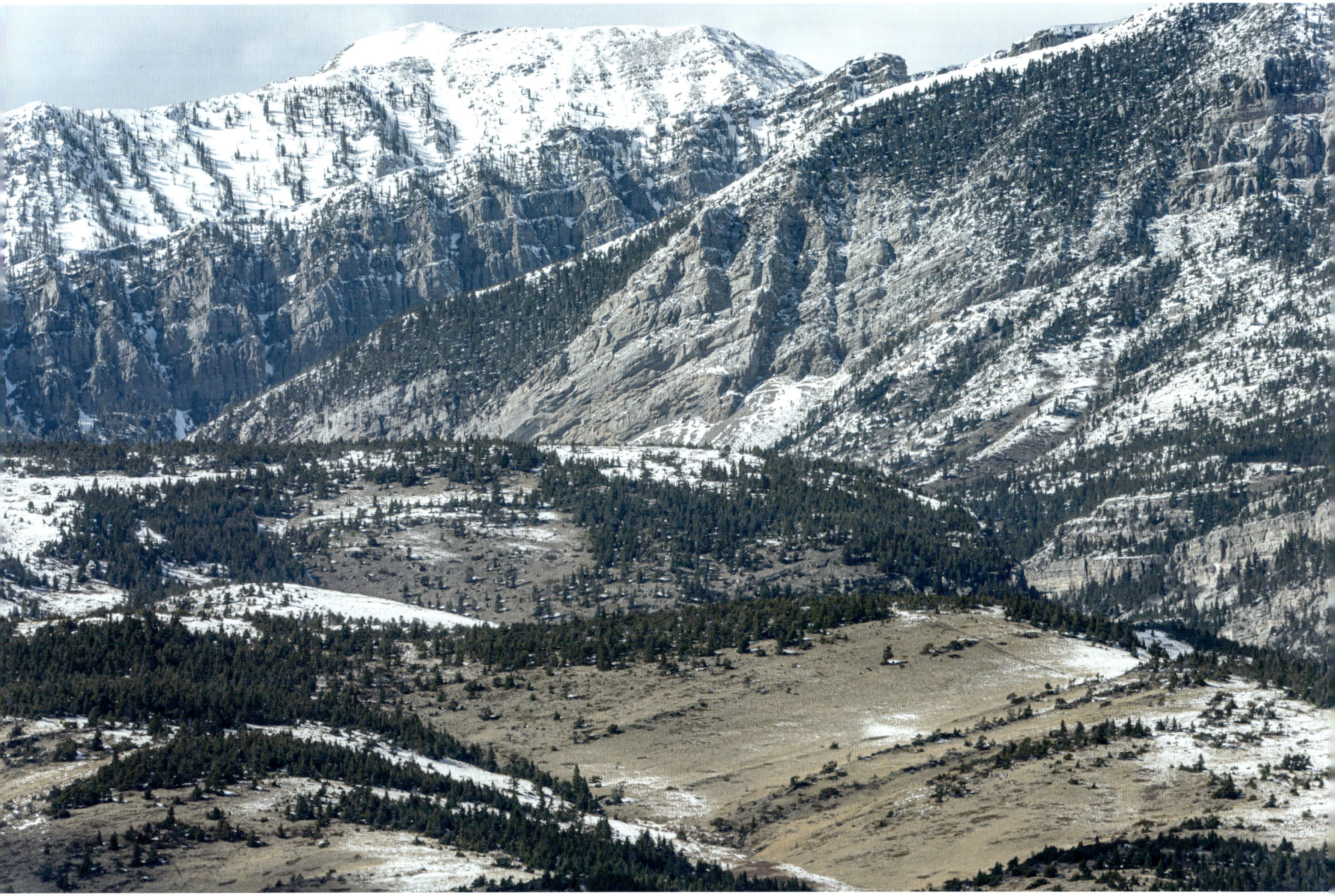

the winter months. (In places prone to deep snow, the animals use their big heads to clear the ground to reach the forage underneath.) The main limitation in ranching bison, says Bechtold, is "you have to have enough landscape. If you don't allow movement, you're actually defeating the whole purpose."

Matt Skoglund also had a deep background in environmental policy and advocacy work in Montana on behalf of wildlife, including bison, when he founded North Bridger Bison and started a bison ranch from scratch with his wife and kids. The Skoglunds' ranch has abundant live water sources and sagebrush-dotted grasslands tucked up against the North Bridger Mountains west of Wilsall, Montana. "I was really craving to do something conservation-based but more tangible," he explains. "I'm a passionate food person; I love to hunt for meat, and garden, and look for morels. When I read about the field harvesting of bison that Dan O'Brien pioneered, I thought, 'I want to do that.'"

The Skoglunds' bison are grass-fed, grass-finished, and field-harvested by Matt right on the ranch. "We try to work with nature as much as possible, to do everything as naturally as possible." To minimize stress, they work the animals only once a

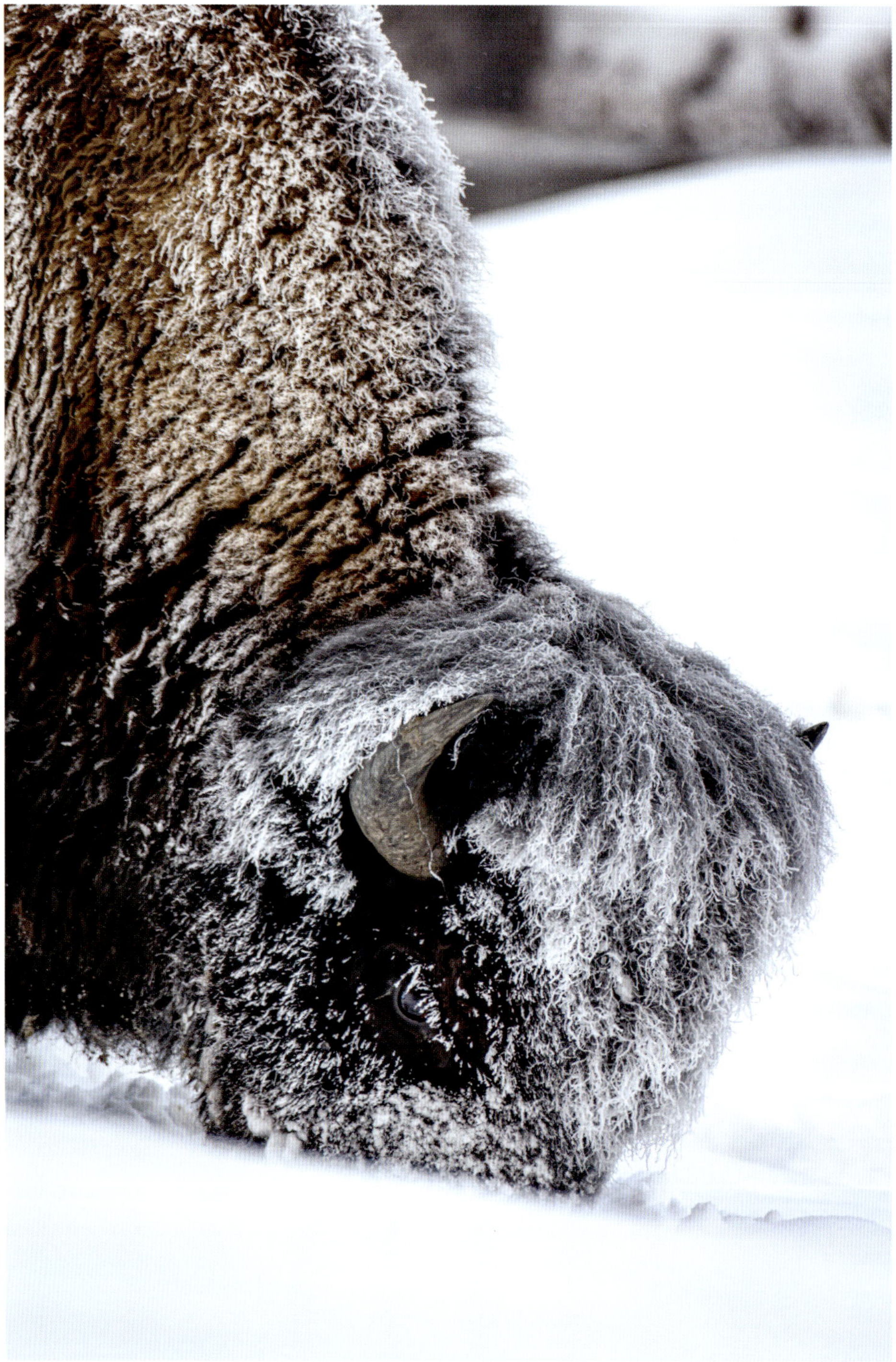

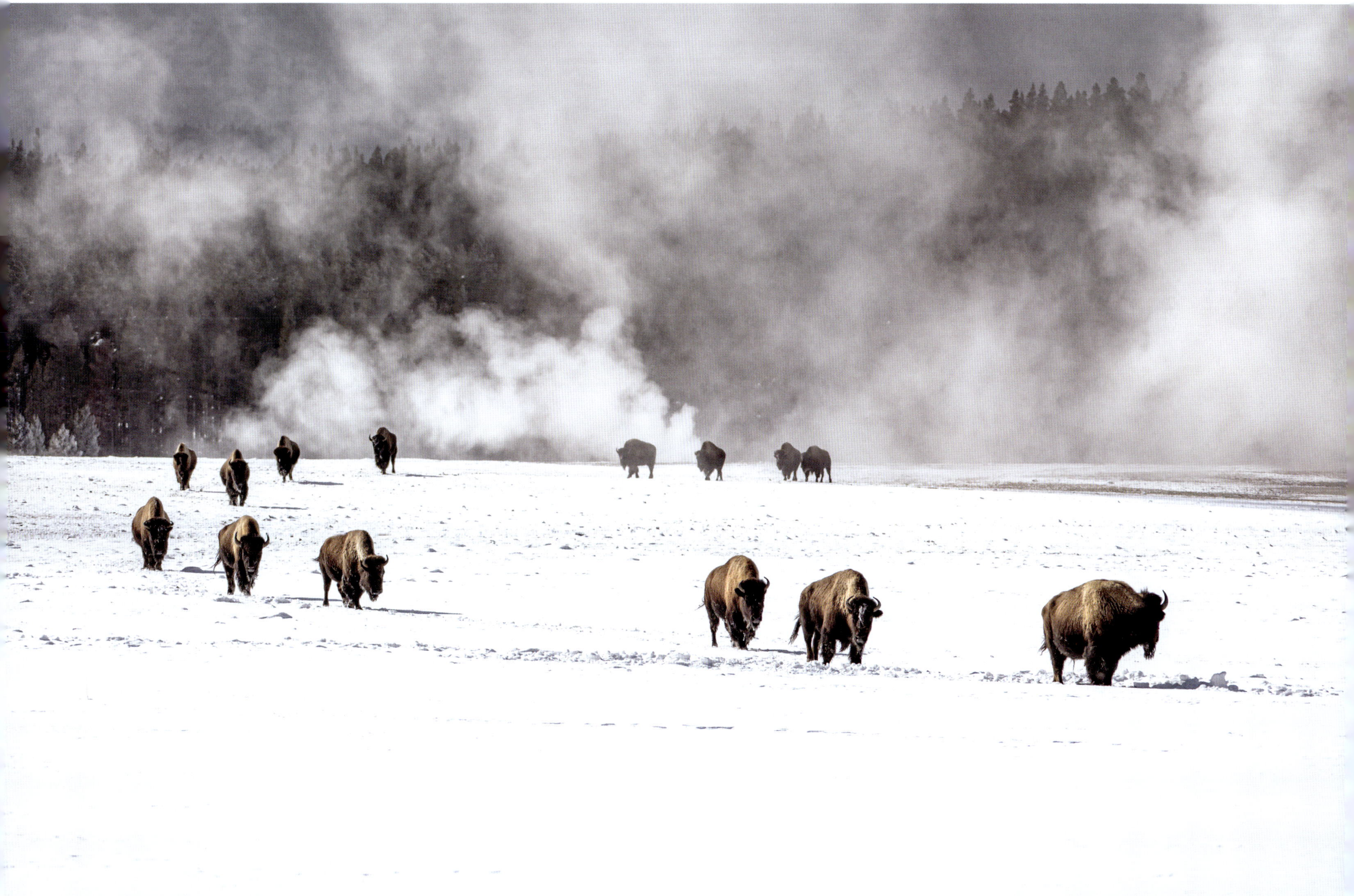

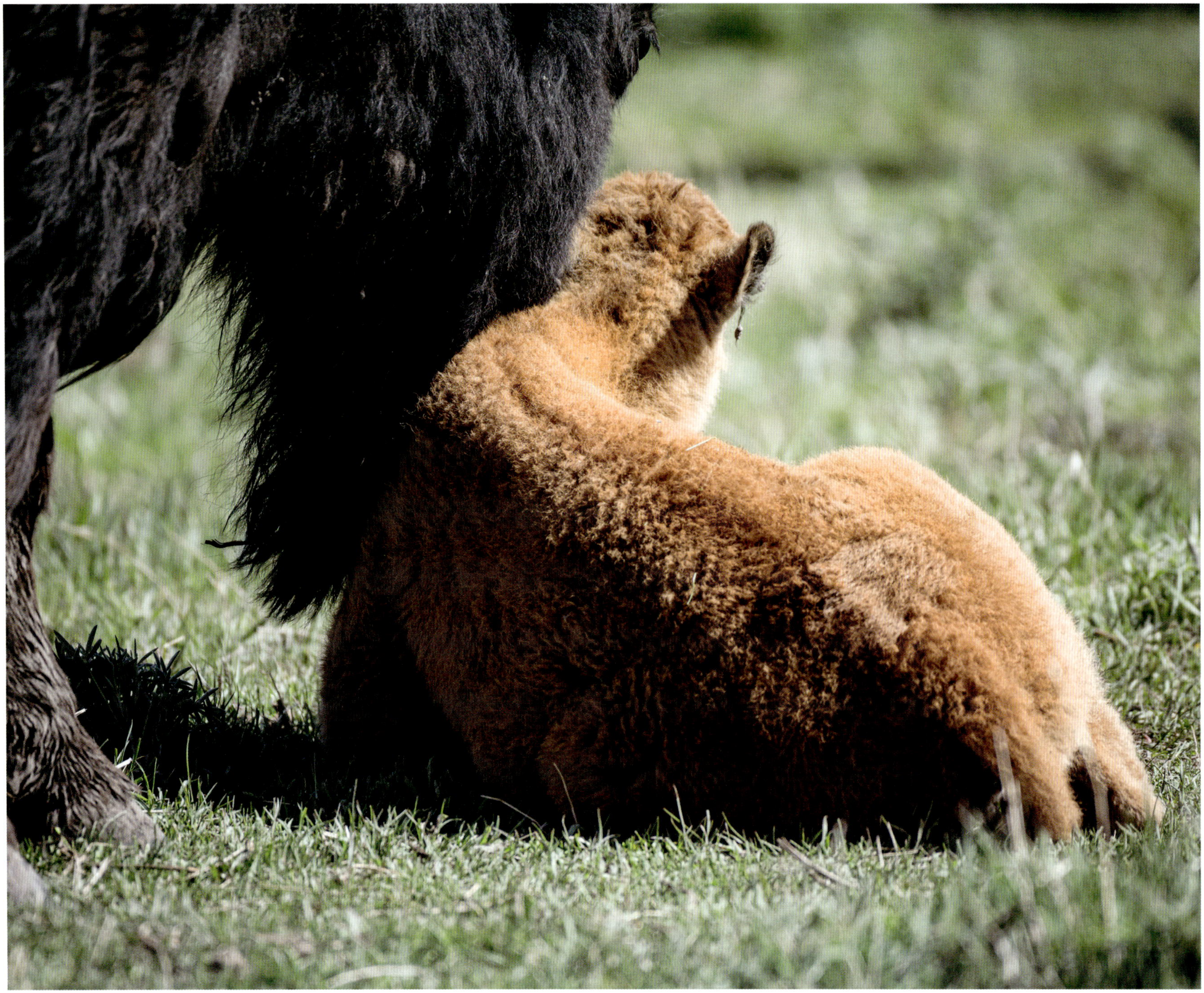

year. Meanwhile, managing the bison in sync with nature helps improve the environment by building topsoil, storing carbon in the ground, and providing habitat for a wide variety of wildlife.

Both North Bridger Bison and Diamond 4D are members of the National Bison Association, a trade association that represents 1,300 members and works with tribal nations, conservation groups, and public herd managers. The NBA collaborated with other stakeholders, most notably the Wildlife Conservation Society (originally founded in 1895 as the New York Zoological Society) on passage of the bill designating bison as the national mammal. The group's purpose, says Assistant Director Jim Matheson, is to restore the species and serve its members, in that order. "We provide resources to help people get started. We're very holistically focused and encourage a sustainable approach." The group's code of ethics is strict, forbidding the use of growth hormones and antibiotics while advocating for the most humane and natural treatment of the animals in its members' care. It also prohibits the crossbreeding of bison with any other species, an important distinction that is often misrepresented but is of utmost importance, not only to the NBA but to the future of the species.

Standing Rock Sioux Indian Reservation straddles the border of North Dakota and South Dakota. There Ron Brownotter, who is of Lakota-Yanktonai descent, runs 600 head of buffalo over 20,000 acres. Brownotter served in the U.S. Marine Corps and was attending college in California while running cattle on his family land, 320 acres that had been passed down from his grandmother to his father. He was commuting back and forth to California State Polytechnic Institute and depending on family and friends to help him with his animals when two things happened that would change the course of his life.

The first was the winter of 1996–97. "It was like a 100-year winter," he recalls. "Everyone had losses that year. Cows are not that hardy. I didn't have a goal of raising buffalo, but I thought, what can I do to prevent this in the future?" The second happened back at school that fall, when he convinced his team for a project in an agricultural entrepreneurship class to choose bison ranching in South Dakota. After researching the land and equipment needed and calculating the costs of the animals and projected income, they presented the report to the class. Out of that experience a backer stepped forward. By 2001, with his wife's encouragement, he'd bought his first six yearlings from the annual Black Hills Auction in Custer State Park. Two decades later, his land is flourishing. "We have birds, insects, snakes, coyotes, badgers, deer, skunk, even wolves and moose, prairie dogs, eagles, hawks . . . lots of wildlife." And his bison ranch is believed to be the largest such enterprise solely owned by an American Indian in the U.S. or Canada. He even hosts his own auction, selling 375 calves and yearlings each year. Brownotter's success has inspired others. The Standing Rock Sioux Indian Reservation now has its own community-owned buffalo herd.

Brownotter finds fulfillment in raising buffalo. "I describe it like putting on a glove; it fits my hand. My culture, the land, the buffalo, are all mixed together, and I feel my mission today is to raise them and bring them back and just be a help to my community, my family, the reservation as a whole—and to know that the buffalo have returned."

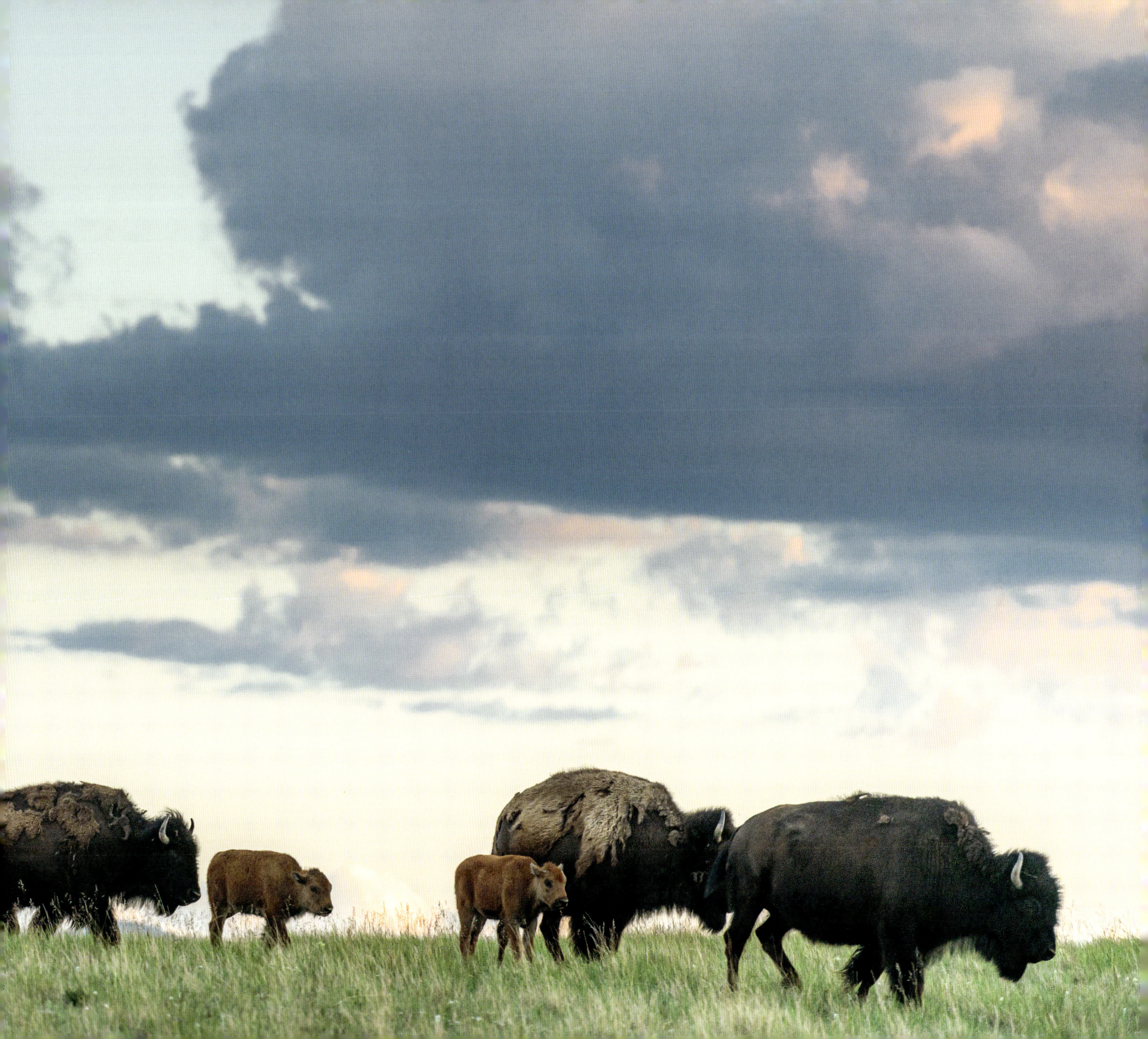

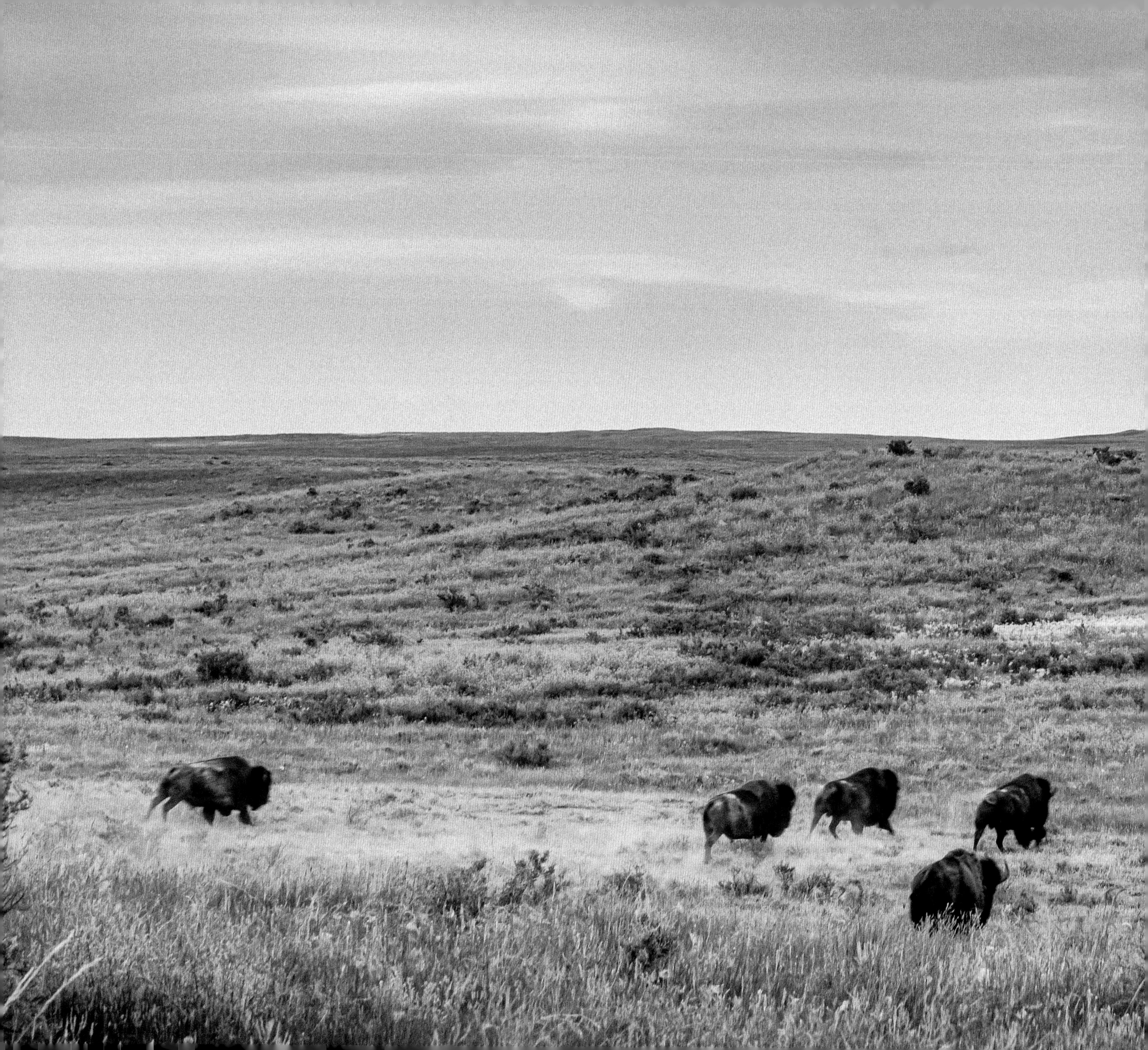

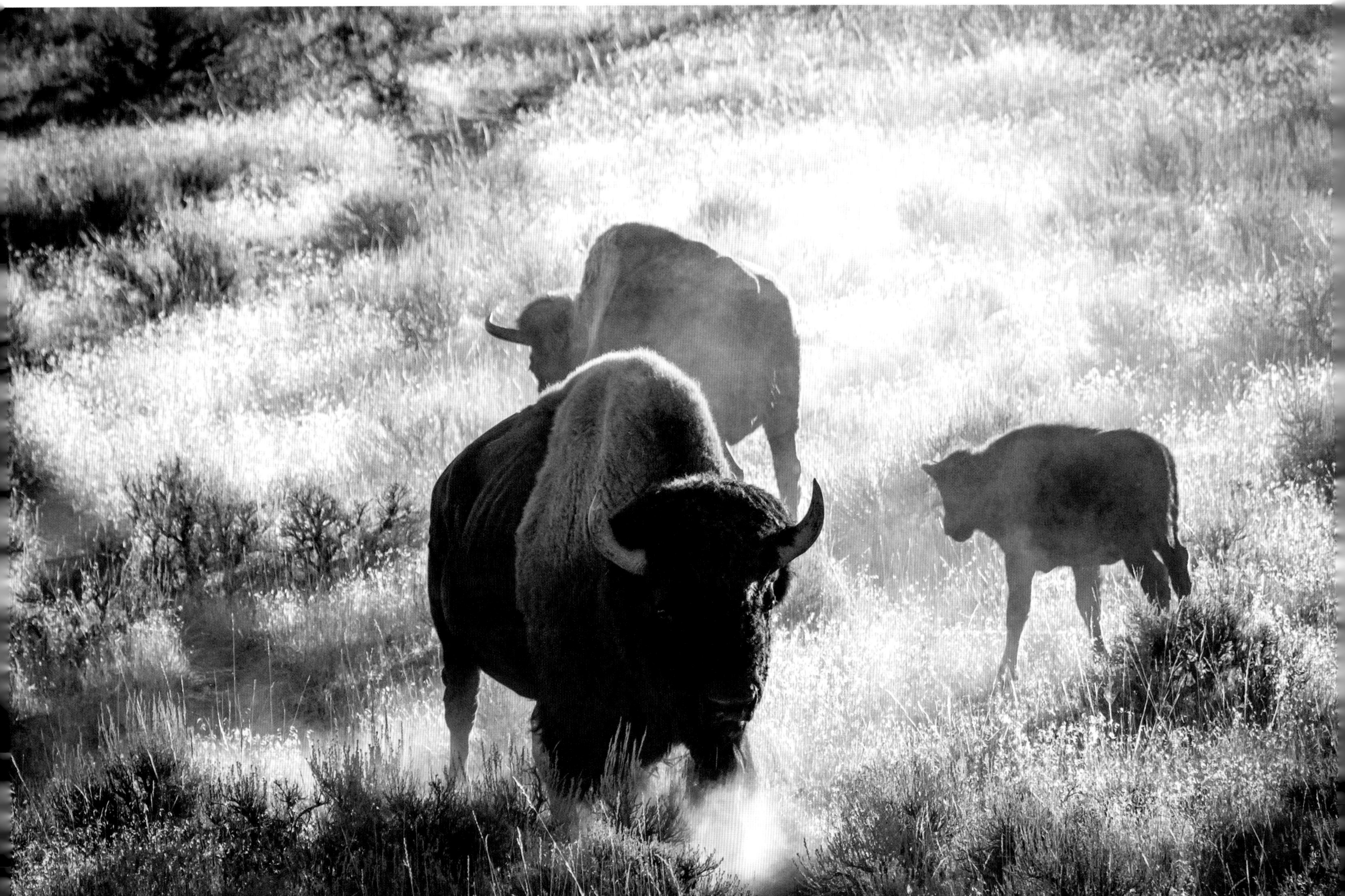

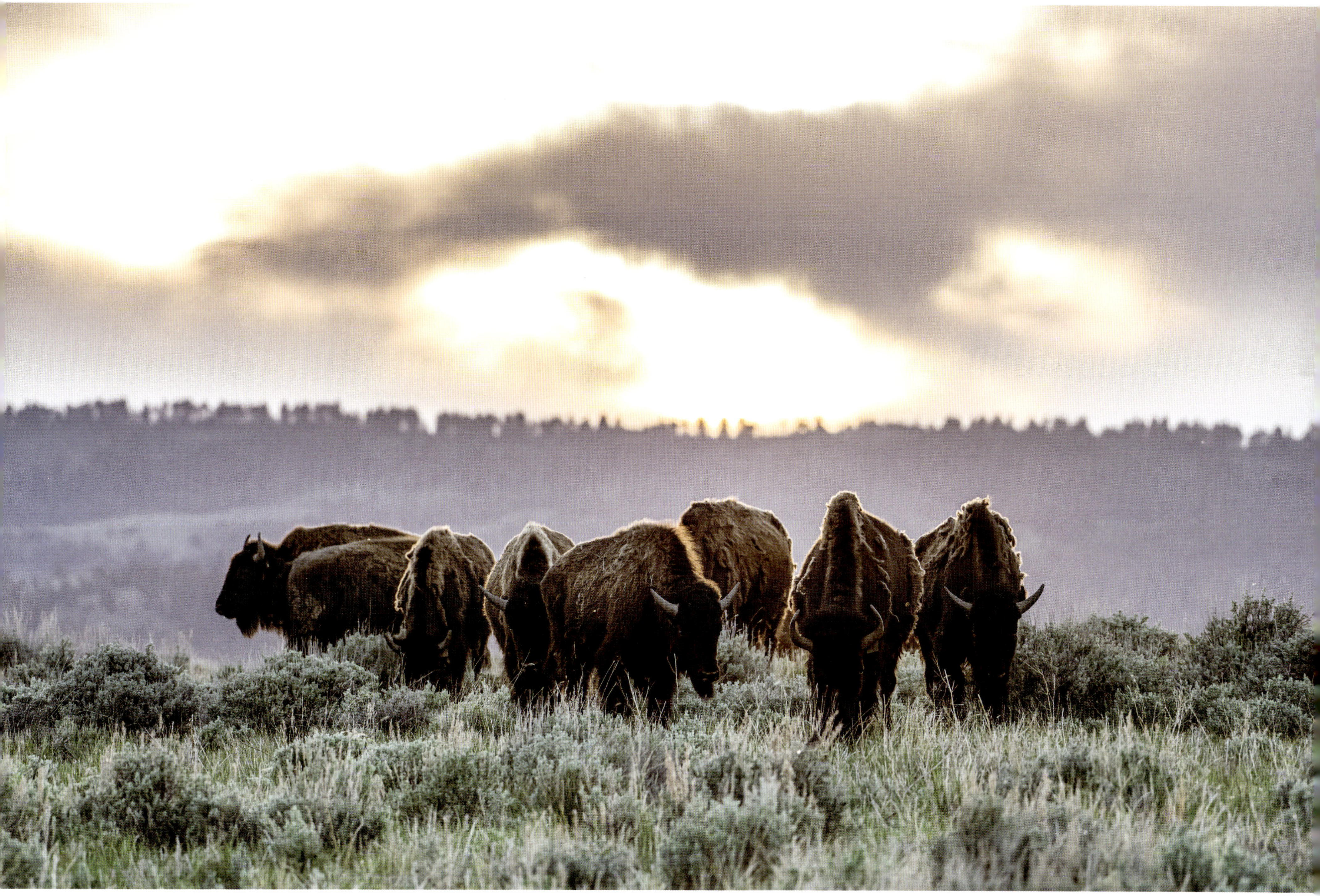

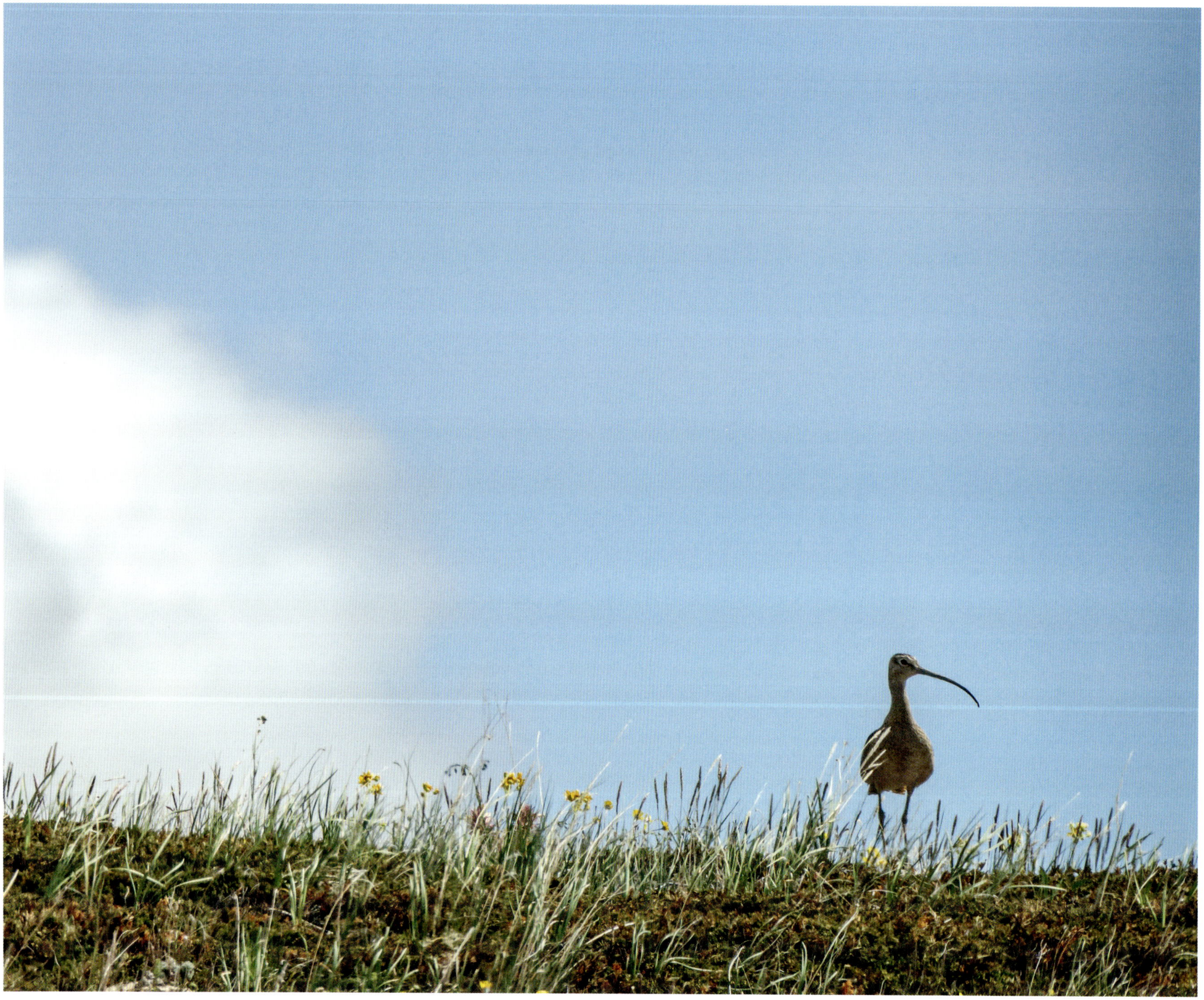

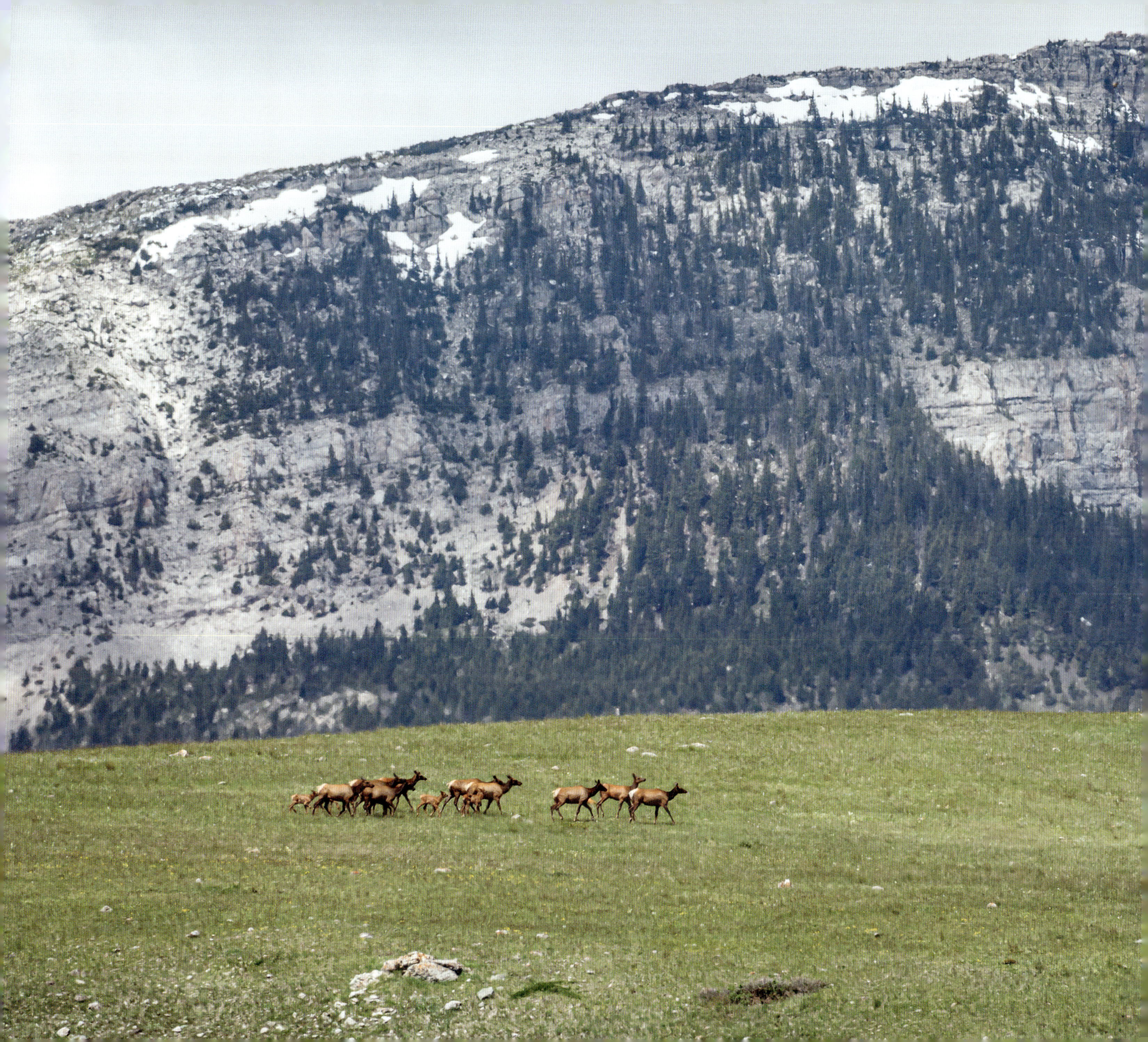

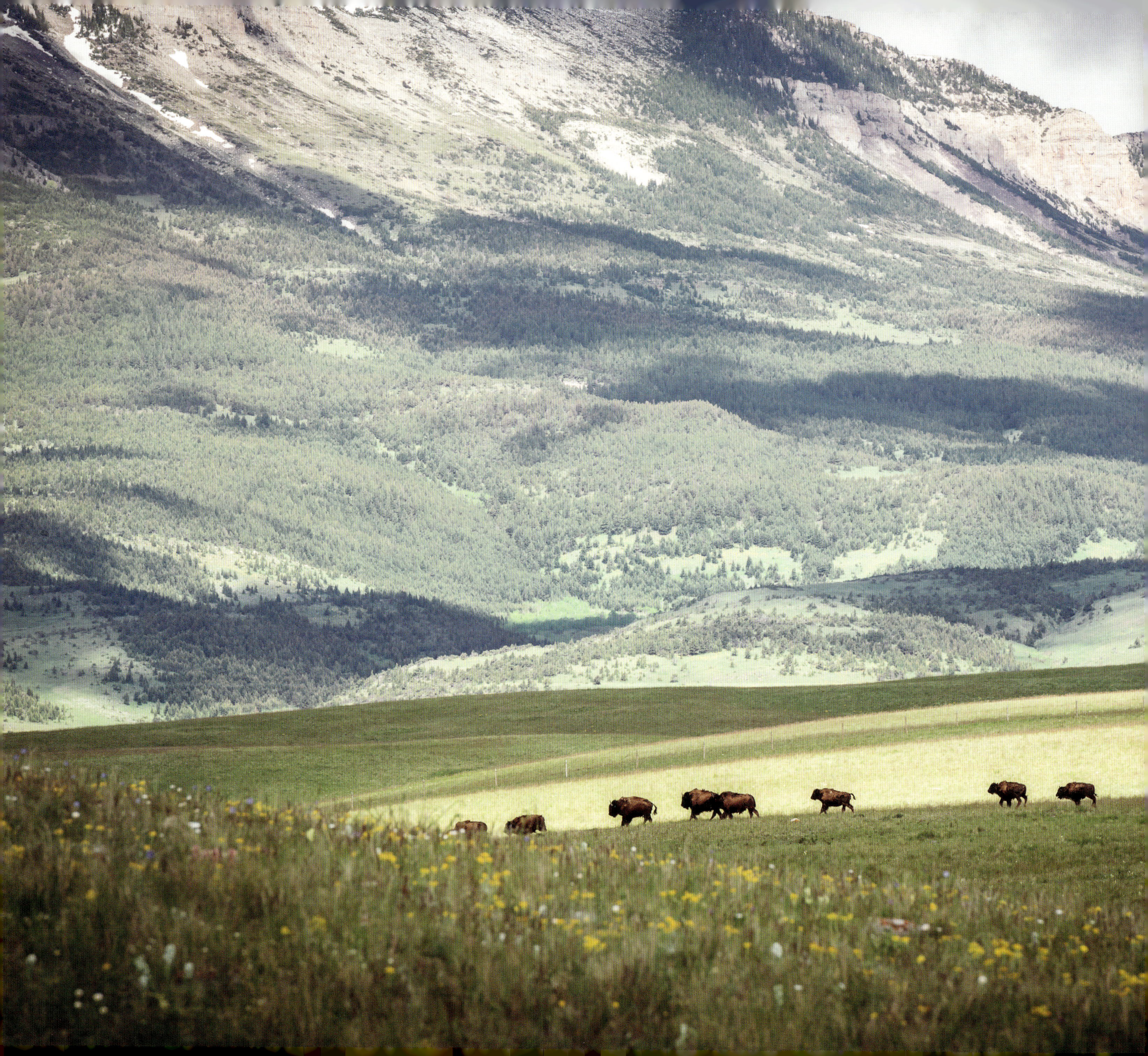

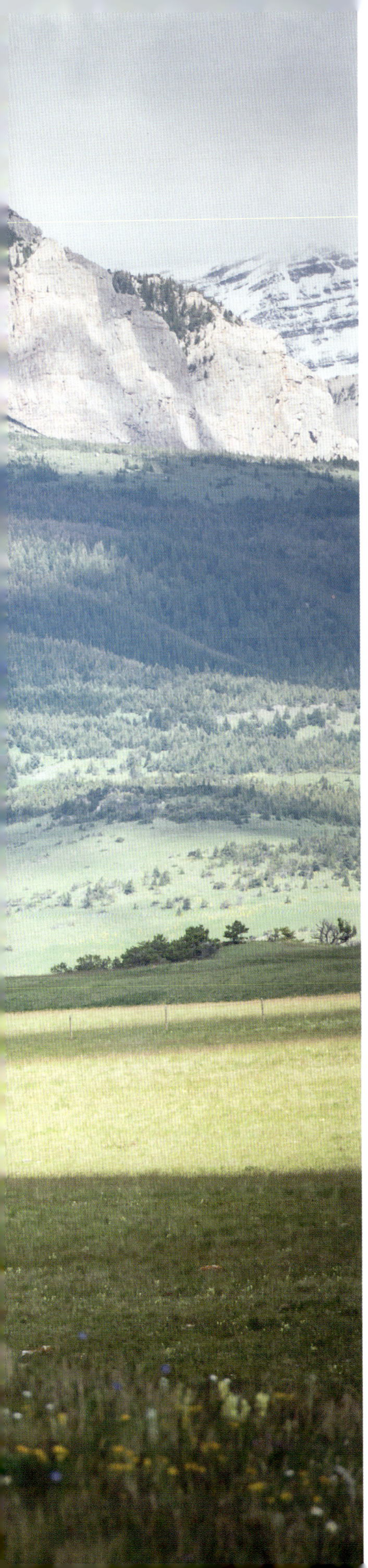

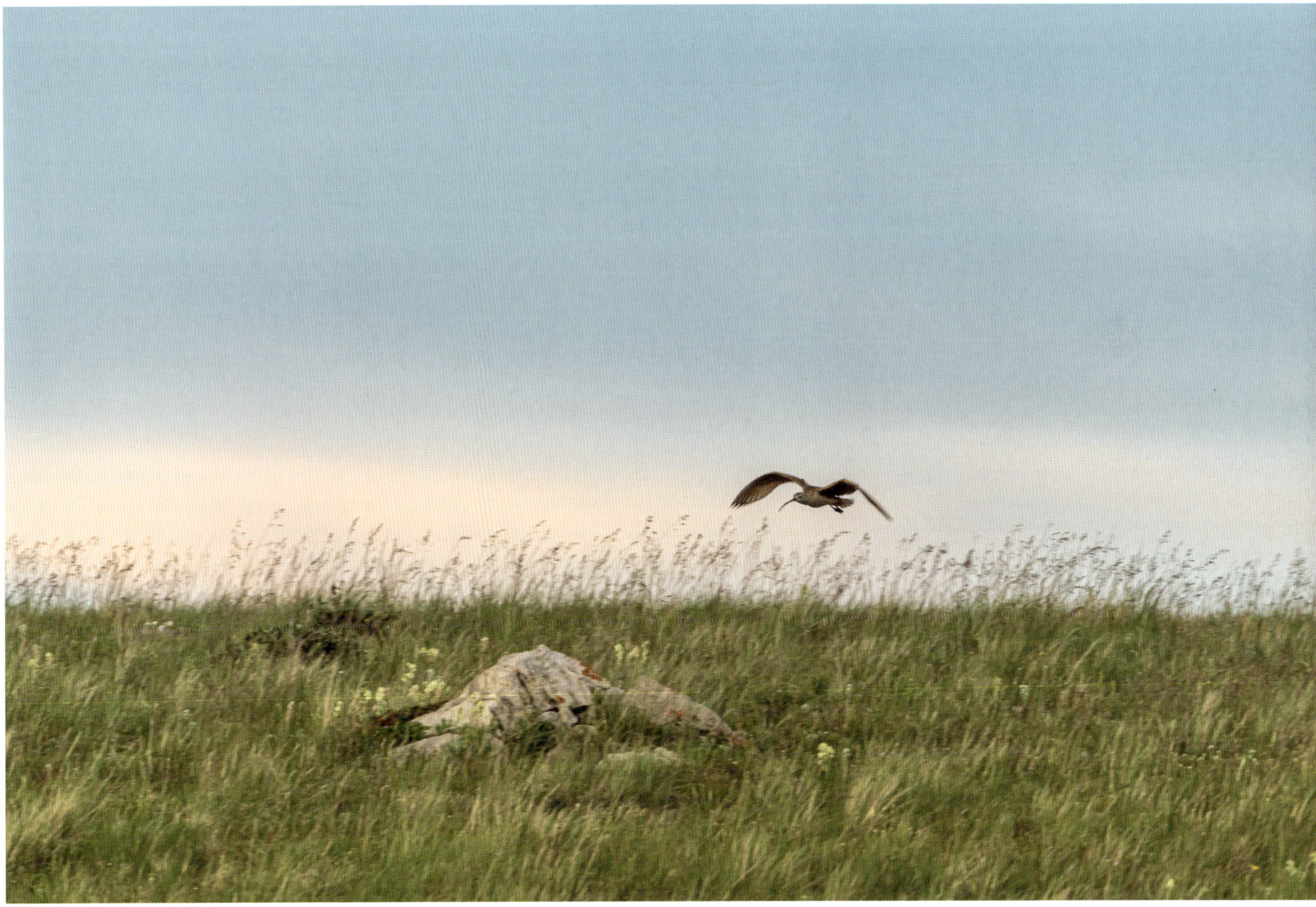

Bison are in fact classified as livestock. Despite their dramatic increase in numbers, they are almost all either literally or virtually fenced.

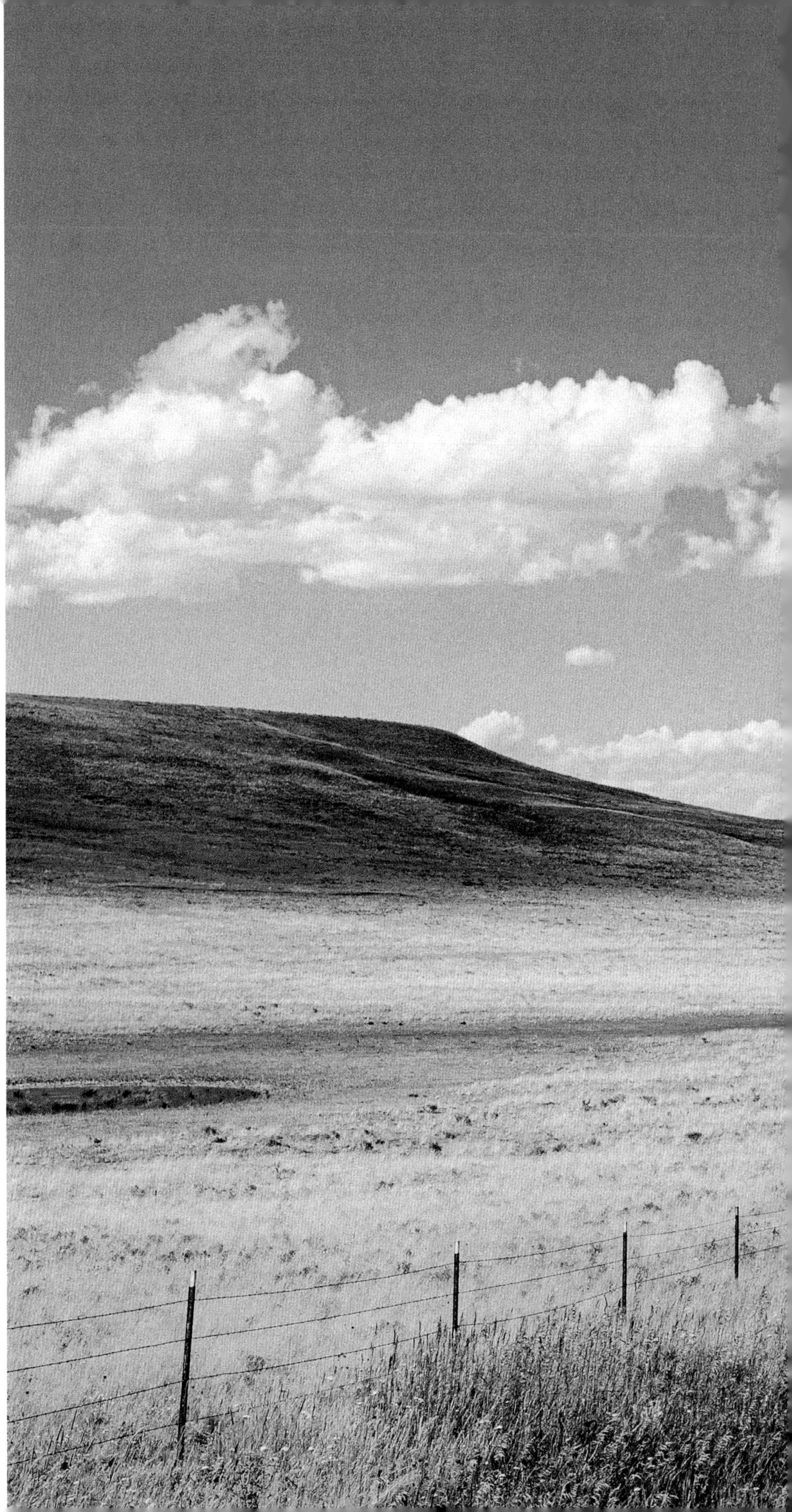

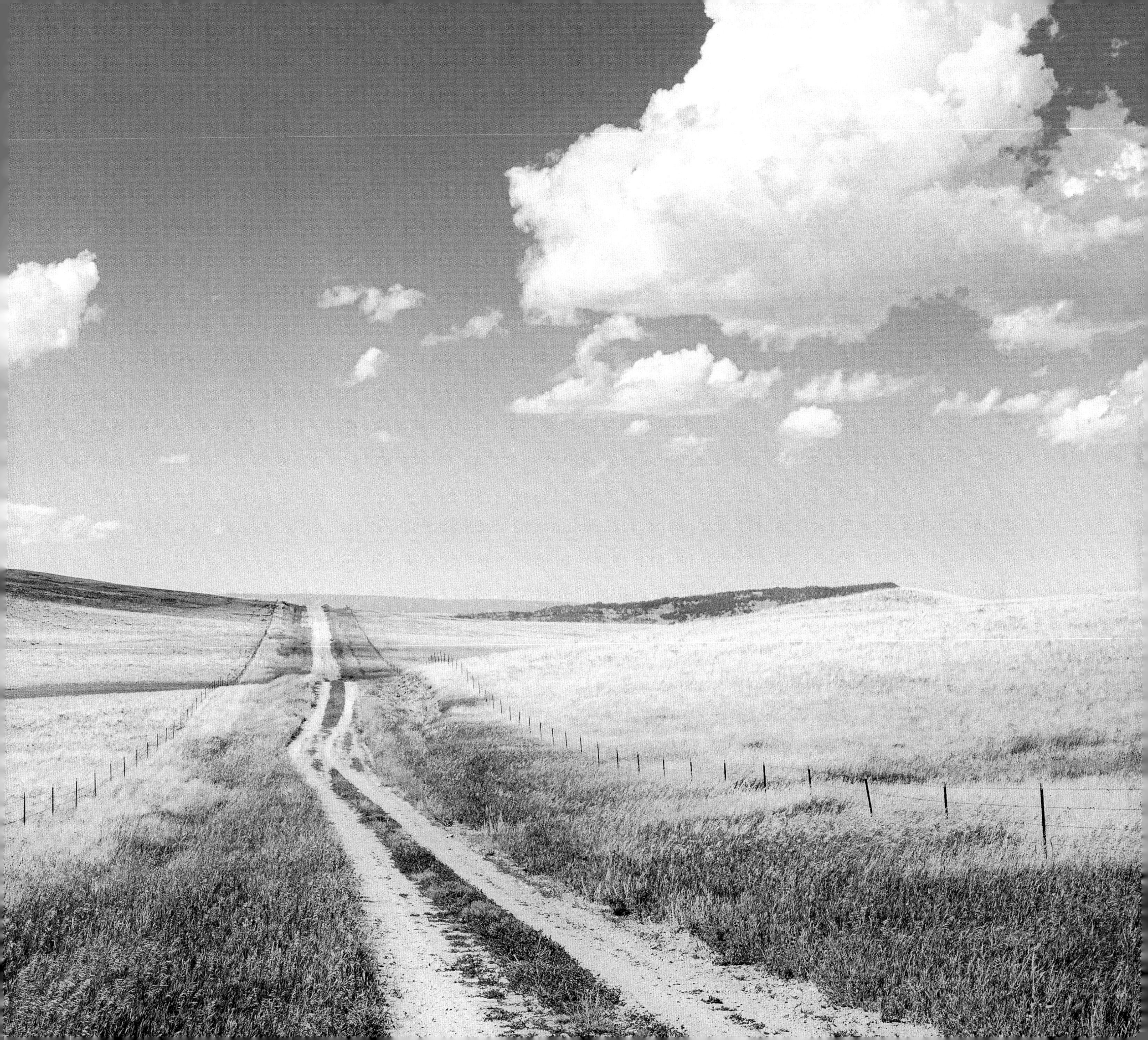

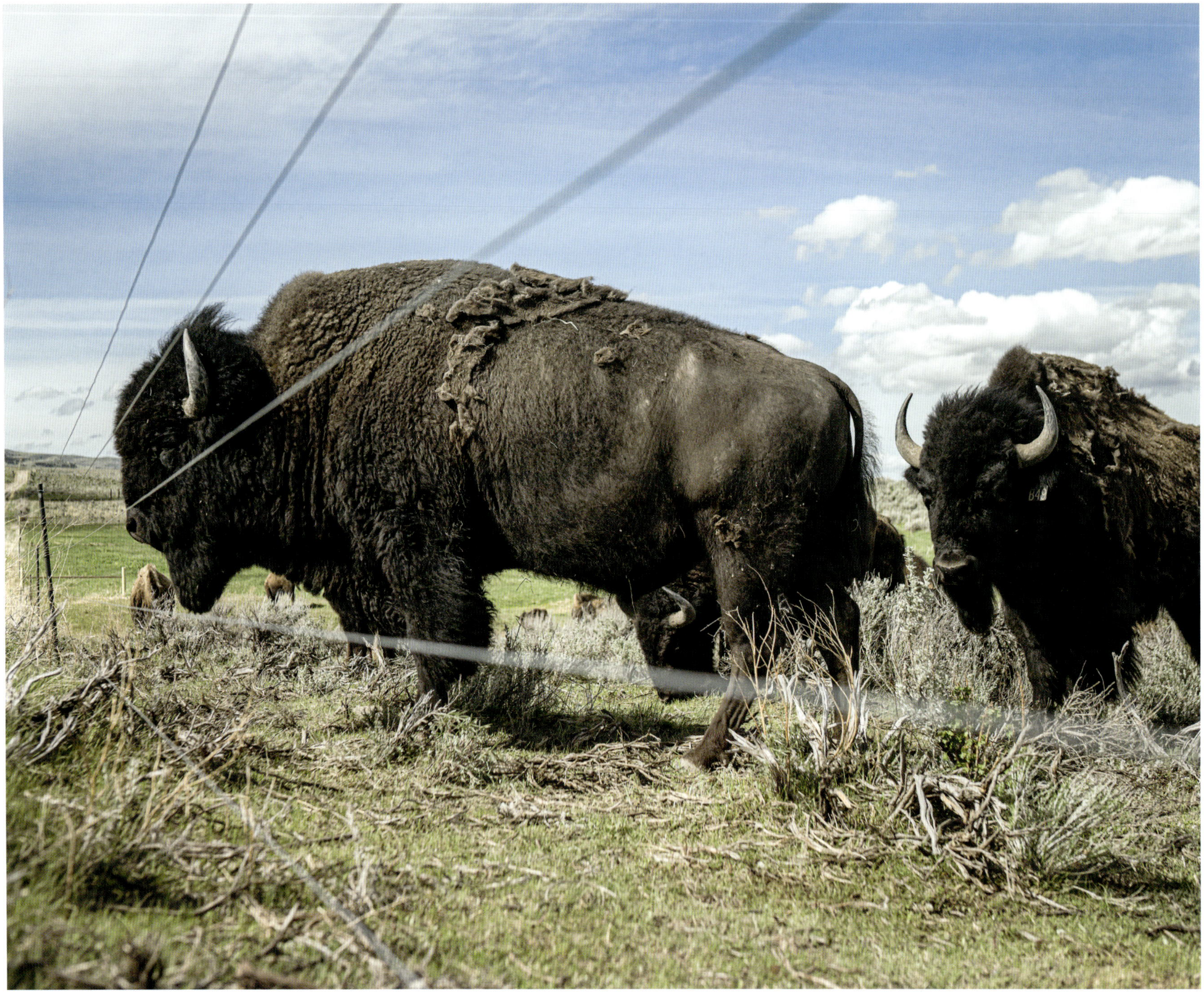

Saved Behind the Fence

In an essay penned in 1987, Frank J. Popper and Deborah Popper advanced the idea of the Buffalo Commons, a proposal that a significant portion of what was originally shortgrass prairie be turned into a vast nature preserve, mainly in order to restore ecological balance to the ecosystem. Considering that the project would have affected ten western, midwestern, and southwestern states and included vast areas that were privately held and already being farmed or ranched, it was untenable on the scale proposed.

One of the many stumbling blocks to any proposal for reintroducing bison on a grand scale is brucellosis, a contagious undulant fever that can cause abortions and infertility in elk, bison, and cattle. Common until the 1950s but now eradicated in all of the United States except the corner where Wyoming, Montana, and Idaho meet, its only known carriers today are elk and bison in the Yellowstone region. Although elk still travel freely, bison are not afforded this privilege, mainly due to concerns from area ranchers, this despite the fact that there have been no documented cases of a Yellowstone bison transmitting brucellosis to cattle.

Yellowstone National Park is the only place in the U.S. continuously occupied by bison since the early days of the country. By the turn of twentieth century fewer than one hundred remained, but within a few decades the herd had propagated so well it was necessary to cull. This practice was stopped in the '60s, the result being that within thirty years there were 4,000 head. ("Keep in mind," notes WWF's Dennis Jorgensen, "that every bison herd, if given enough land, is going to exceed the carrying capacity of the land.") They would travel out of park, especially in the winter, but Montana ranchers objected. The winter of 1996–97 was especially harsh. Thousands of bison left the park searching for food and more than 1,000 were slaughtered at the border. Hundreds more had to turn back into the park, where they starved. Since 2005 the Montana Department of Fish, Wildlife & Parks has experimented with public hunts, allowing 25–50 animals to be hunted by lottery near park borders—much to the dismay of groups like the Buffalo Field Campaign, whose motto is Protecting the Last Wild Bison.

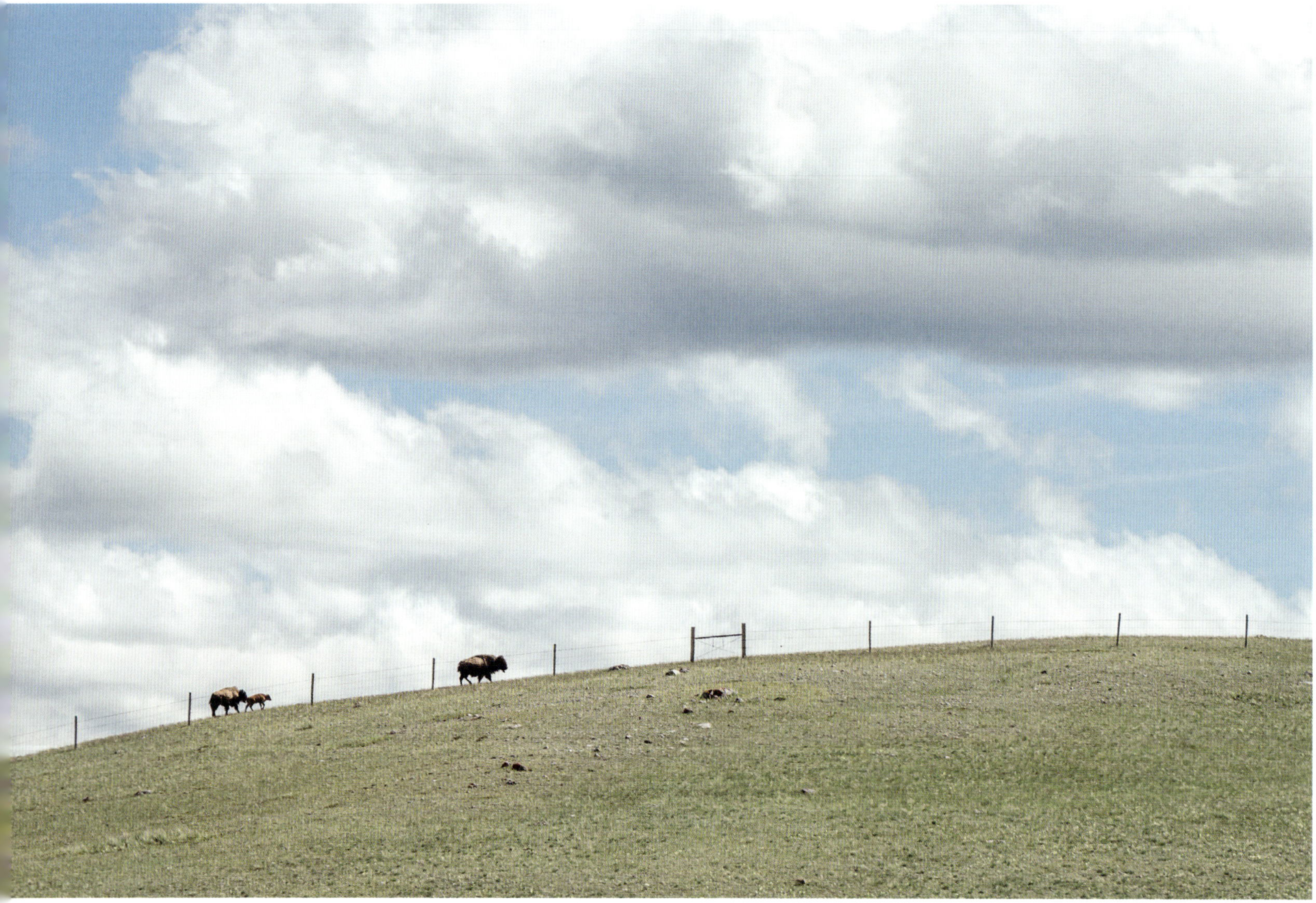

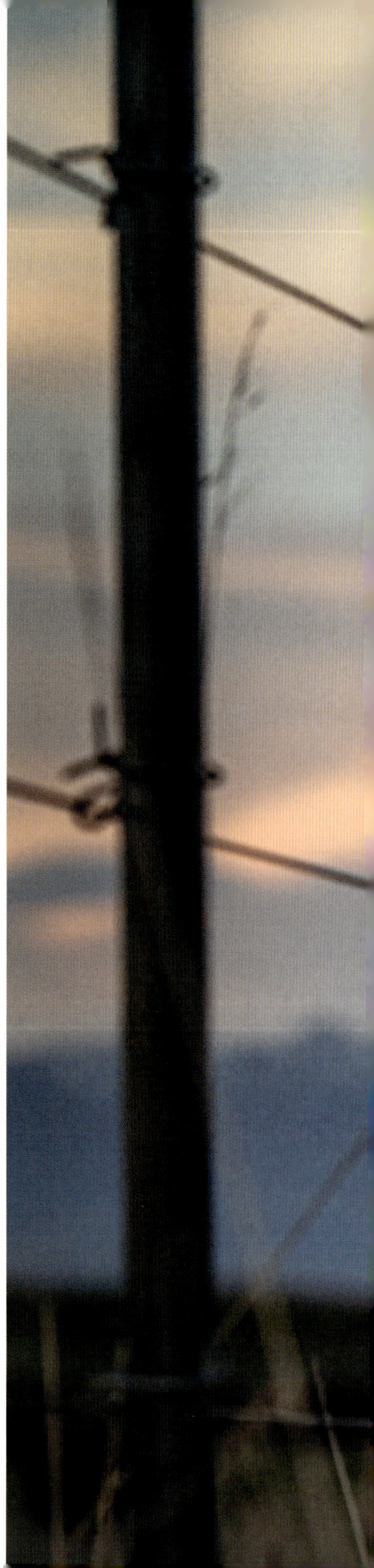

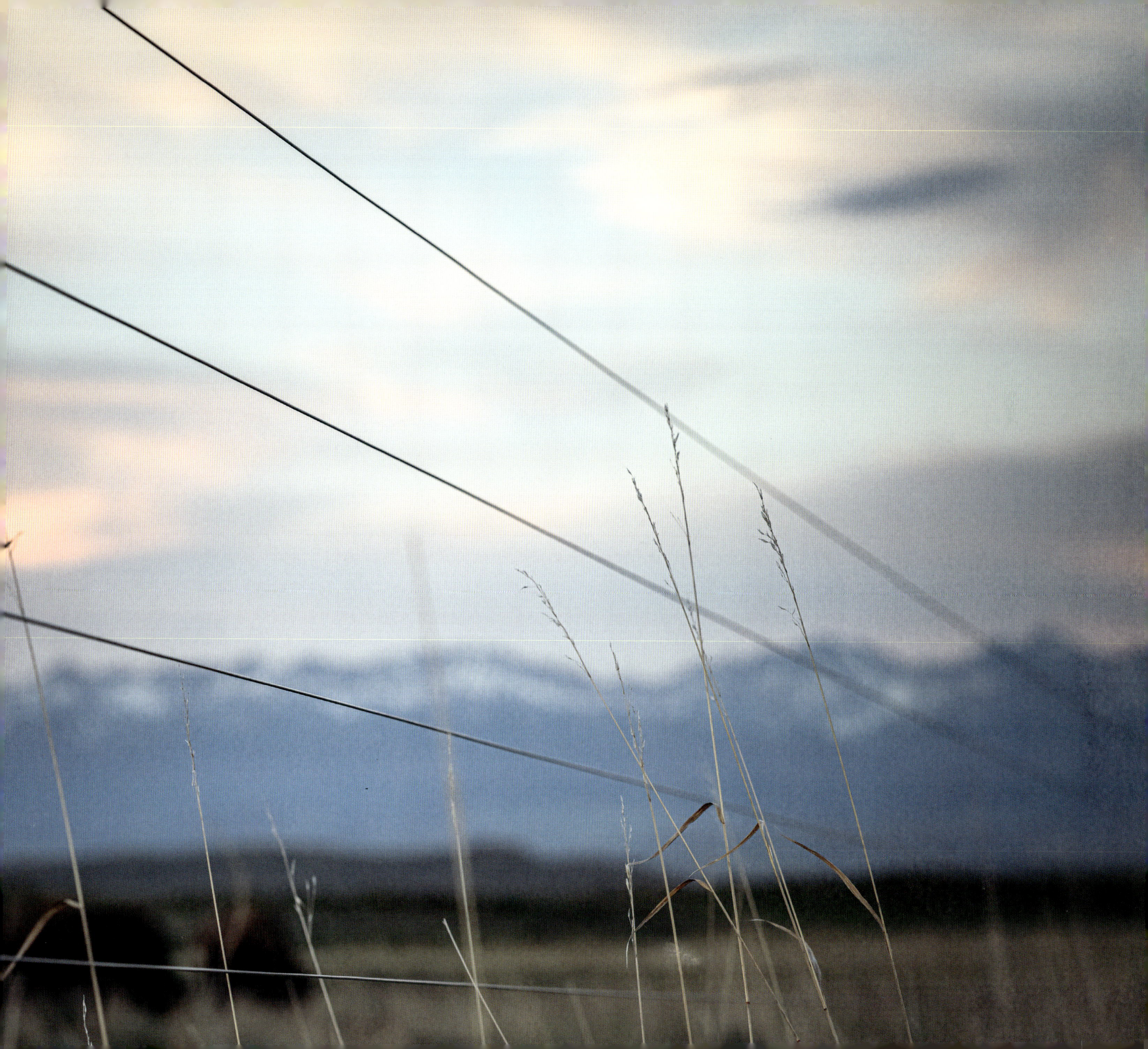

Under the terms of the Interagency Bison Management Plan, drafted by federal and state agencies and Native American groups, there's also an annual capture within the Park. "Yellowstone buffalo are not free roaming," notes George Horse Capture, Jr. "It's like Indians in the past; you left your reservation and that could be a death sentence."

Bison are in fact classified as livestock. This has resulted in a conundrum that some experts refer to as "being saved behind a fence." Grizzly bears, for instance, have been repopulated and wolves reintroduced in the American West; today their populations are both free-roaming and successfully propagating. Bison, however, despite their dramatic increase in numbers, are almost all either literally or virtually fenced. Unlike moose and elk, bison as a wildlife species don't yet exist—though many are hoping to change that.

The thesis of the Buffalo Commons—that reintroducing bison to prairie environments is the ecologically sound path—is being advanced on multiple fronts and there are efforts underway to restore bison as a wildlife species, a particular focus of yet another group, the National Wildlife Federation.

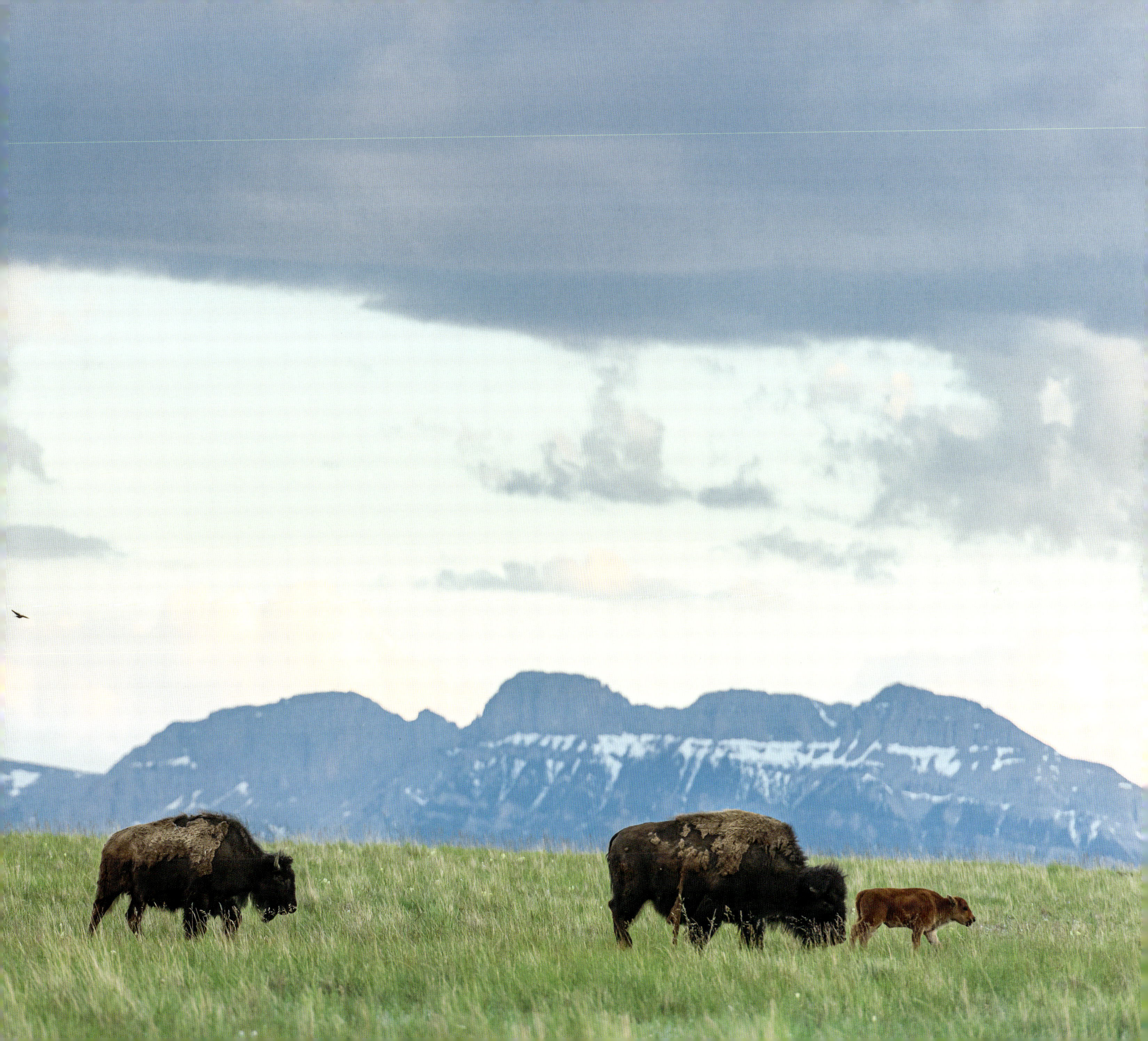

Jason Baldes is Tribal Buffalo Coordinator for the National Wildlife Federation and serves as Secretary of the InterTribal Buffalo Council. As a member of the Eastern Shoshone tribe, he also manages the Shoshone tribe's buffalo herd on the Wind River Reservation in Wyoming. "The tribes on the Wind River Reservation have a history of conservation success," he says, pointing to the establishment of 180,000-acre roadless area in the 1930s, several decades prior to the passage of the 1964 Wilderness Act. More recently, strict game regulations established in the 1980s have resulted in the reestablishment of robust populations of species such as pronghorn antelope and bighorn sheep.

Baldes' father was a wildlife biologist with the Fish and Wildlife Service. After father and son traveled to East Africa to witness the wildebeest migration, Baldes returned home with a vision to restore buffalo to the Wind River Reservation where they hadn't been seen since 1885. Beginning with ten animals in 2016, and adding ten more the following year, the herd now numbers about thirty essentially free-ranging buffalo on 67,000 acres. The eventual goal is a herd of one thousand, but the process will take time. "It's as much about managing the animals as it is about managing people," Baldes says. "The challenge is to step out of the paradigm of treating them like cattle. Initially they have to be behind a fence because no one can remember how to see them as animals on the landscape."

In 2020 the U.S. Department of the Interior announced the Bison Conservation Initiative, a ten-year directive to pursue the goals of wild, healthy bison herds; genetic conservation; ecological restoration; cultural restoration; and shared stewardship. Around the same time, the Rosebud Sioux Tribe announced the creation of the Wolakota Buffalo Range, almost 28,000 acres of native grassland with the capacity to support a herd of 1,500. WWF's Dennis Jorgensen also points to an effort by the Blackfeet to restore a semi-free-ranging herd, one that would be allowed to move freely across the border between Glacier National Park and Canada's Waterton Lakes National Park. Perhaps one day the bison-as-livestock versus wild-and-fenced versus free-ranging question might ultimately arrive at a comfortable resolution. In the meantime, increasing numbers of buffalo are ranging more and more unfenced acres in as close to a wild state as possible to the benefit of the land, the people, and other wildlife species.

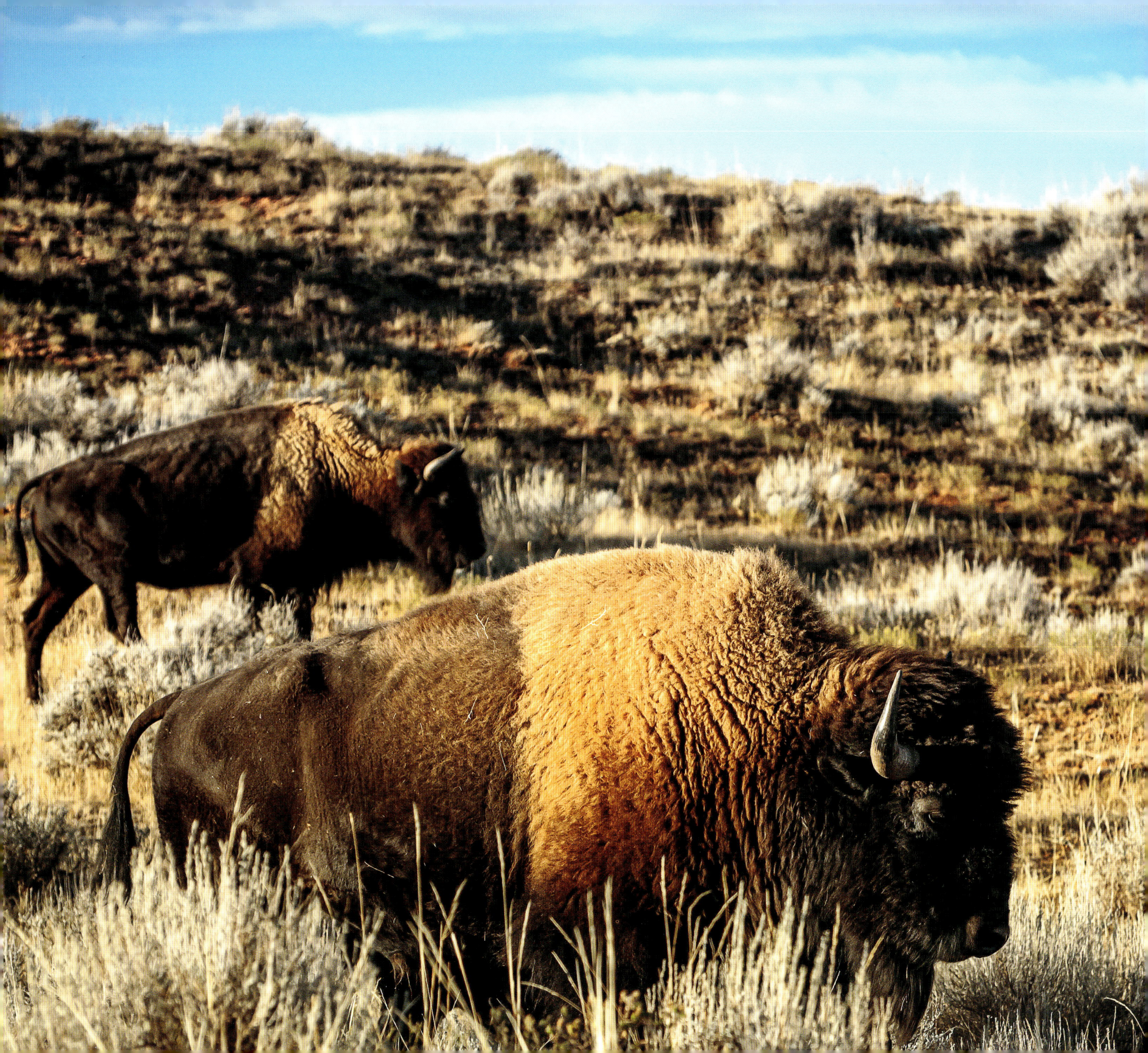

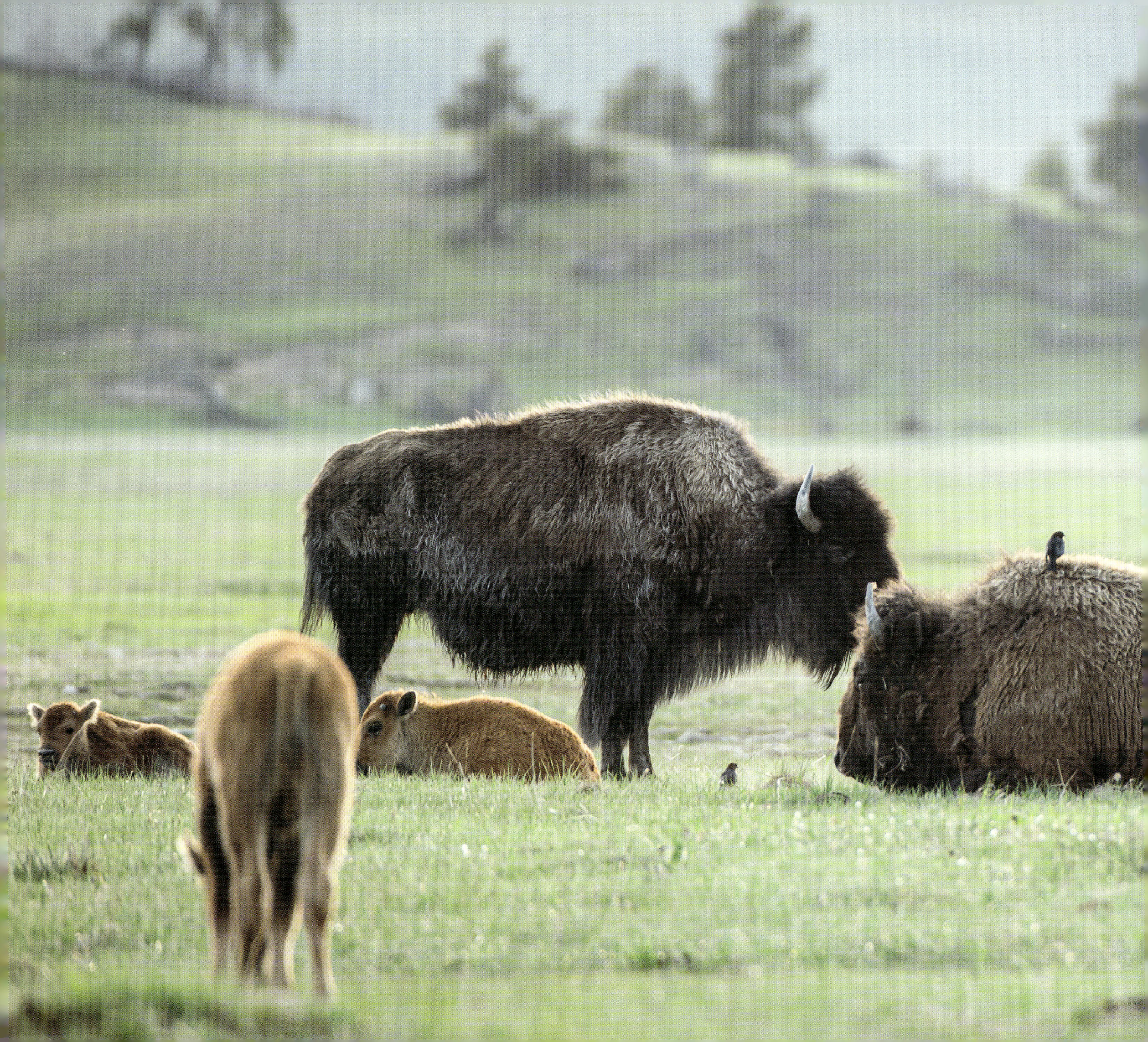

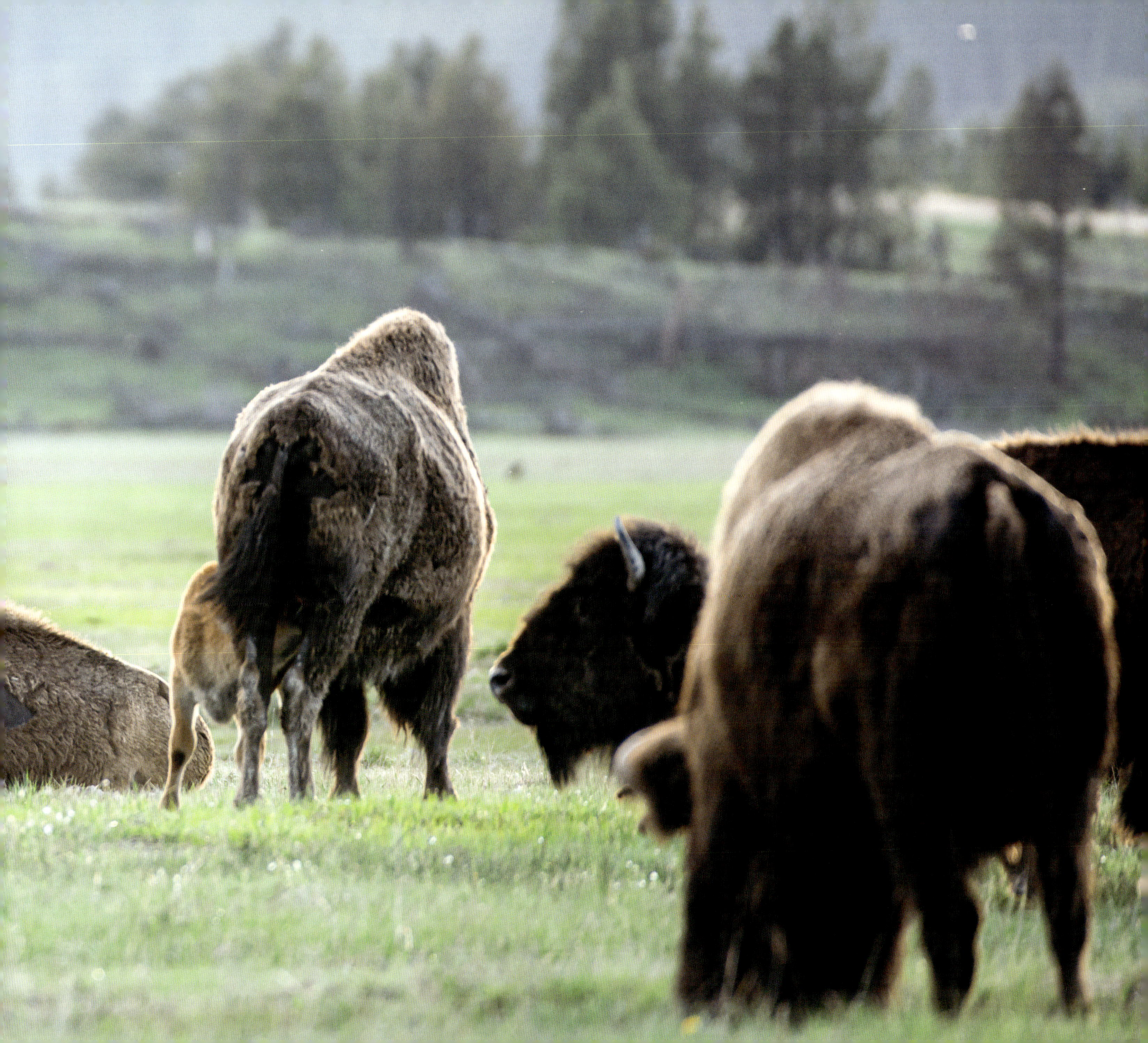

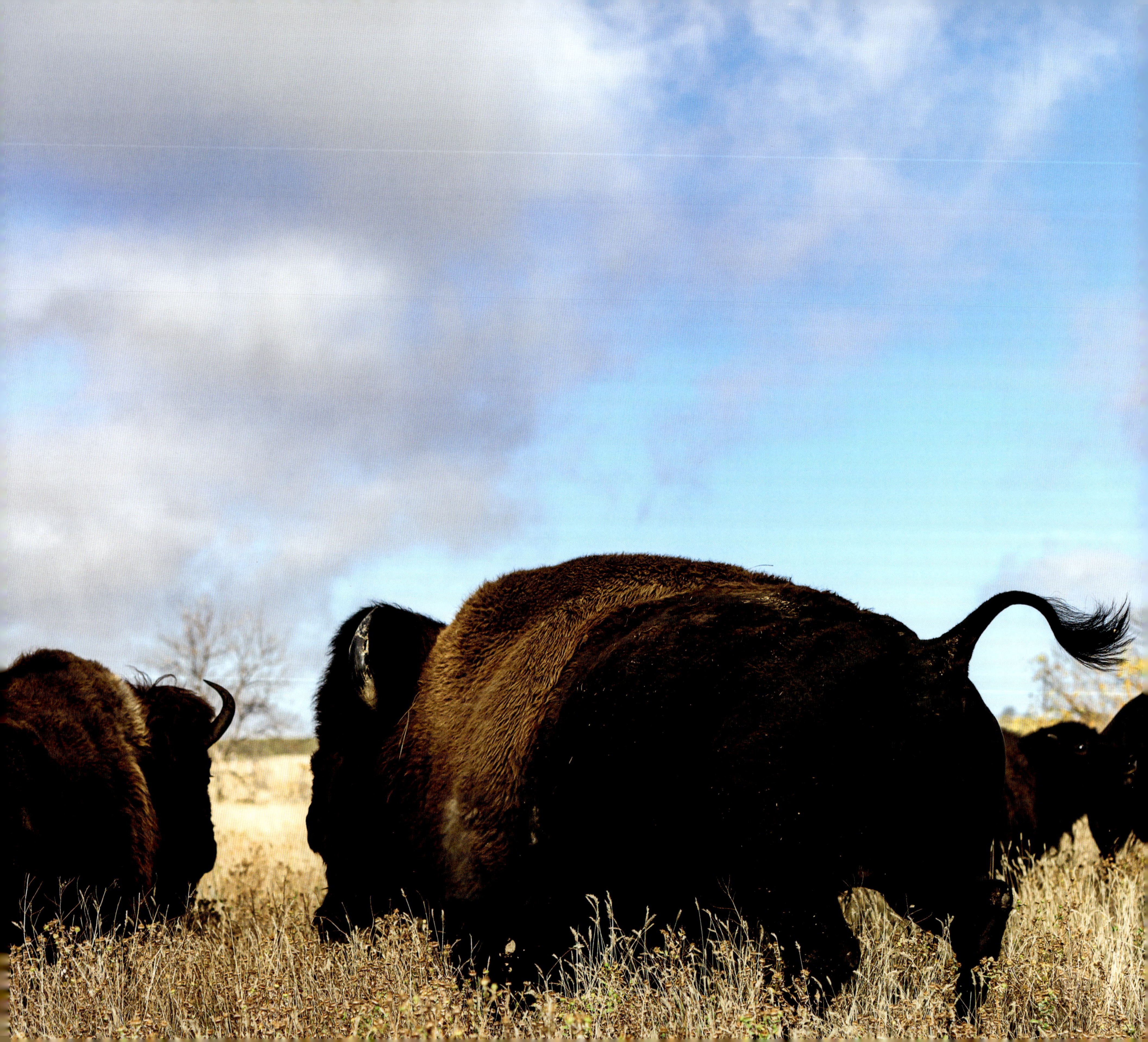

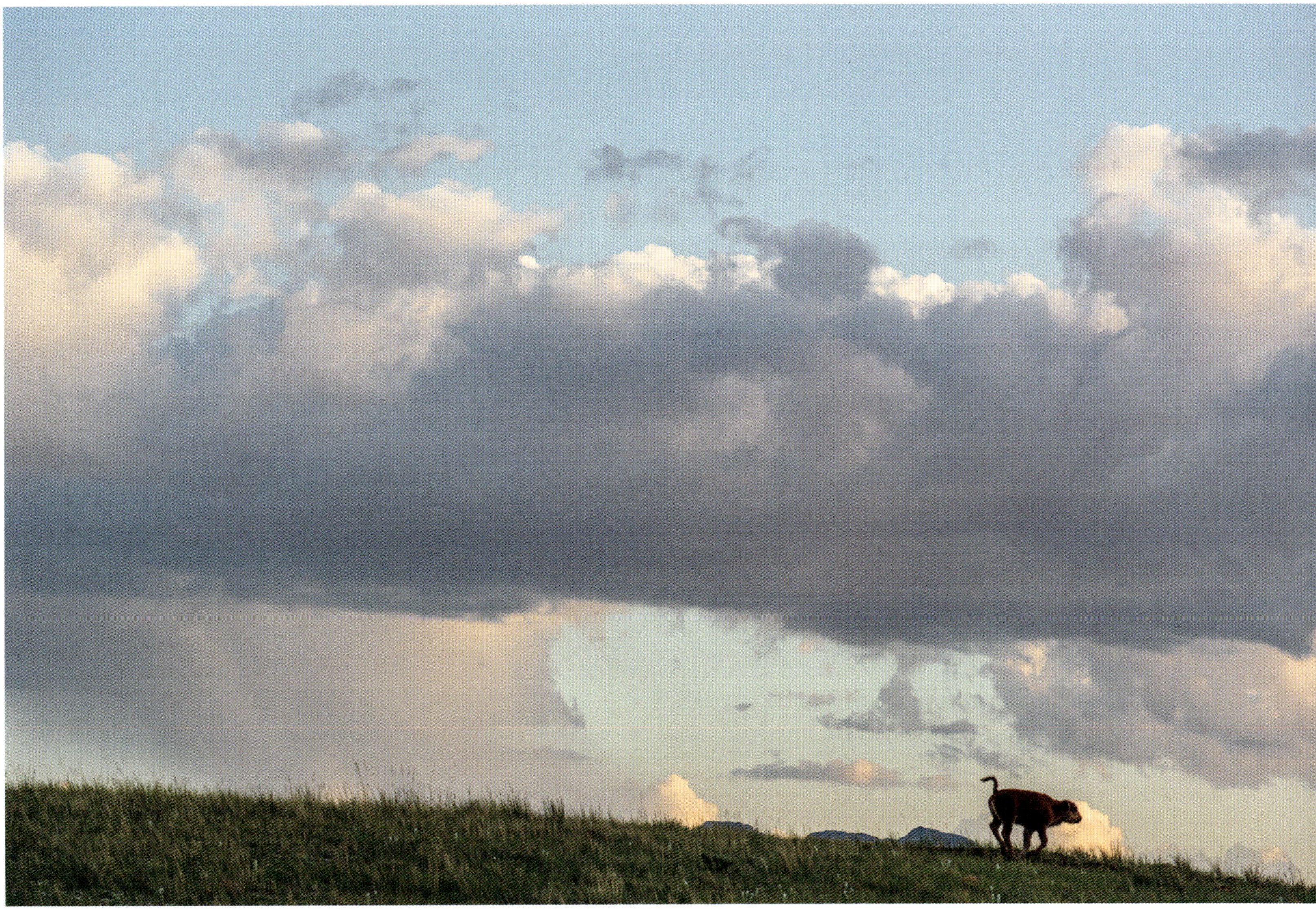

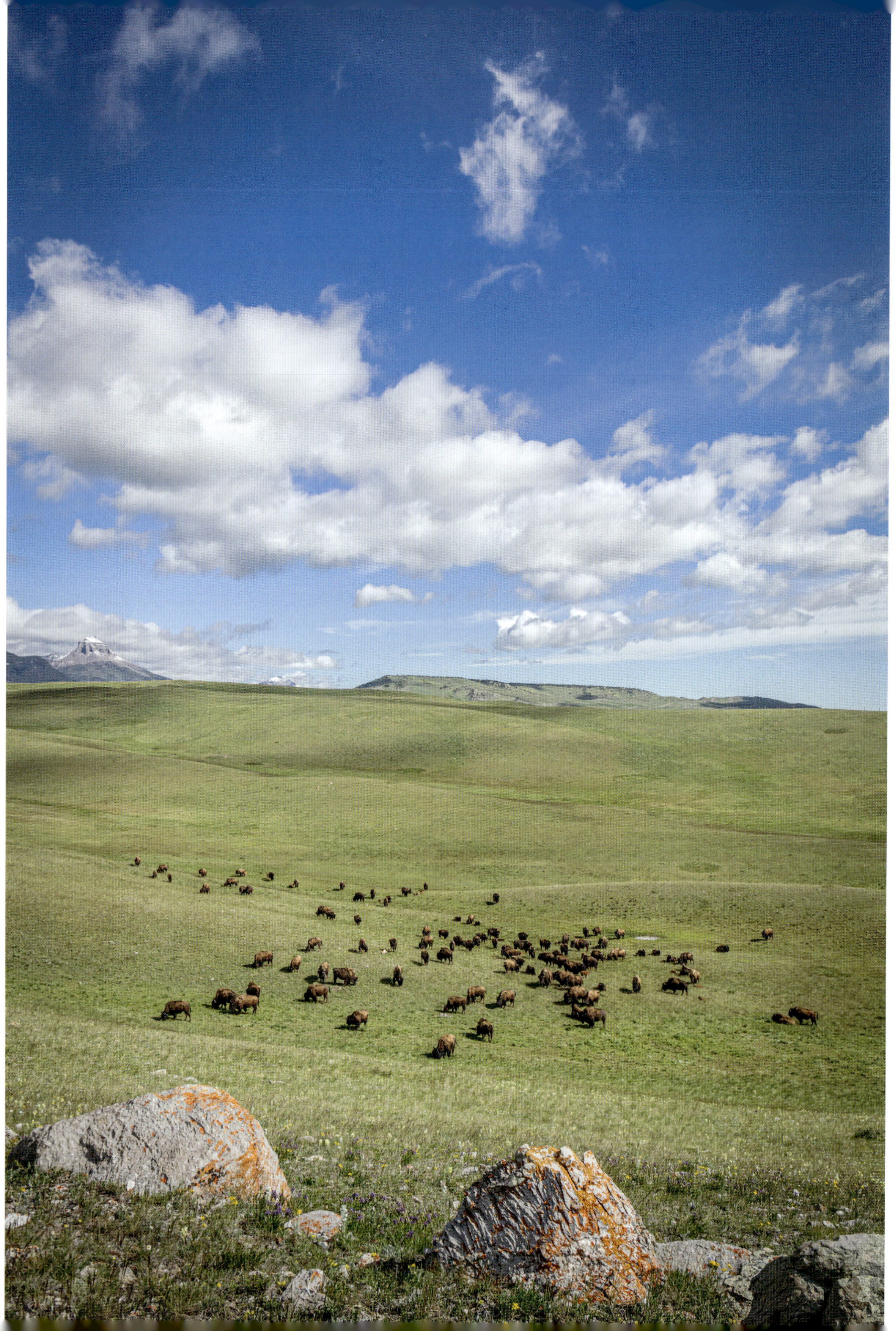

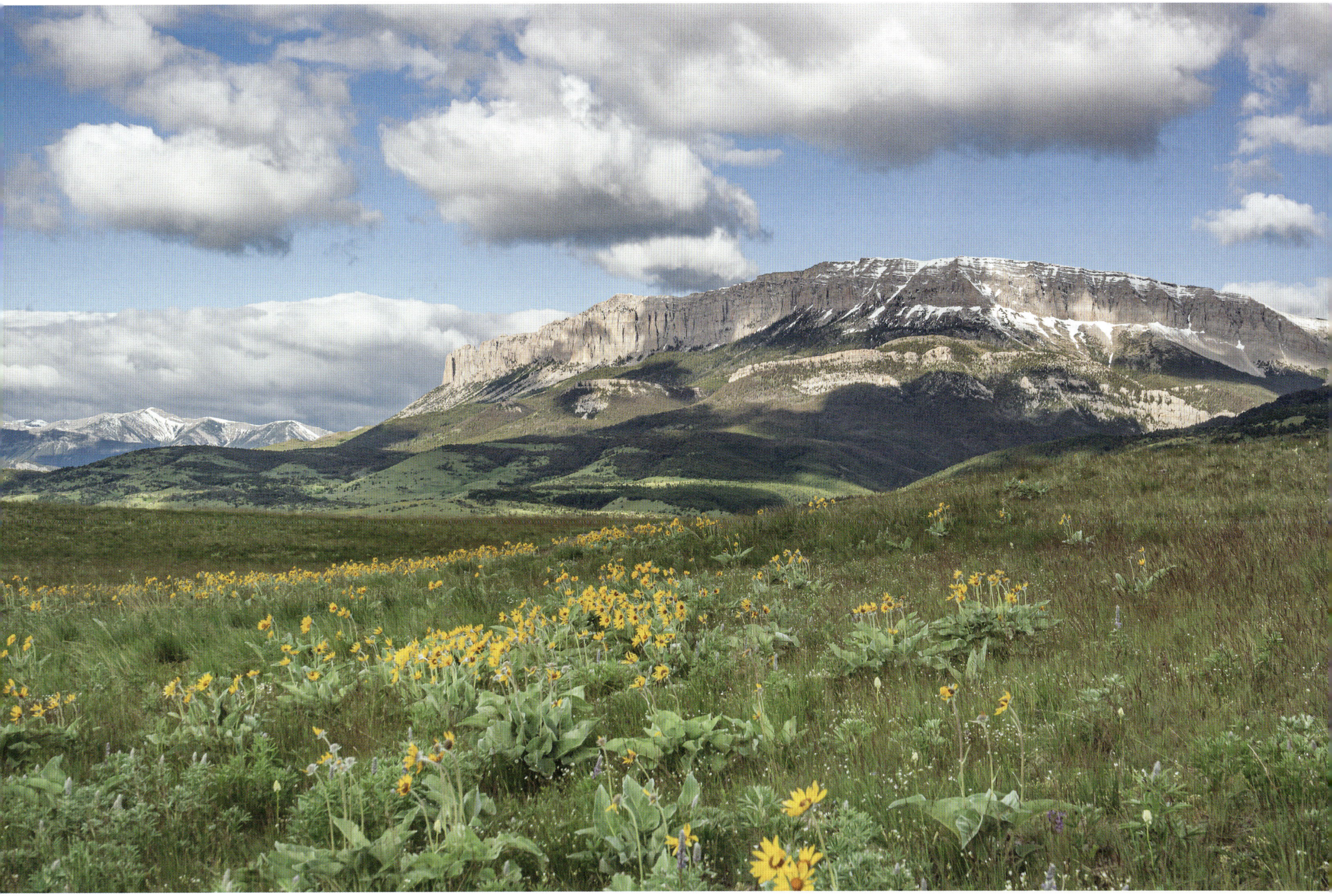

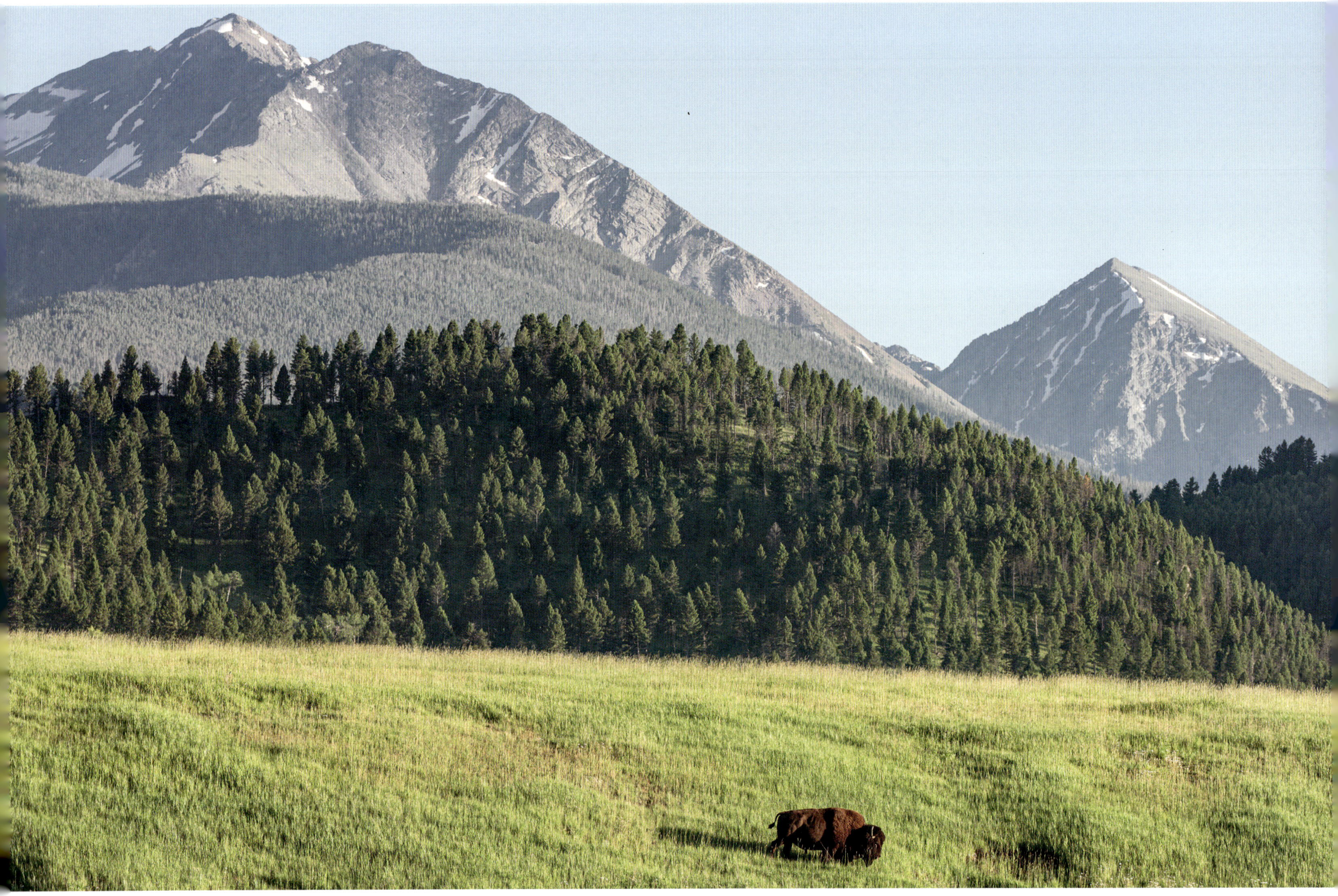

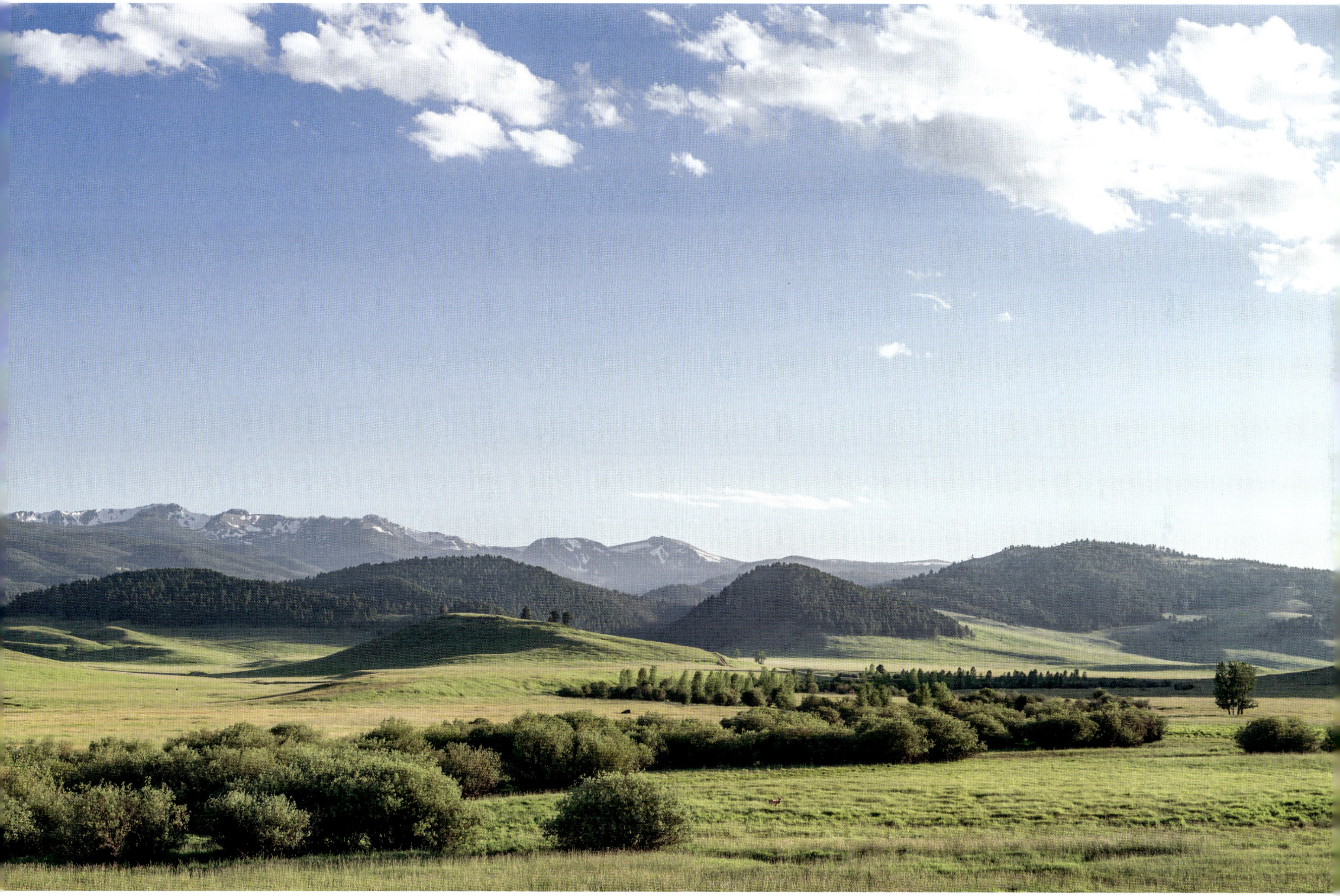

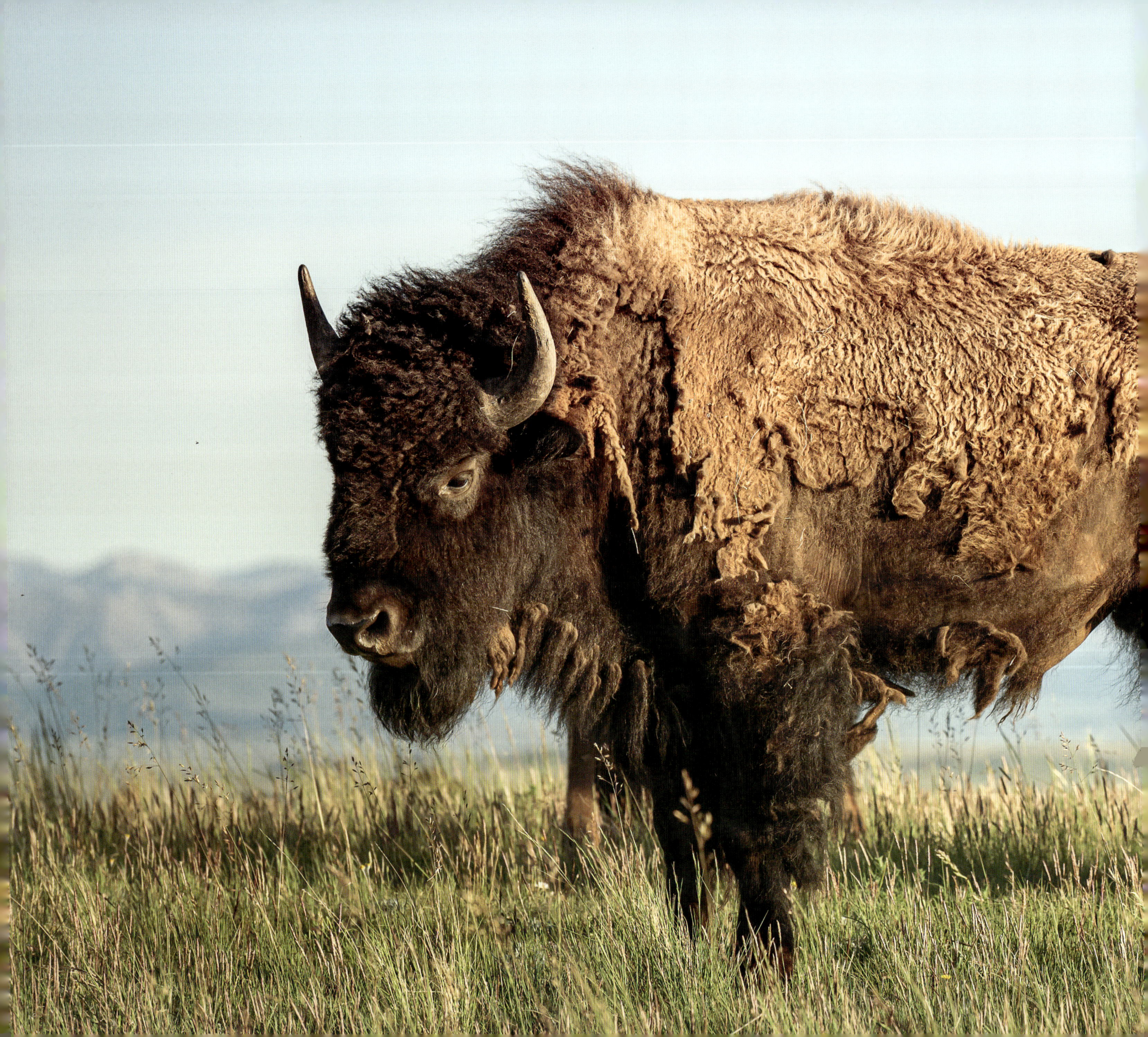

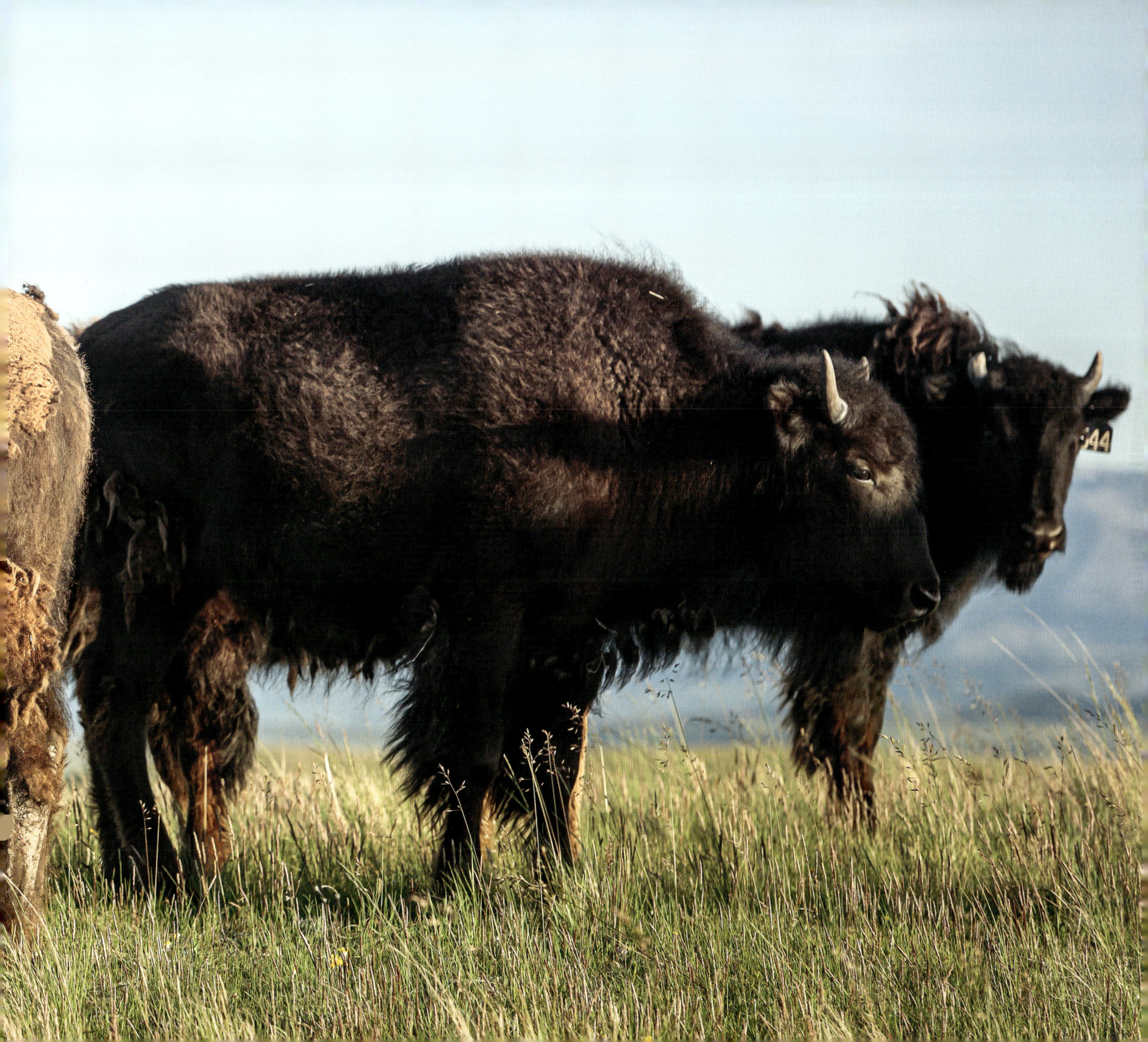

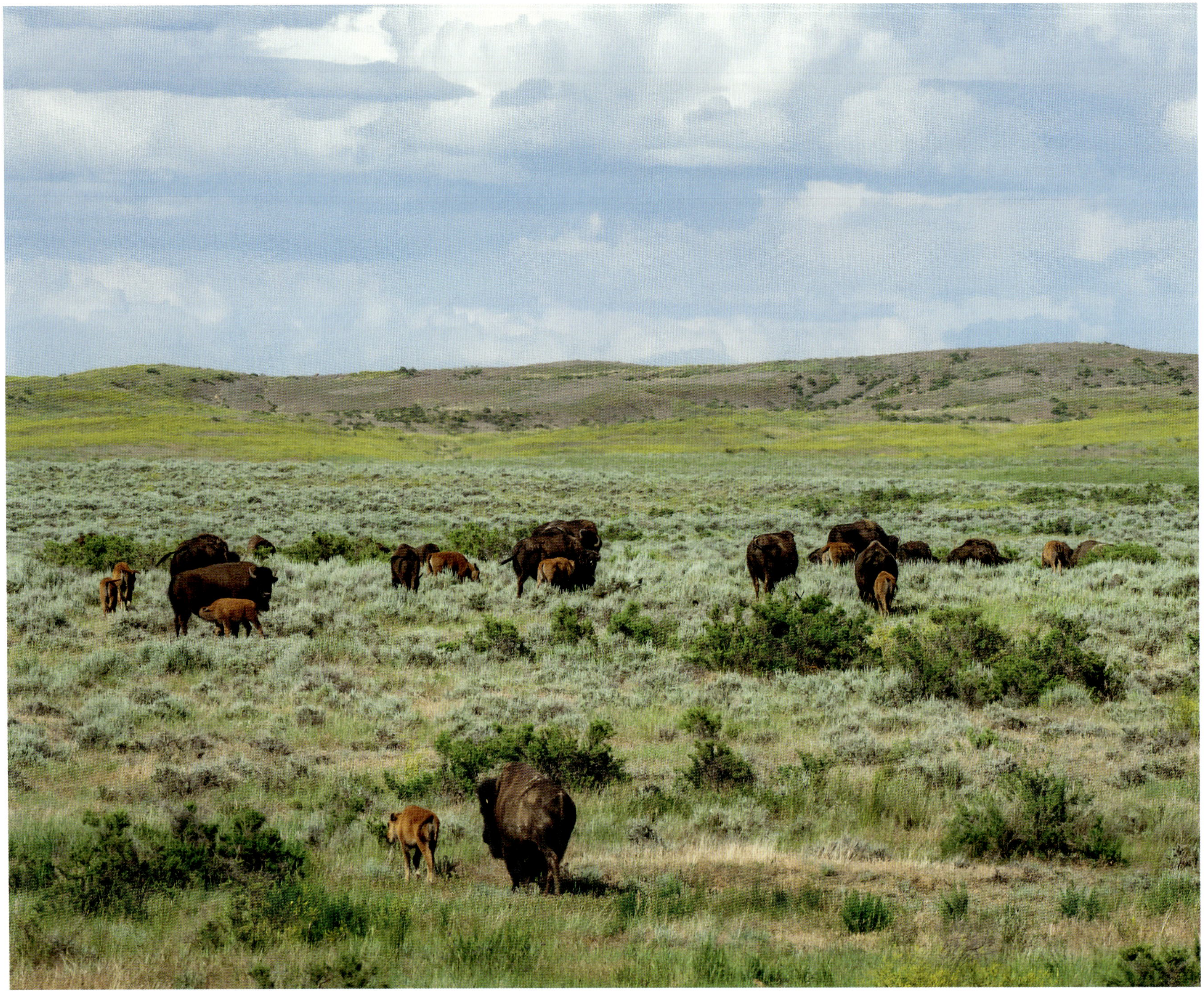

Eternal Spring

The American Prairie Reserve's Kestrel Camp in Montana is an hour from the nearest town (Malta, population 1,950) and accessed by dirt roads. When it rains, they become slick with bentonite, a substance that causes slow-moving cars on level surfaces to slide sideways right off the road. During a six-month winter, blizzards are a fact of life. Although snow, hail, and downpours can happen anytime, summer and fall are when the prairie comes alive with wildlife, when the green grasses ripple in the breeze, when sunflowers bob their heads along the roadsides, and meadowlarks sing on barbed-wire fences. It's when the big skies of Montana are at their most expansive.

Most of us think Montana and visualize mountains, trout streams, high-end ski areas, or the spectacular Going-to-the-Sun Road in Glacier National Park. But in the north-central section of the state, where the Missouri River flows for more than 700 miles, it's open country, sparsely settled, with nary a wi-fi cafe in sight. It's a land of farms and ranches, combines, and grain silos. It's also a land of natural grassland prairie, tens of thousands of acres of which have never been tilled.

When in the spring of 1805 the Lewis and Clark Expedition first passed through the spectacular Missouri Breaks region, they encountered vast herds of bison. As settlers subsequently moved onto the Plains and bison were virtually exterminated, other charismatic megafauna—grizzlies, mountain lions, elk—were displaced to the mountains while nuisance animals and predators such as prairie dogs, coyotes, and wolves were enthusiastically poisoned and shot. Some of the land was given over to grazing, and some to farming, but in much of the region its native grasslands are untilled.

Two decades ago, a group of conservationists conceived of a project to protect a large swath of this ecosystem, one of the world's last intact grassland ecosystems, by creating a vast wildlife preserve to restore the ecosystem and the species that evolved there. Their vision can be imagined on a map as a 3.2-million-acre oval through which the Missouri River runs from west to east. Up to 500,000 acres of private land would be combined with a patchwork of public lands (the Charles M. Russell Wildlife Preserve, the Upper Missouri River Breaks National Monument, national forest, and BLM lands) and the Fort Belknap Indian Reservation to create the largest wildlife preserve in the continental U.S. Fences would be removed, pronghorn antelope and prairie dogs protected, and bison herds would be reestablished. Ultimately, other species would follow. This would be possible because, unlike most of the United States, "The foundation of our ecosystem remains largely intact," explains Dr. Daniel Kinka, APR's Wildlife Restoration Manager. "It's changed drastically over 150 years, but by and large the grasslands are still there and have tons of capacity to support large amounts of wildlife."

Bison, he explains, are "ecosystem engineers," partly due to their manner of grazing, and the ways in which they move across the landscape, and partly because of their habit of wallowing. Bison create depressions in the earth that they then roll in, to cool off, shed fur, rid themselves of parasites, scratch their backs on a treeless plain, and mark their territory (they often urinate in the wallows, especially during fall rut). They'll continue to use the same spots year after year, which creates permanent intermittent wetlands. The inadvertent effects of this behavior on the ecosystem are enormous, Kinka explains, from carrying seeds picked up in the animals' coats from one location and transplanting them to another, to literally creating habitat:

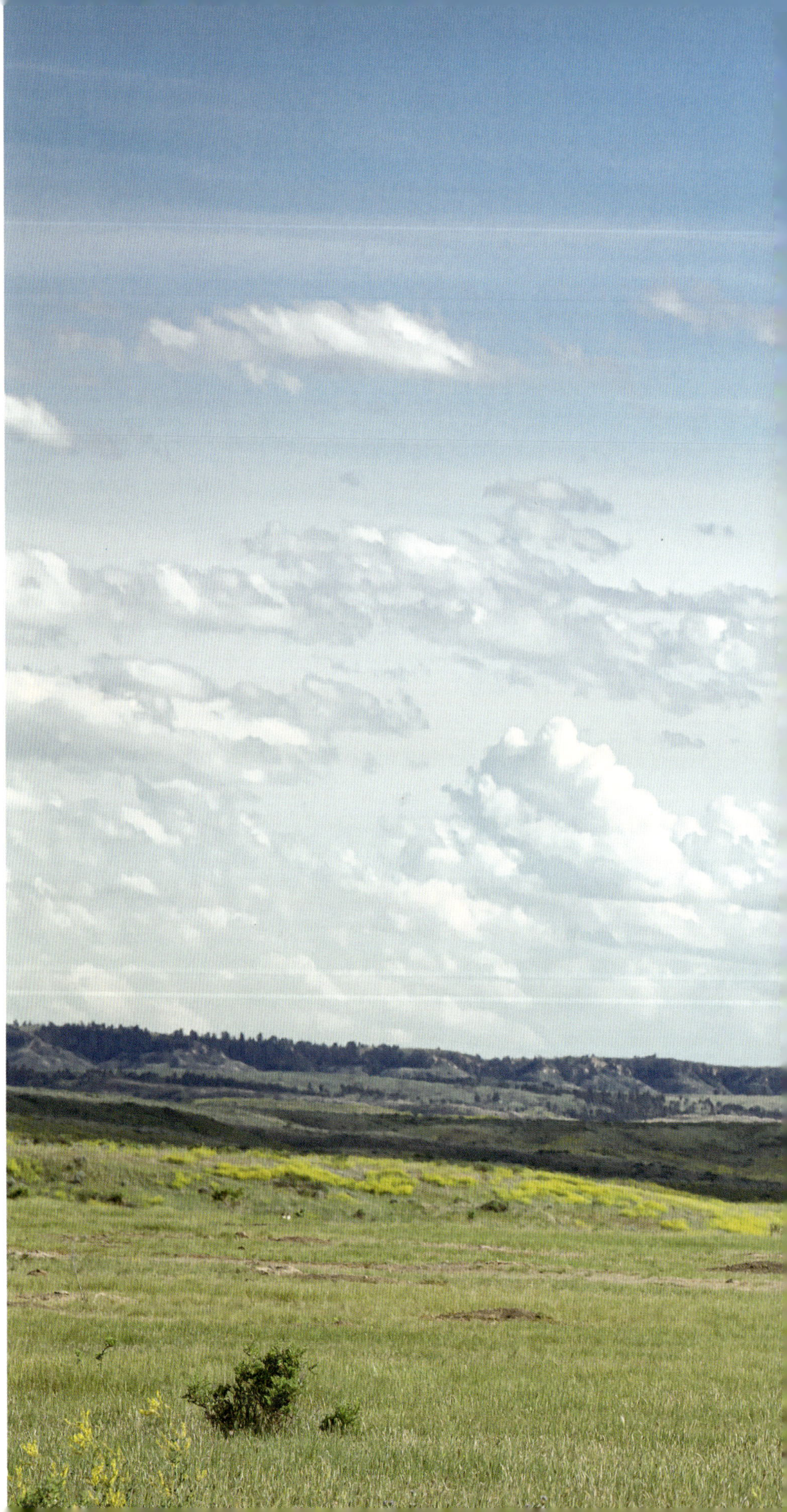

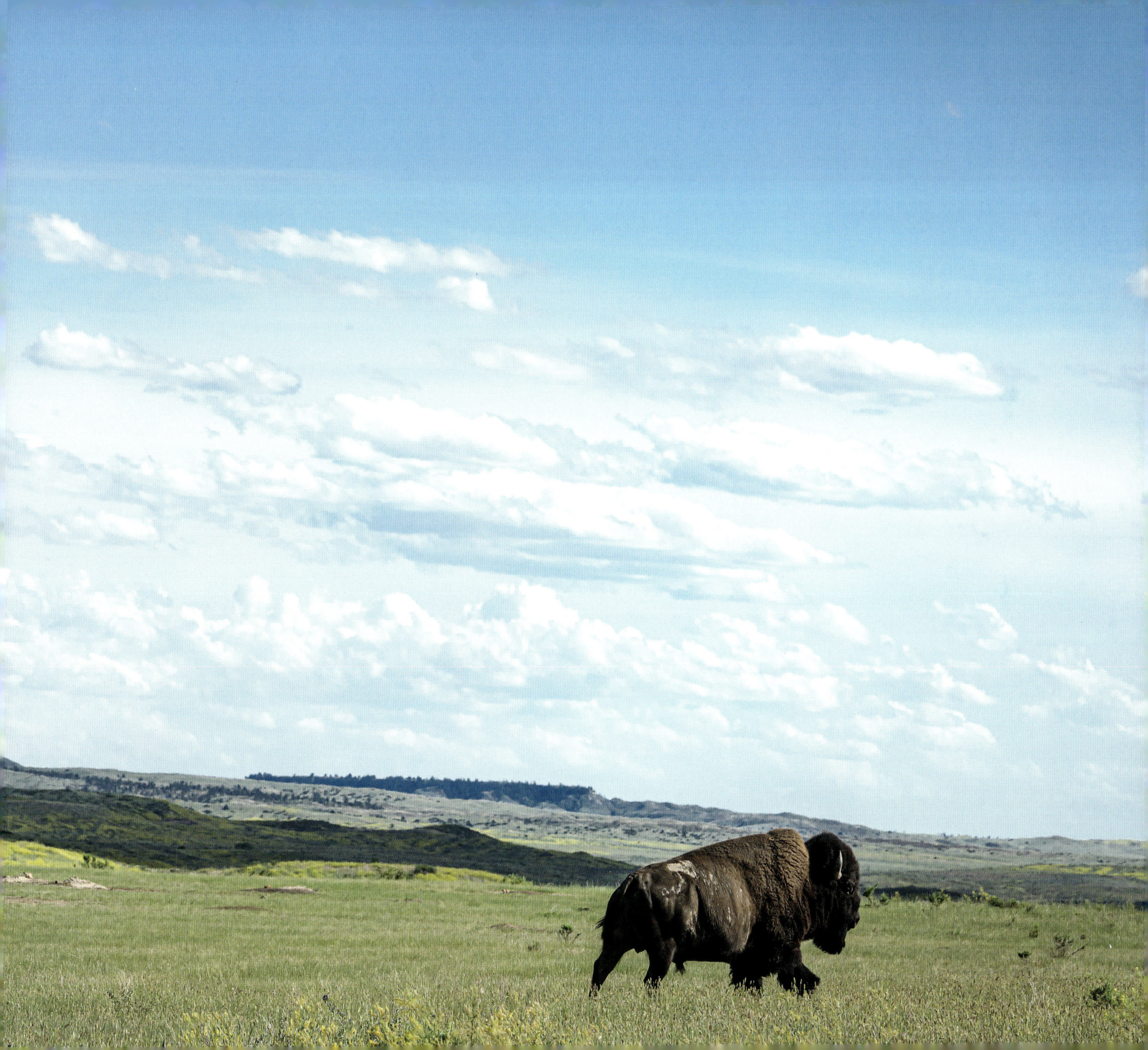

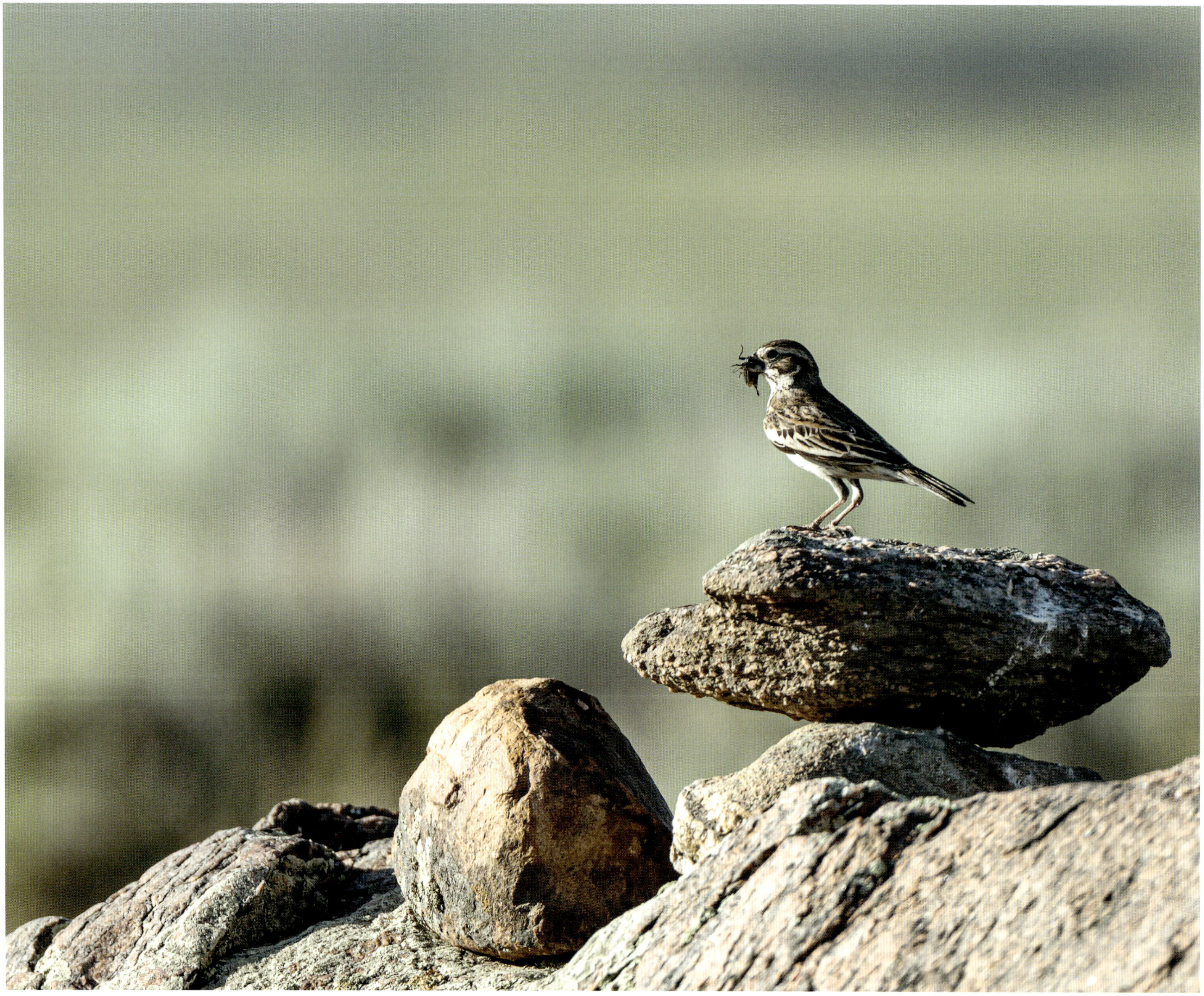

Sage grouse need soggier ground; it's very important for this iconic species. Bugs and amphibians spawn in the wallows. Even if they live in a dry environment, they need a wet environment for spawning. We have preliminary evidence that even the plants at the edge of a wallow are different because there the soil is more compacted by the bison. The edges of the wallows become their own kind of habitat, like a tidepool. It might only be six inches wide, but these kind of halos become important for a whole series of grazers.

And unlike cattle, he adds, bison don't tend to hang out in riparian corridors. They drink and move on, allowing flourishing willows, healthy stream edges, and plants that haven't been trampled to sustain elk, mink, beaver, and those species of fish that need shade.

Bison earn their title as a keystone species by "mowing the North American lawn," Kinka continues:

Some areas are grazed where they're taking half and leaving half. Others areas are grazed to the ground. Still others are missed and can go ungrazed for years. This creates a mosaic of grass types and habitats that create important habitat for birds. {Until the bison returned}, the birds that require the grass to be half grazed down were fine, but birds requiring super low or ungrazed grass were in peril. A prairie system experiences regular booms and busts with ice and fire and it has to have a mosaic. As soon as you smooth out that mosaic, you lose biodiversity.

New research based on observation of Yellowstone's bison herds conducted by scientists from the U.S. Geological Survey, the National Parks Service, the University of Montana in Missoula, and the University of Wyoming suggests an amendment is needed for the Green Wave Hypothesis, the idea that as spring returns a green wave of new plant life creeps up in elevation. "Most herbivores follow that green wave because the spring grasses are the most nutritious," says Kinka. "But bison break the mold. The way they graze locks the grassland into spring because every time the plants start to grow, the bison nibble them down just enough [to promote regrowth]. Most creatures have to chase the seasons. Bison have this ability to freeze spring in place."

Since 2004, the APR's starter herd of bison has grown from sixteen to about 850, managed in three groups. The Reserve has completed twenty-nine transactions for a protected network for wildlife of more than 350,000 acres. APR is also working with local ranchers to implement antelope-friendly fencing and install wildlife cameras so landowners can be reimbursed for providing habitat for key species. They've set up campgrounds, restored a historic schoolhouse, built an education center, and launched a hut-to-hut system for hikers. Campgrounds are maintained for public use free of charge. Despite the fact that ranching continues in certain areas of the preserve and some controlled hunting is allowed, some neighbors regard its activities with suspicion, concerned that the return of bison might somehow impact their land, their livestock, or their way of life.

A traverse of the Reserve in search of bison involves hours of driving on dirt roads. A routine early summer foray might include braking for a family of sage grouse, glimpsing baby marbled godwits tumbling amongst sunflowers, and seeing a flock of pelicans take flight from a lake ringed with rushes. There's an ancient buffalo jump, used by generations of Indians to herd large numbers of bison to their deaths. The buffalo jump represents one way of life on the Plains, while a restored schoolhouse represents another—the latter poignantly symbolizing the demise of the former.

When visitors locate a herd—perhaps about 100 bison, mostly females with their bright red calves, plus a few young bulls—they might take photos and observe through binoculars. But even at 300 yards distant, the cows are uneasy. One by one the animals get to their feet. Slowly the group starts moving away. Soon they are trotting.

They disappear down into a little wash lined with vegetation and reappear on the far side as a sizable herd moving across the prairie. This isn't like Yellowstone, where the last direct descendants of the millions that roamed the Great Plains, although technically wild, nonchalantly share the winter roadways with snowmobiles. Wary of humans, driven by instinct, they churn the ground with their hooves, gathering momentum, and merge into a single dark line across the vast brown and sage plain.

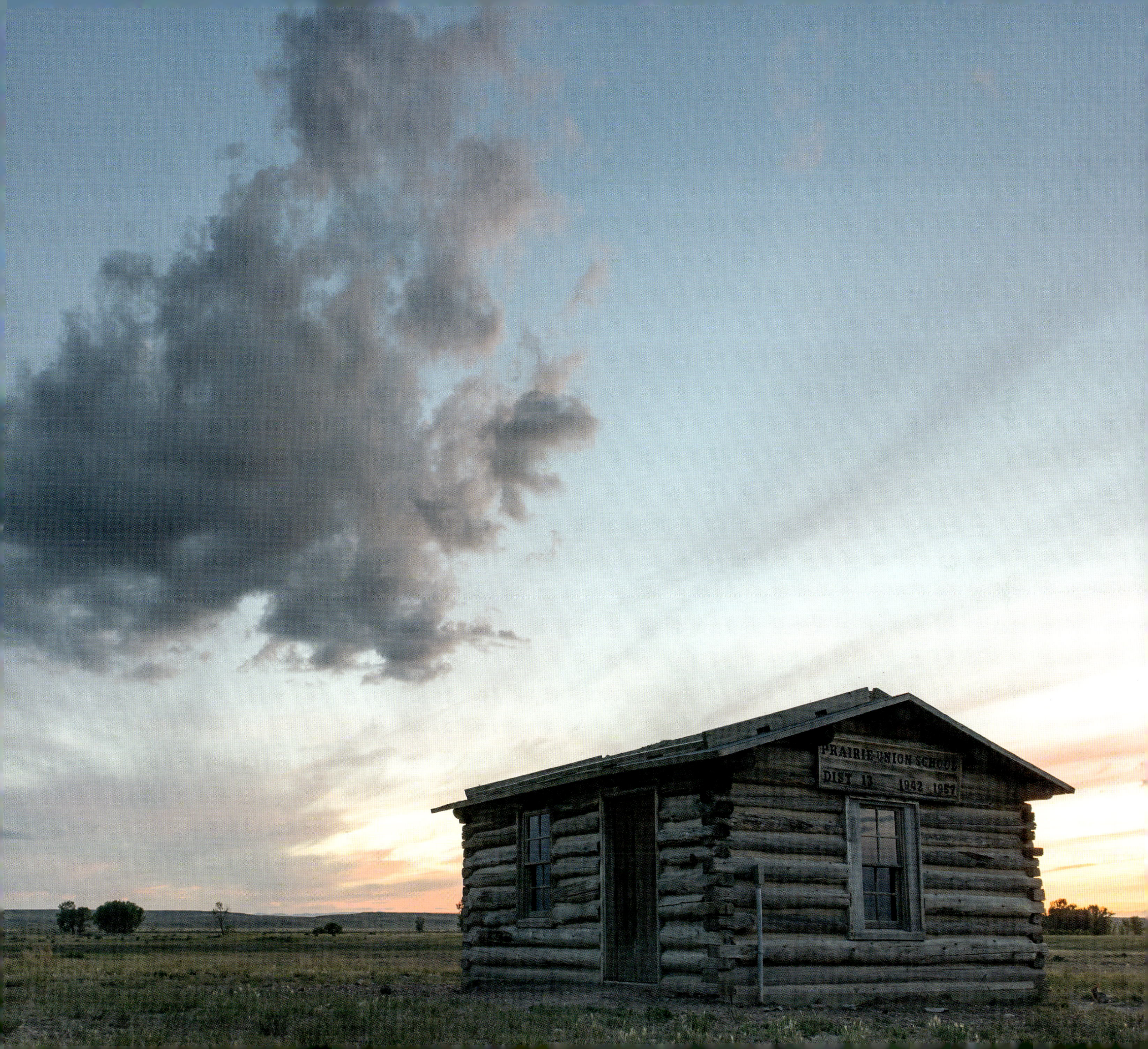
PRAIRIE UNION SCHOOL
DIST 13 1942 1957

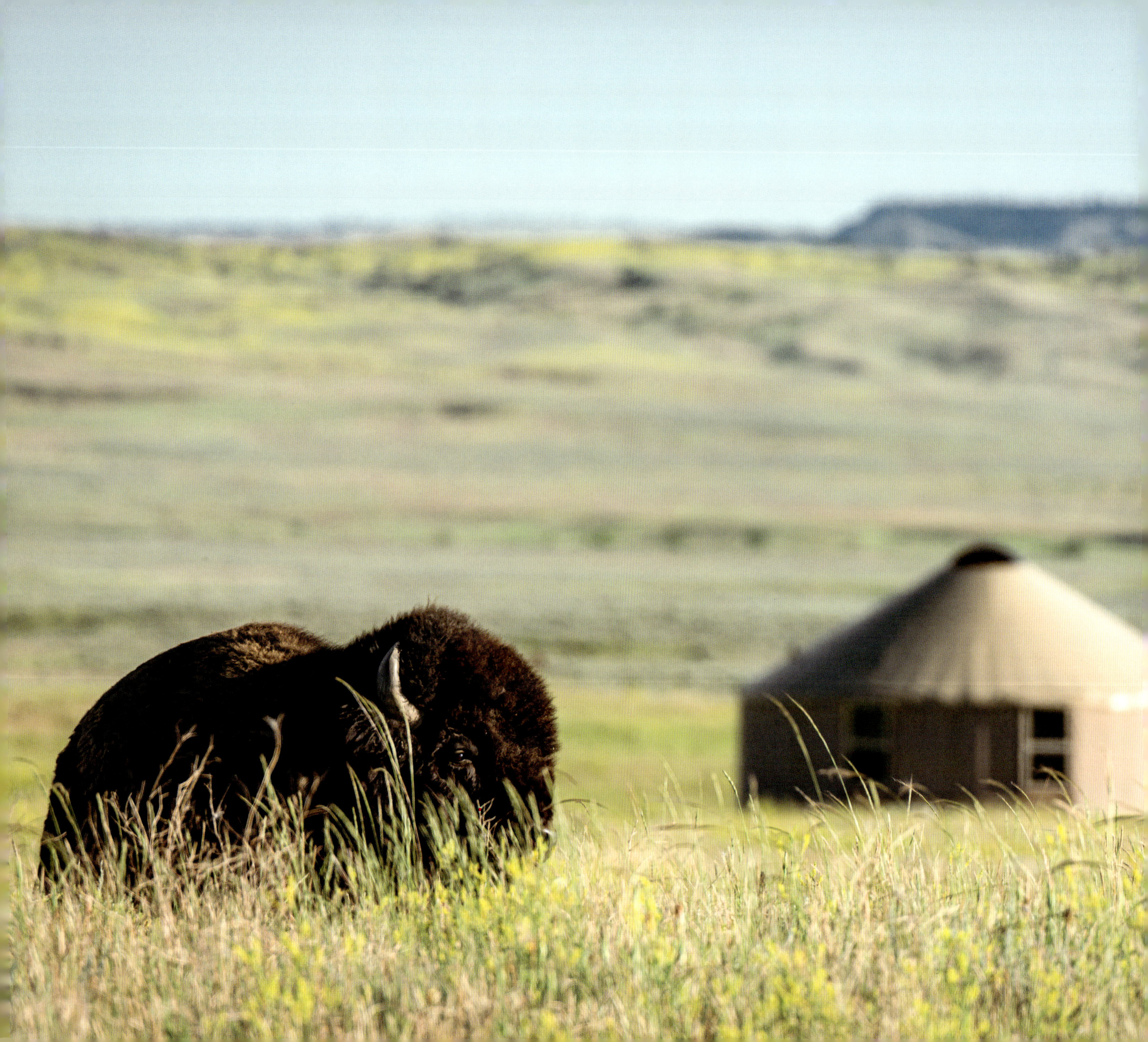

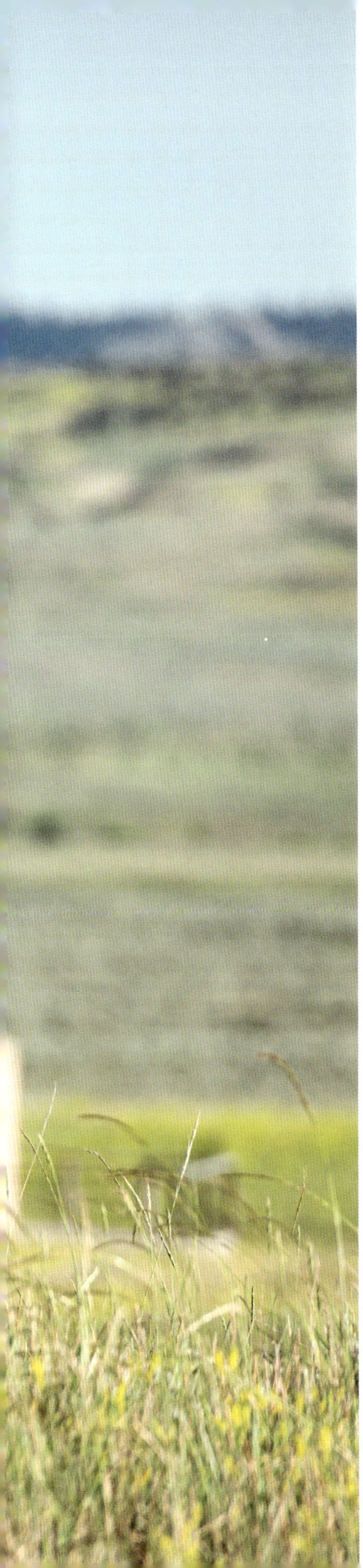

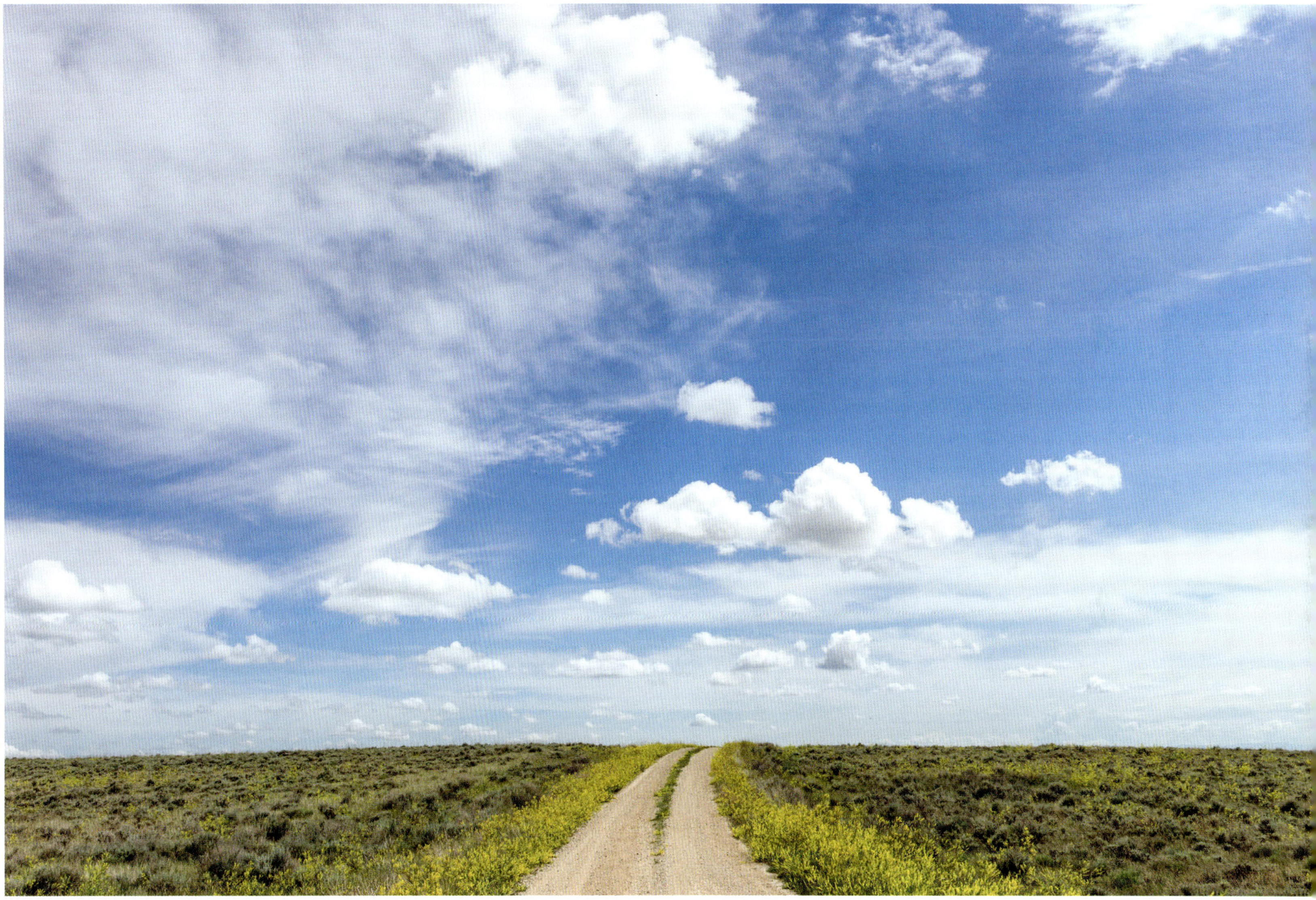

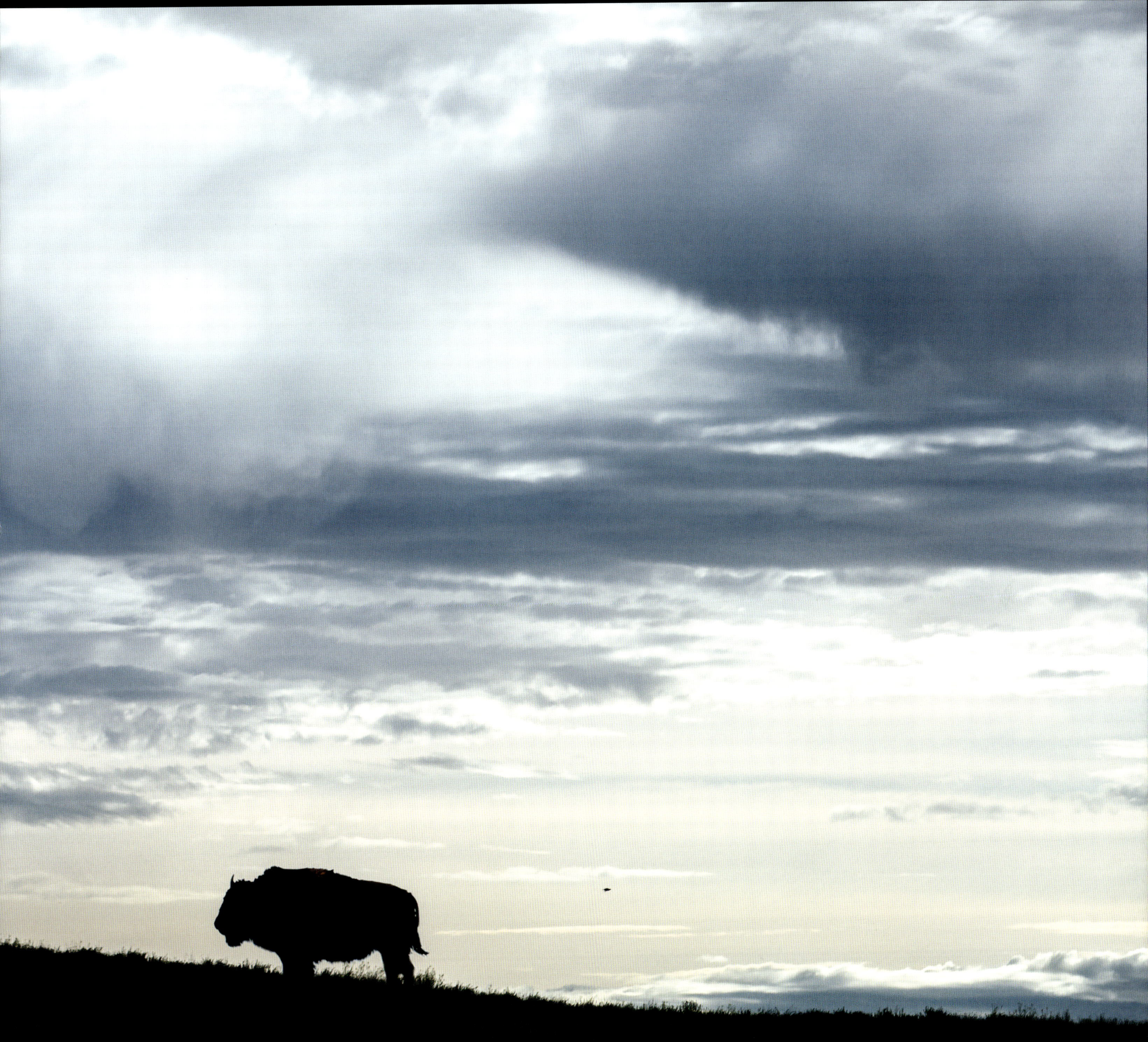

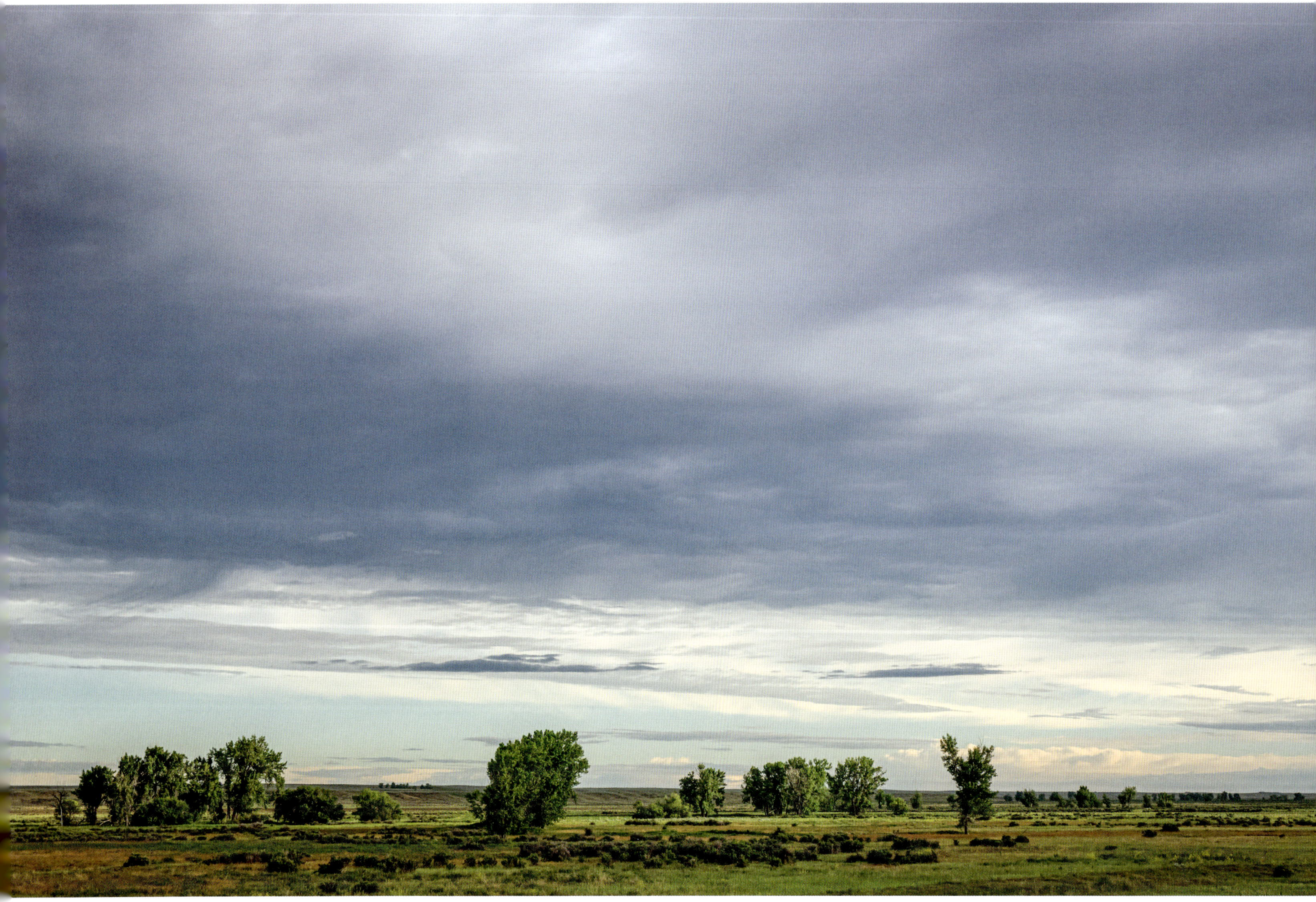

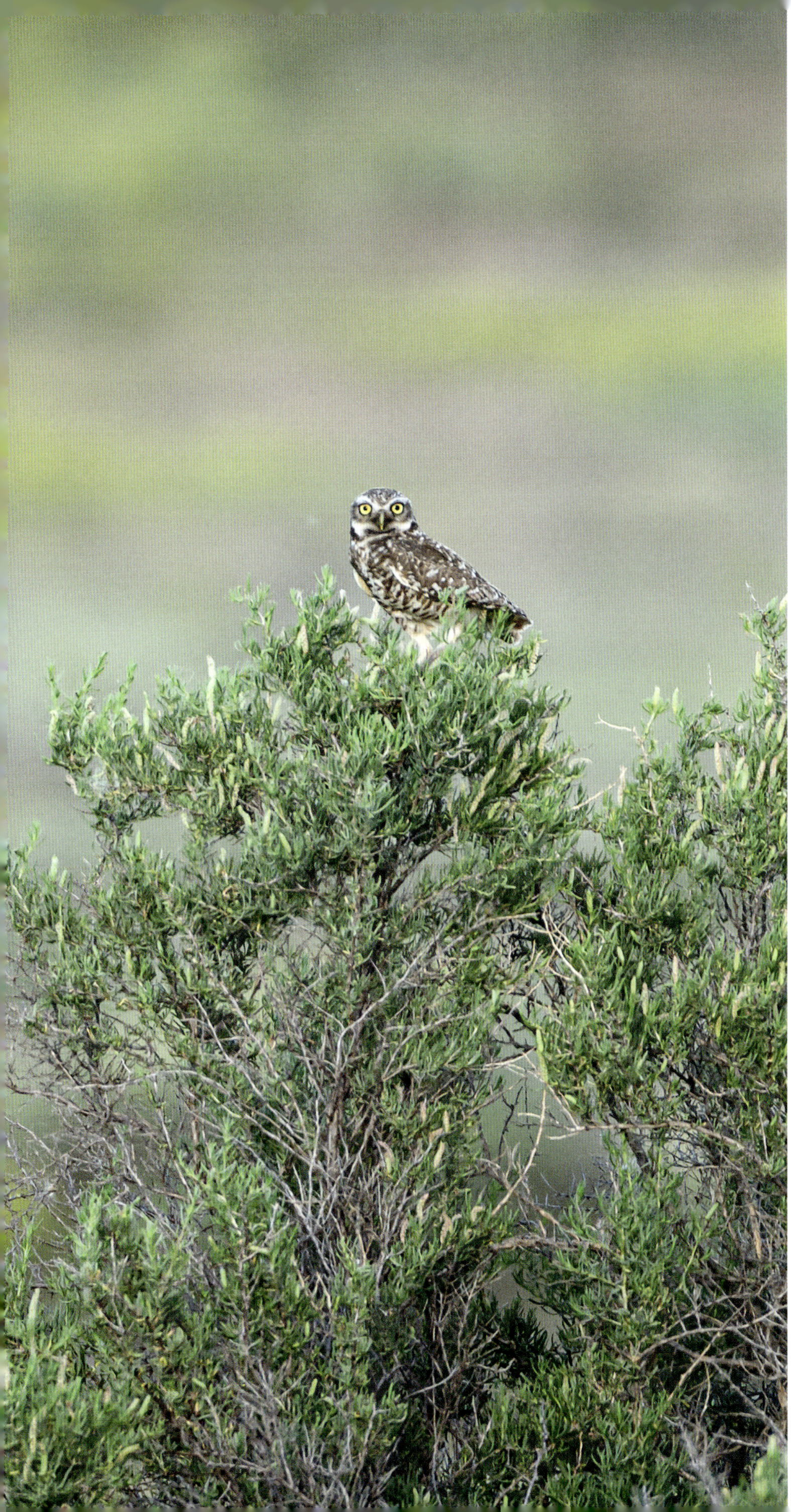

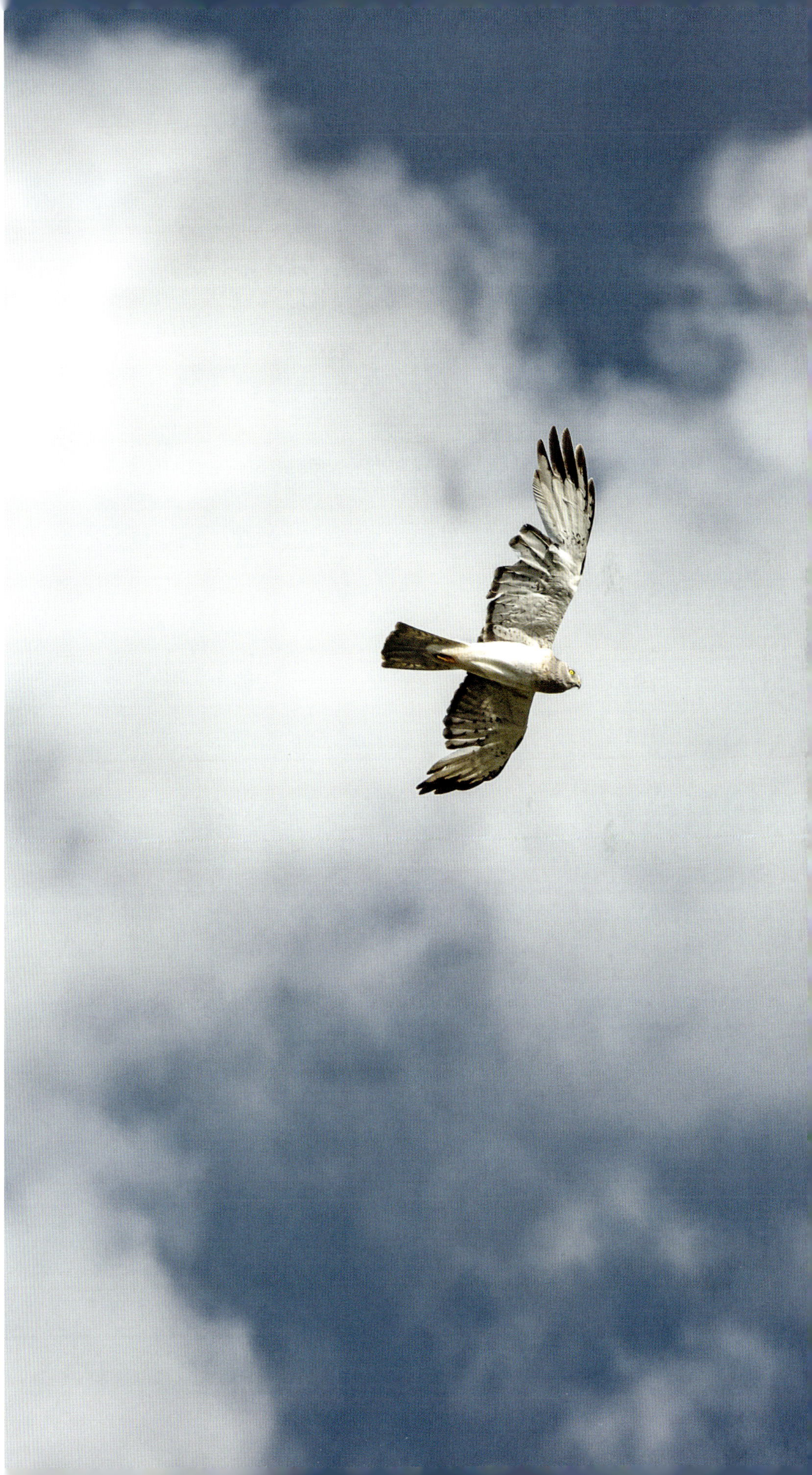

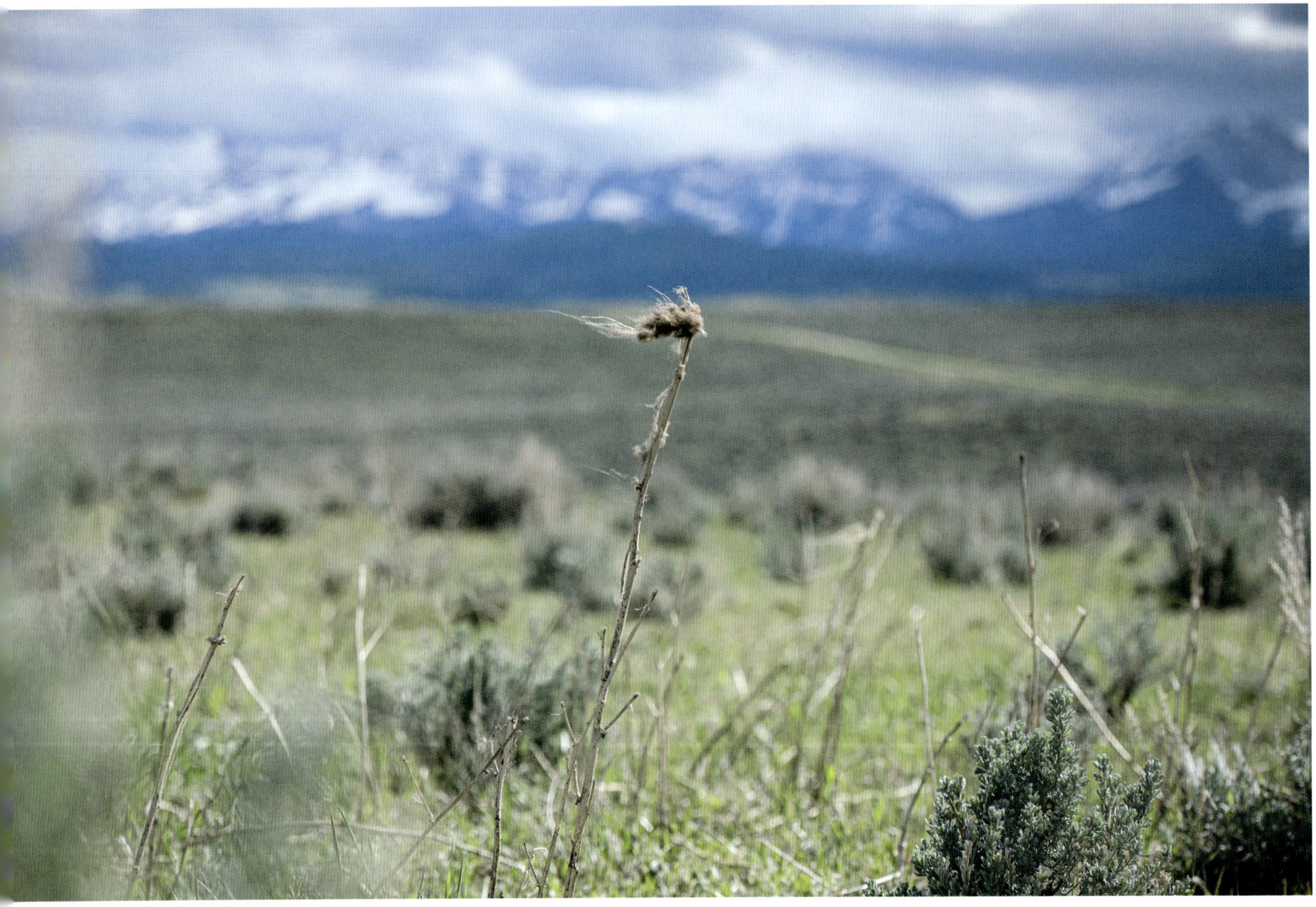

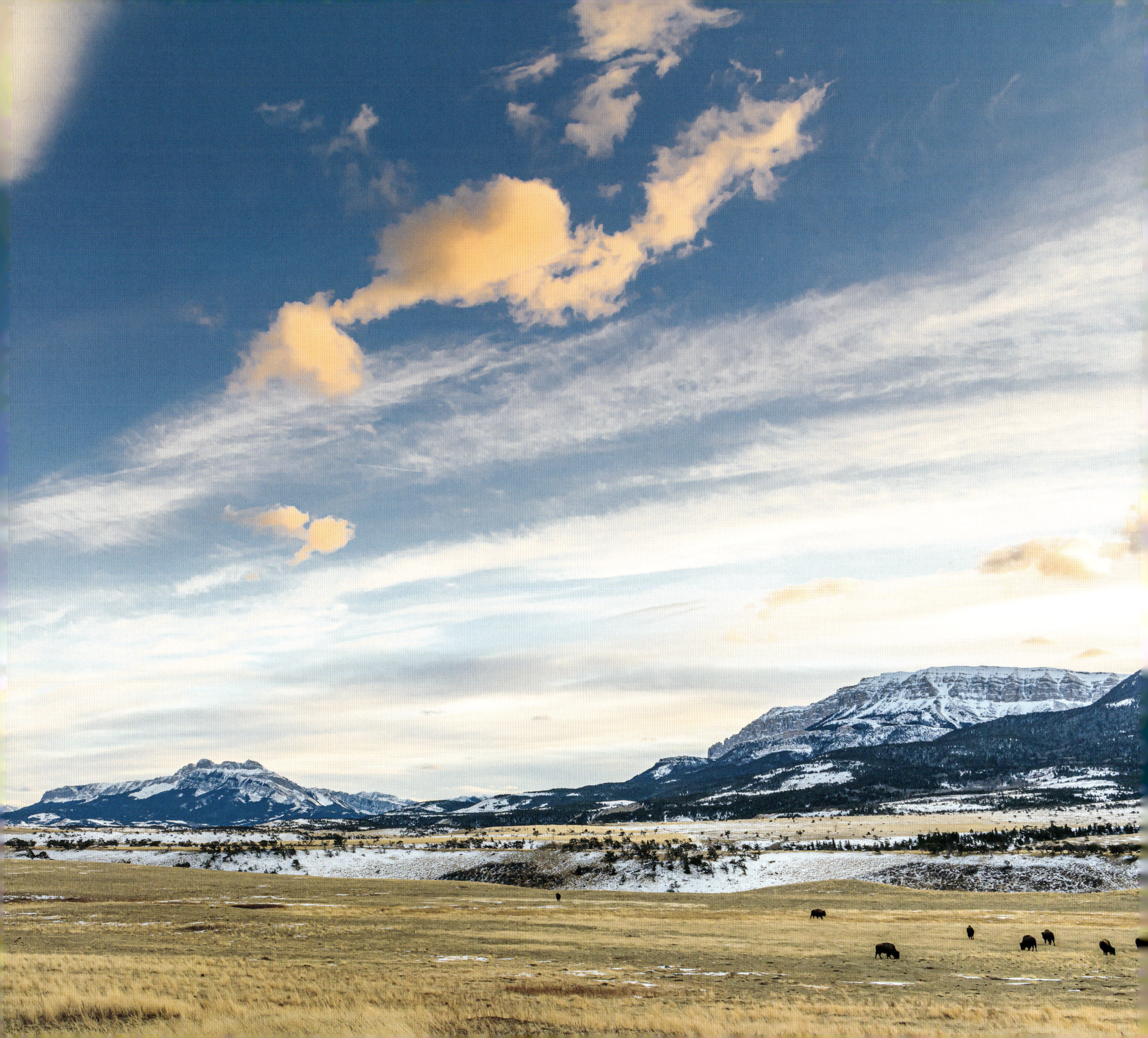

A spirit voice then said to Plenty Coups, "There it is: the buffalo will be no more, and the spotted buffalo will now eat the buffalo grass. The many lodges {teepees} moving camp on the trail of the buffalo will be gone forevermore."

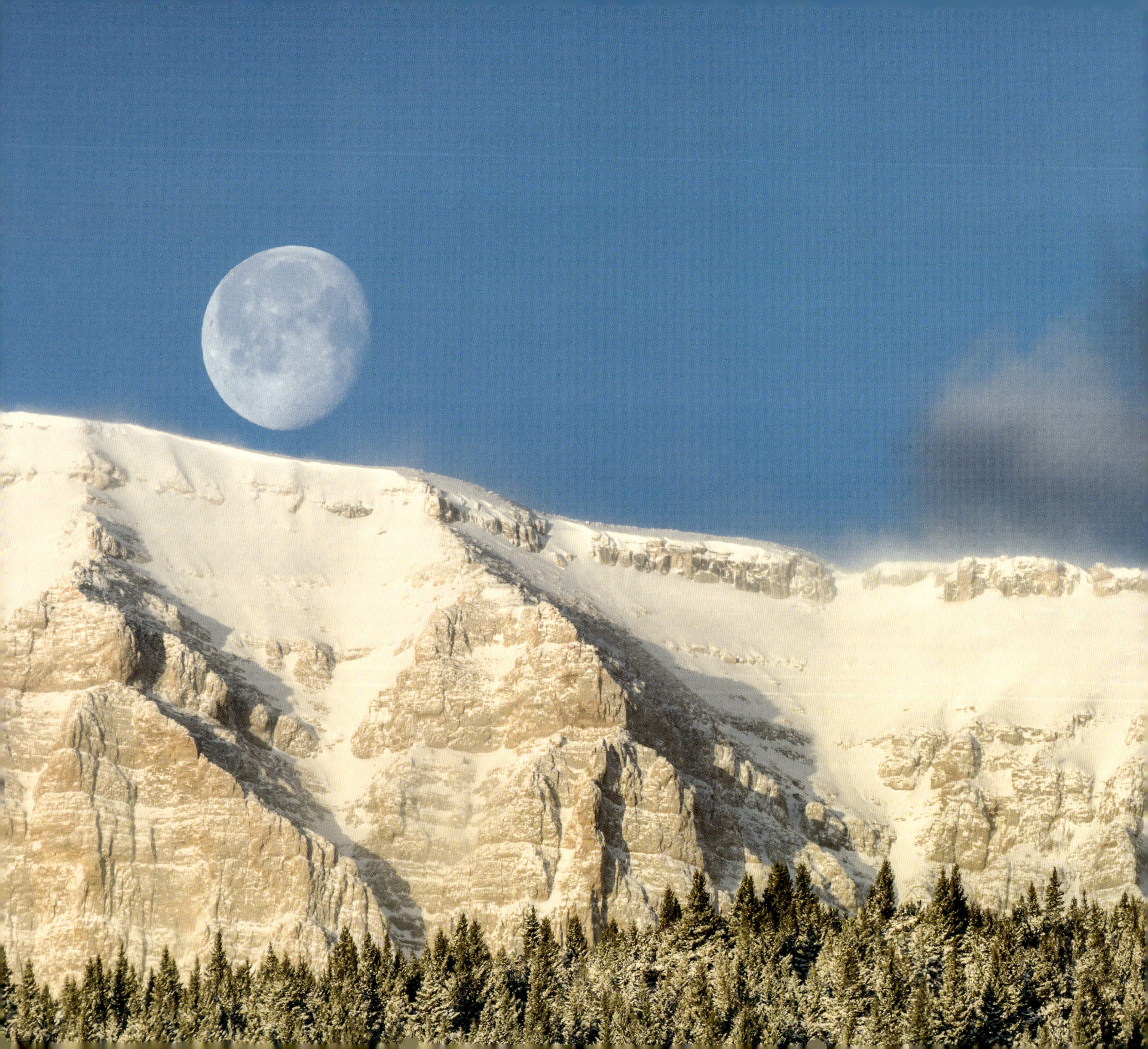

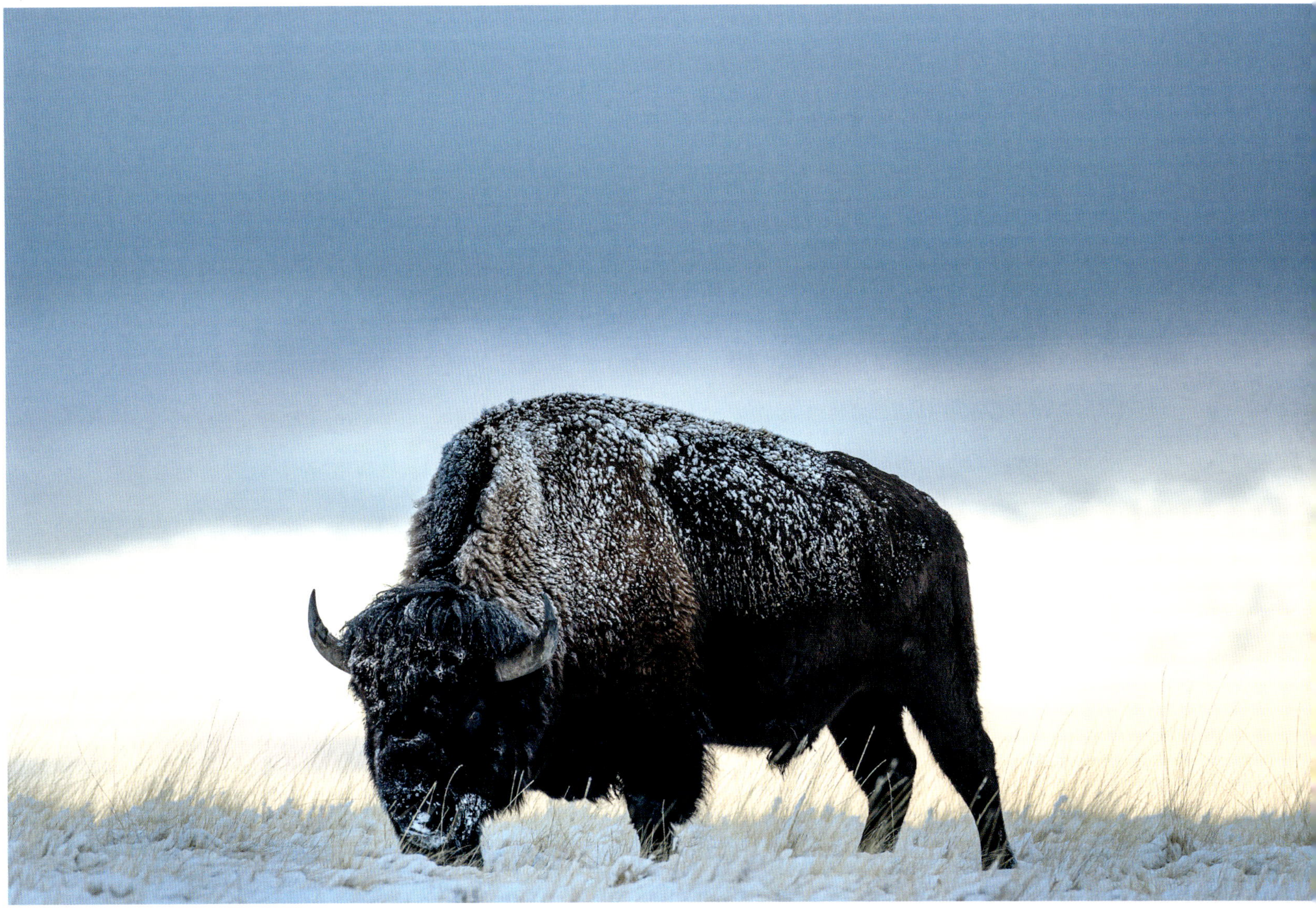

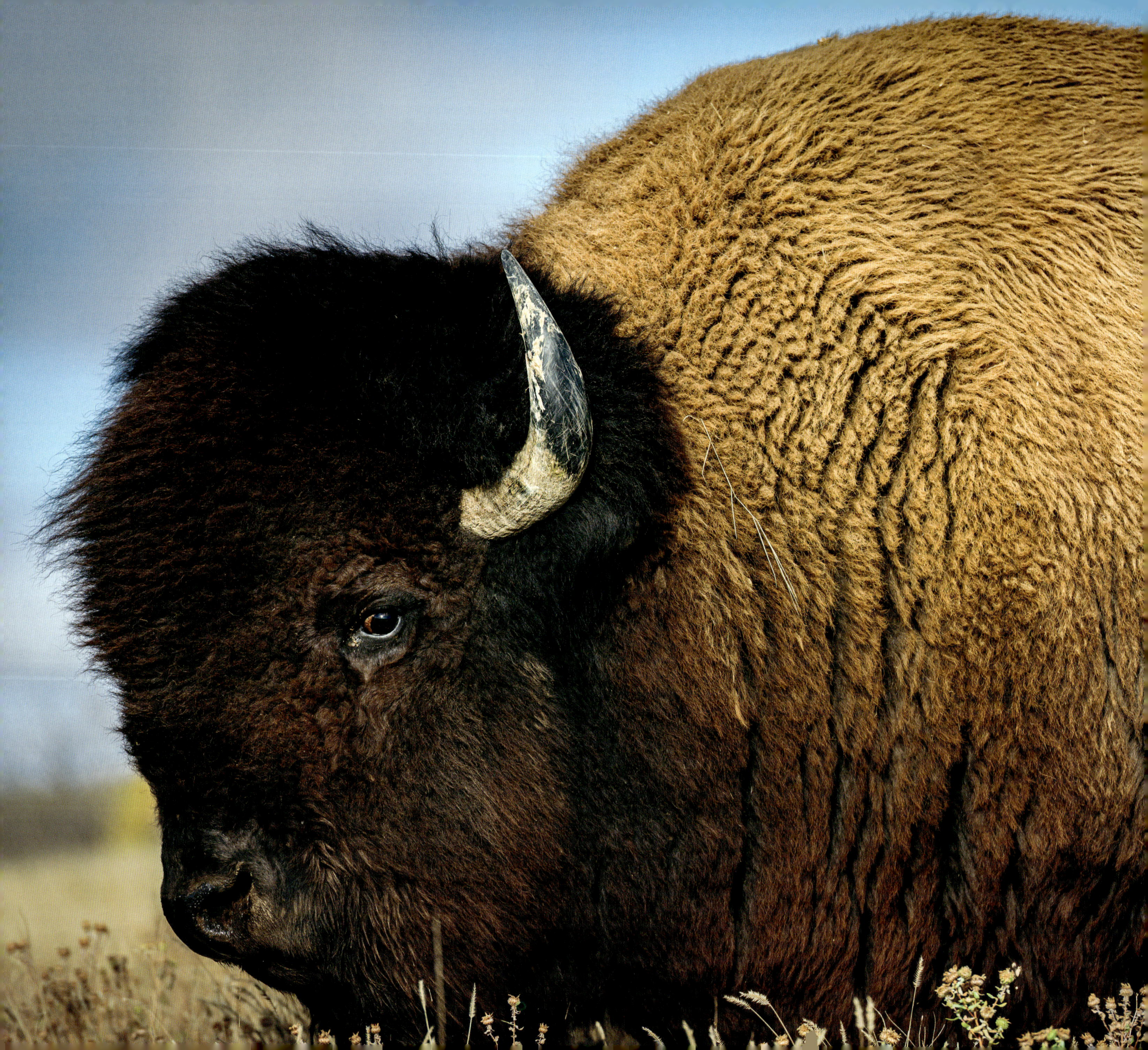

The Disappearance and Return of the Buffalo

The Upsauloeka *(Crow Indians), the Ones Who Put Rocks At The Base Of The Teepee, or the Ones Who Follow The Buffalo, expected to always walk with the wolf on the trail of the buffalo to operate the buffalo jump. But the disappearance of the buffalo was shown through Plenty Coups' supernatural dream, which he had when fasting on a peak in the Crazy Mountains (in present-day Montana) during the summer of 1868.*

In the dream, Plenty Coups went to the head of Arrow (Pryor) Creek Canyon (also in present-day Montana), where to the rising sun he looked as a massive herd of buffalo thundering hooves shook and trembled the ground, Mother Earth. While his heartbeat was with Mother Earth's in the rhythm of buffalo hoofbeats, the buffalo emerged from over the luscious, green horizon, below the gold turquoises, the sun-barely-rising-sky. Many buffalo engulfed the gentle slope, racing toward where it is winter all of the time, turning to where the sun sets, then toward where it is warm all of the time, and finally heading toward where the sun rises. When they headed toward where it is cold all of the time again, they all vanished, disappearing into a massive hole in the ground.

From out of that hole in the ground emerged the spotted buffalo (cattle), many to eat the buffalo grass. As the sun was rising, and from the center of the sun, came so many light-eyed people using spotted buffalo to pull wagons. Wagons entered the plains, wandering toward the rivers and mountains.

The dream moved on to a cluster of pine trees, where many different kinds of birds were fluttering around, chattering and whistling. There were magpies, meadowlarks, hawks, ravens, robins, sage grouse, and many more. As they were busy making a lot of noise and flying around, a chickadee was quietly sitting on the branch of a pine tree when a big gust of wind knocked down all of the trees the many birds were sitting on, except for the pine tree with the little chickadee.

A spirit voice then said to Plenty Coups, "There it is: the buffalo will be no more, and the spotted buffalo will now eat the buffalo grass. The many lodges {teepees} moving camp on the trail of the buffalo will be gone forevermore. The big wind that knocked down all of the pine trees, except for the one pine tree the chickadee was on, is representative of the many light-eyed people who emerged from the morning sacred sunrise. If you get rid of some of these people, many more will come. You have to learn to live with them, be like the chickadee, sit quietly and observe and learn from these light-eyed people. There it is: tell your people the buffalo will disappear."

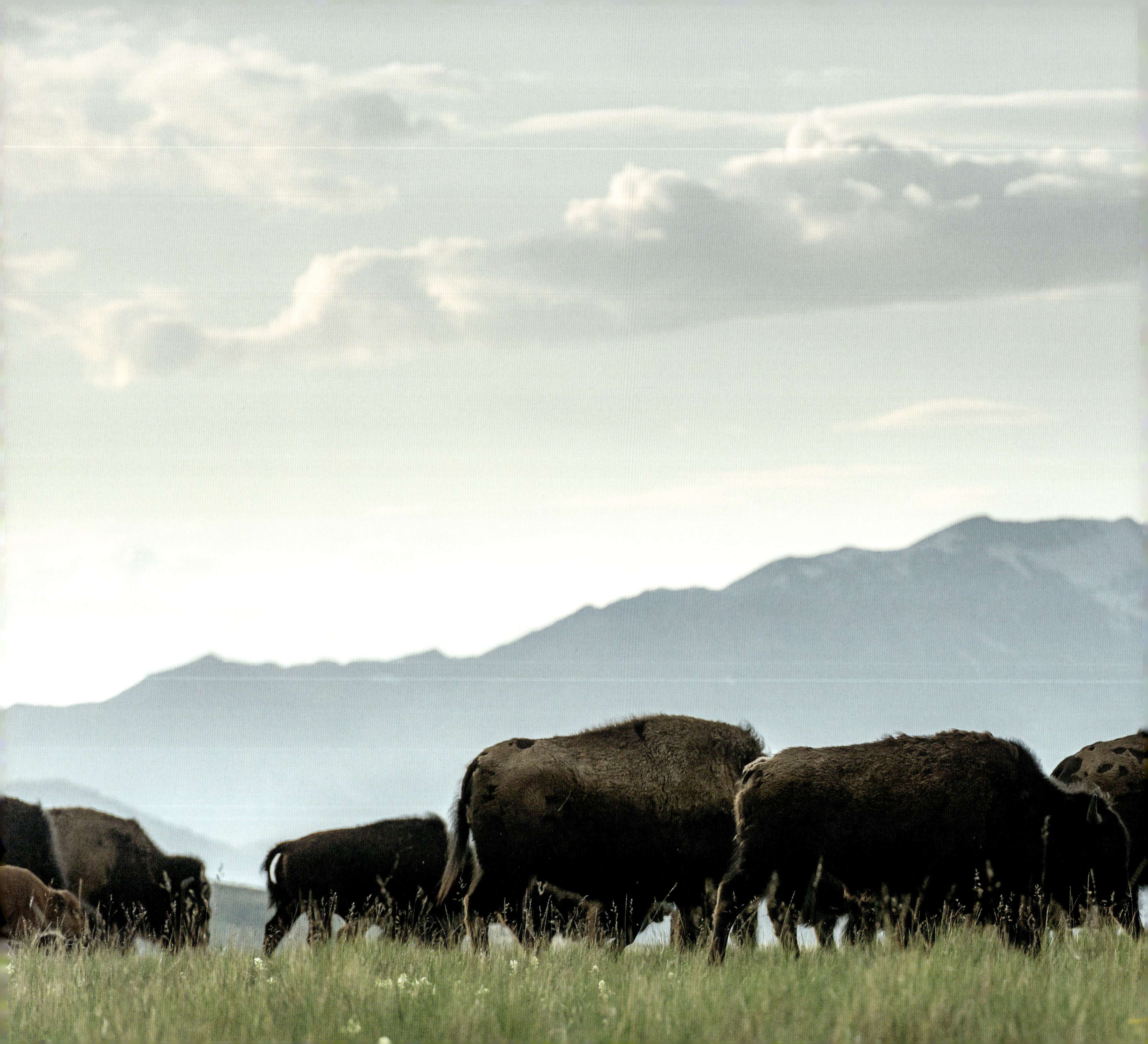

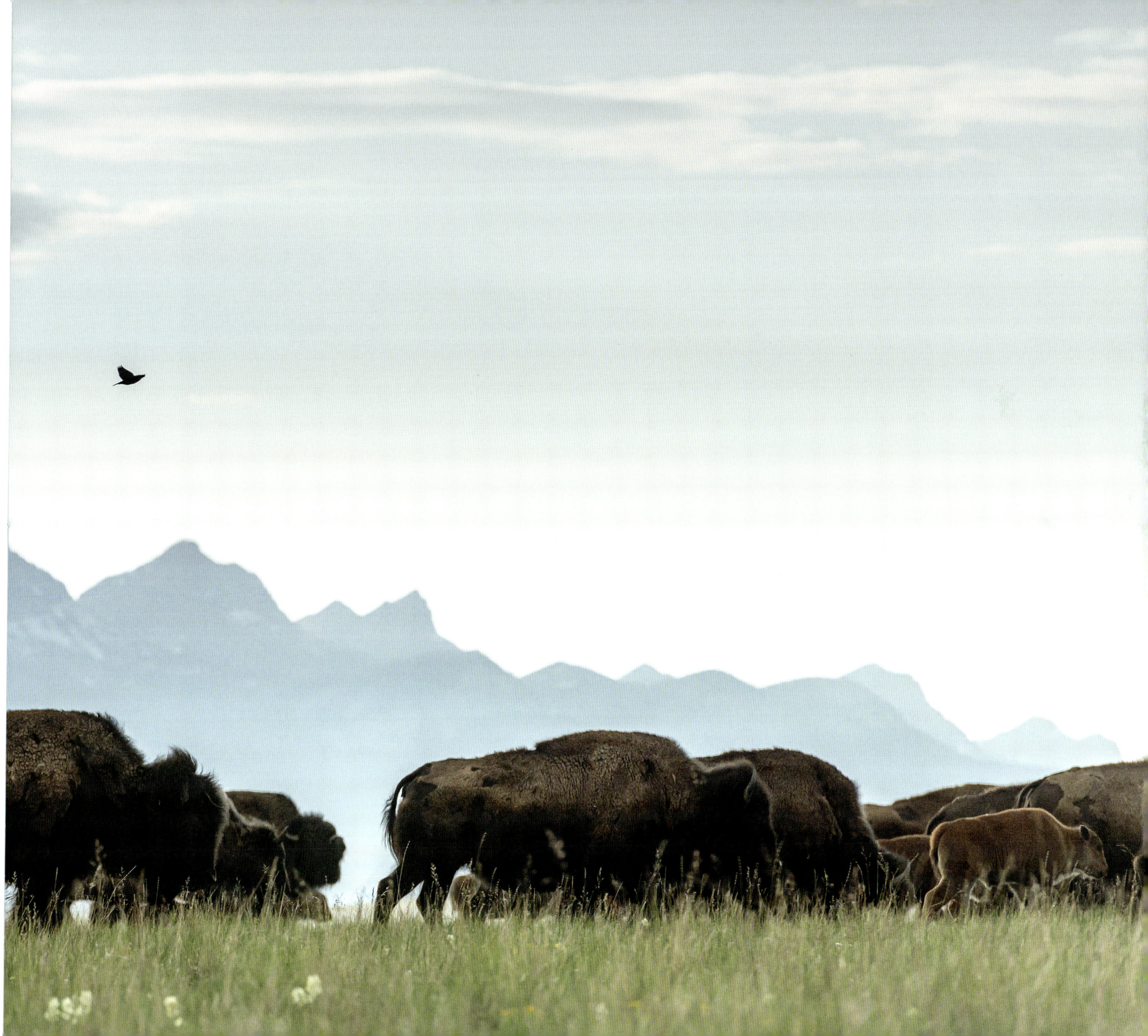

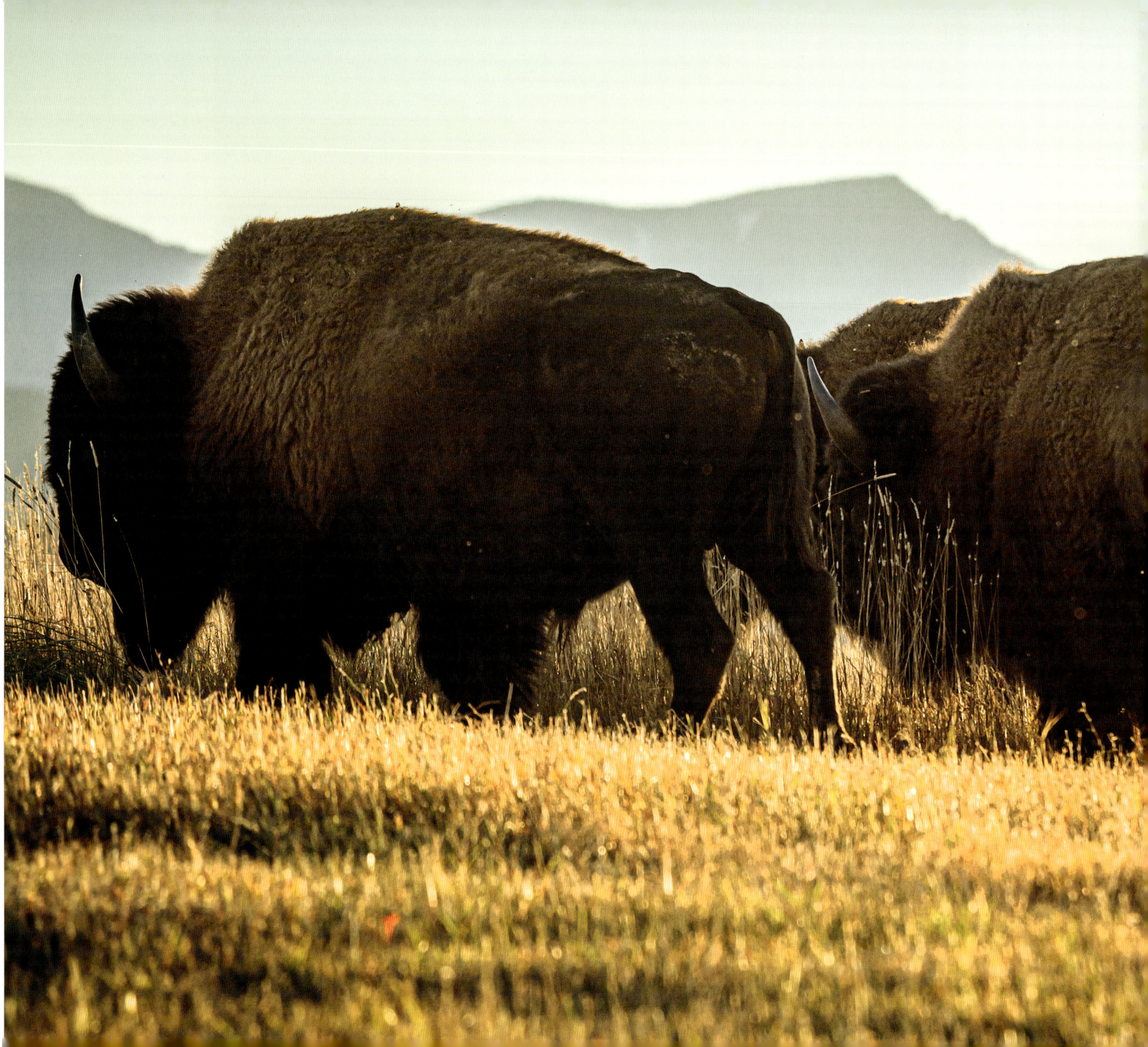

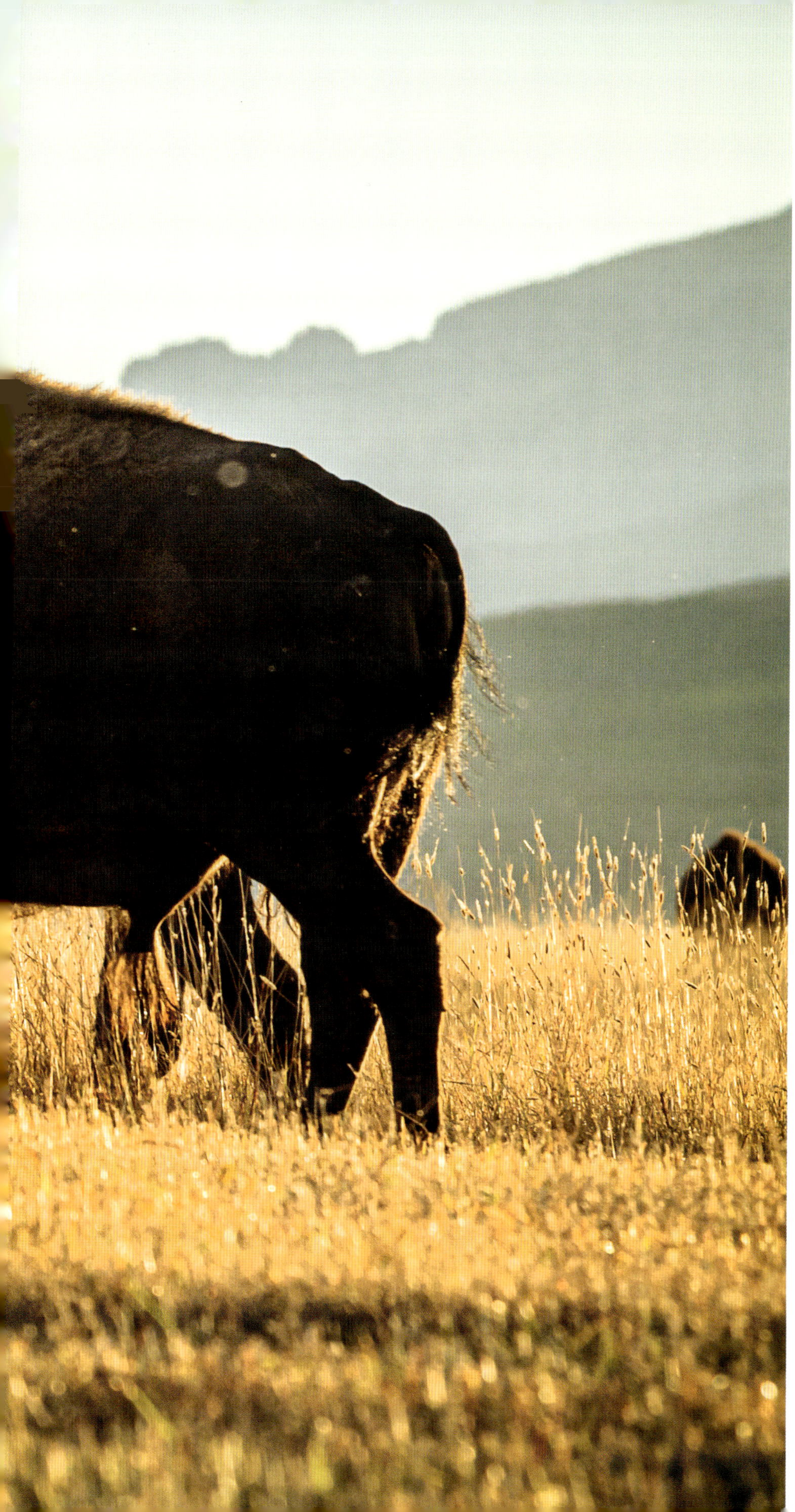

at Bear All Of The Time's home, and in the quiet silence of the main smoke somewhere in the night's middle, an announcement was heard. Bear All Of The Time was praying with tobacco in a corn shuck as the prayer and smoke entered the sacred fire smoke to ascend toward the many stars of the peaceful night sky, and a spirit voice said, "Bisha kun dauk kook *{The buffalo are returning}.*"

Everyone heard the voice come from behind the meeting's lodge, off toward where the sun sets, and all were happy, but stunned, because the buffalo had been gone for over 80 winters. And how could they return? There were many light-eyed people and many spotted buffalo. How could it be?

But the spirit voice was trustworthy: In the Moon When The Chokecherries Ripen (August), and the nights become cool, just before the Moon When The Sun Is The Same (September), Robbie Yellow Tail, of the Whistling Water clan and Superintendent of Crow Agency, would announce his decision to bring some buffalo from Yellowstone National Park (in Wyoming, Montana, and Idaho) to turn loose upon the Crow Indian Reservation's/Montana's Big Horn Mountains.

Although these buffalo (reintroduced in 1934) would eventually be killed off because of greedy ranchers, who wanted the buffalo grass and labeled our herd with brucellosis, the Upsauloeka would not be deterred from bringing back the buffalo, and in 1972, a replacement herd was brought from Theodore Roosevelt National Park in North Dakota to the Big Horn Mountains (and in 1980, I myself helped transplant 27 more buffalo from the park to the mountains).

Today, due to the above restoration efforts, and others, the buffalo eat the buffalo grass in the Upsauloeka Buffalo Pasture, and we grease the edges of our mouths with their meat. The tongues of these mighty buffalo are given to respectable women of the Upsauloeka, as in the best winters on the trail of the buffalo. Many Upsauloeka use hearts that are good because we have felt the soul of life that returned to our people with the reintroduction of the buffalo. Our hearts beat as one with the buffalo and Mother Earth.

—HENRY REAL BIRD

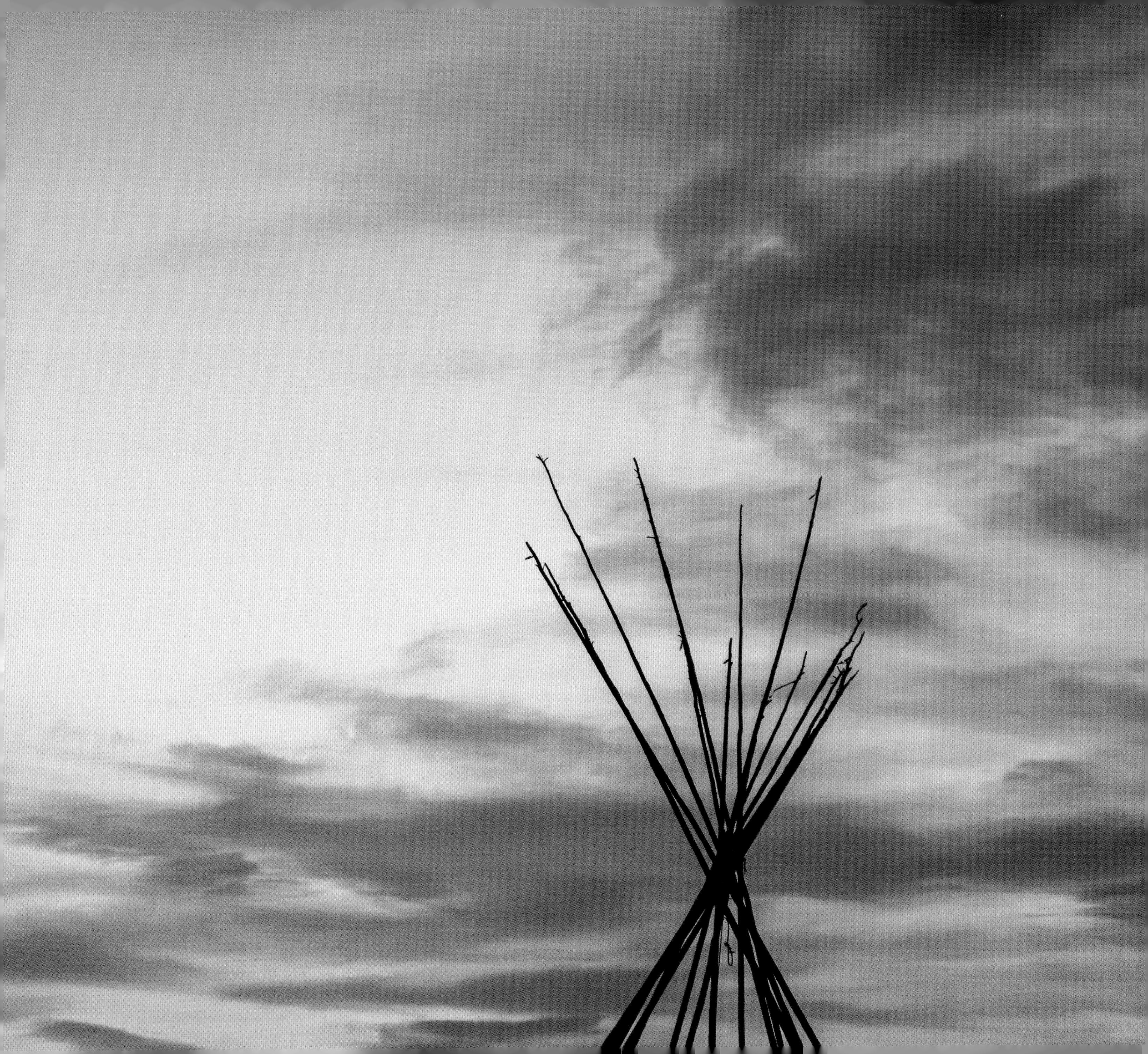

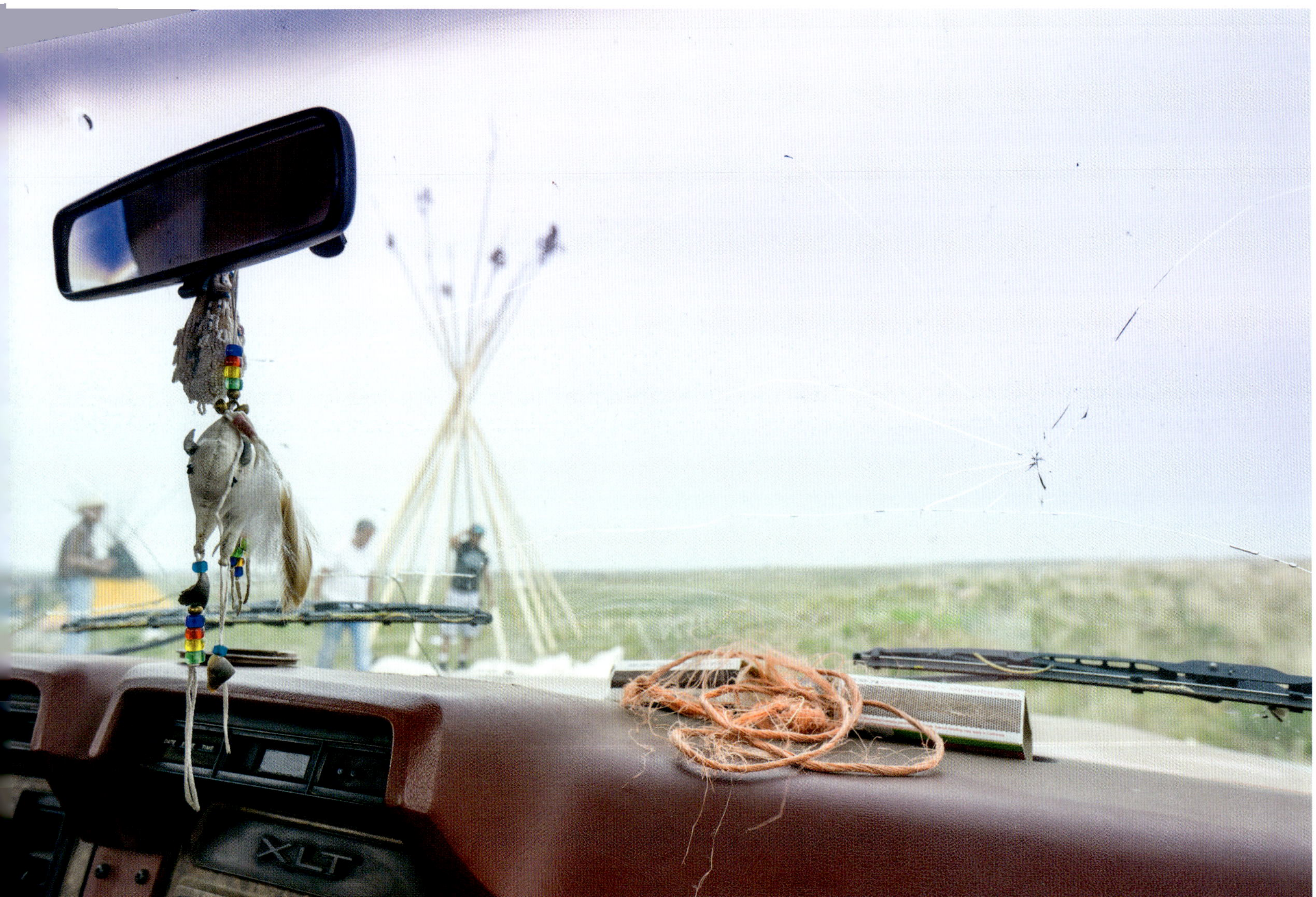
XLT

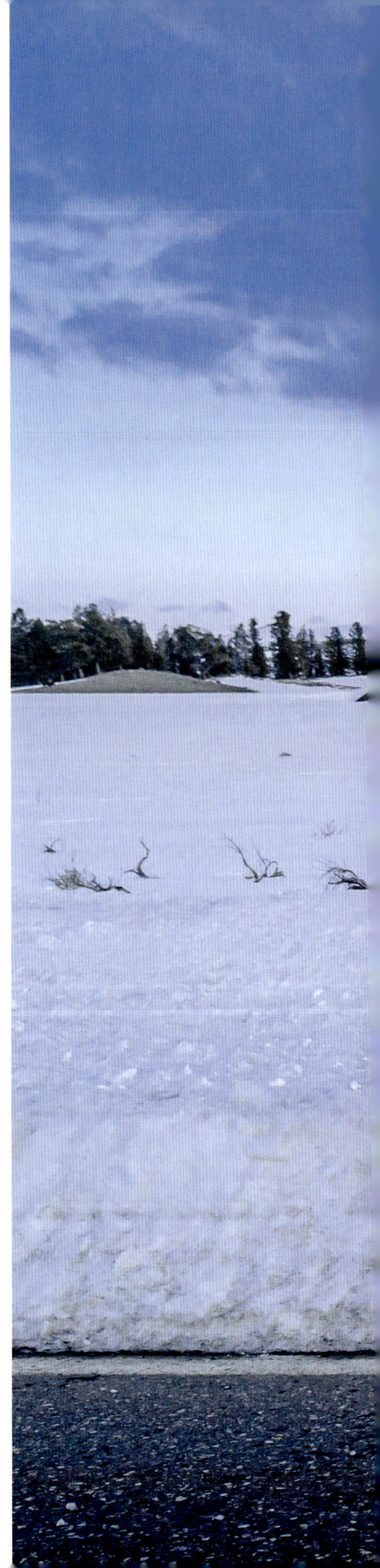

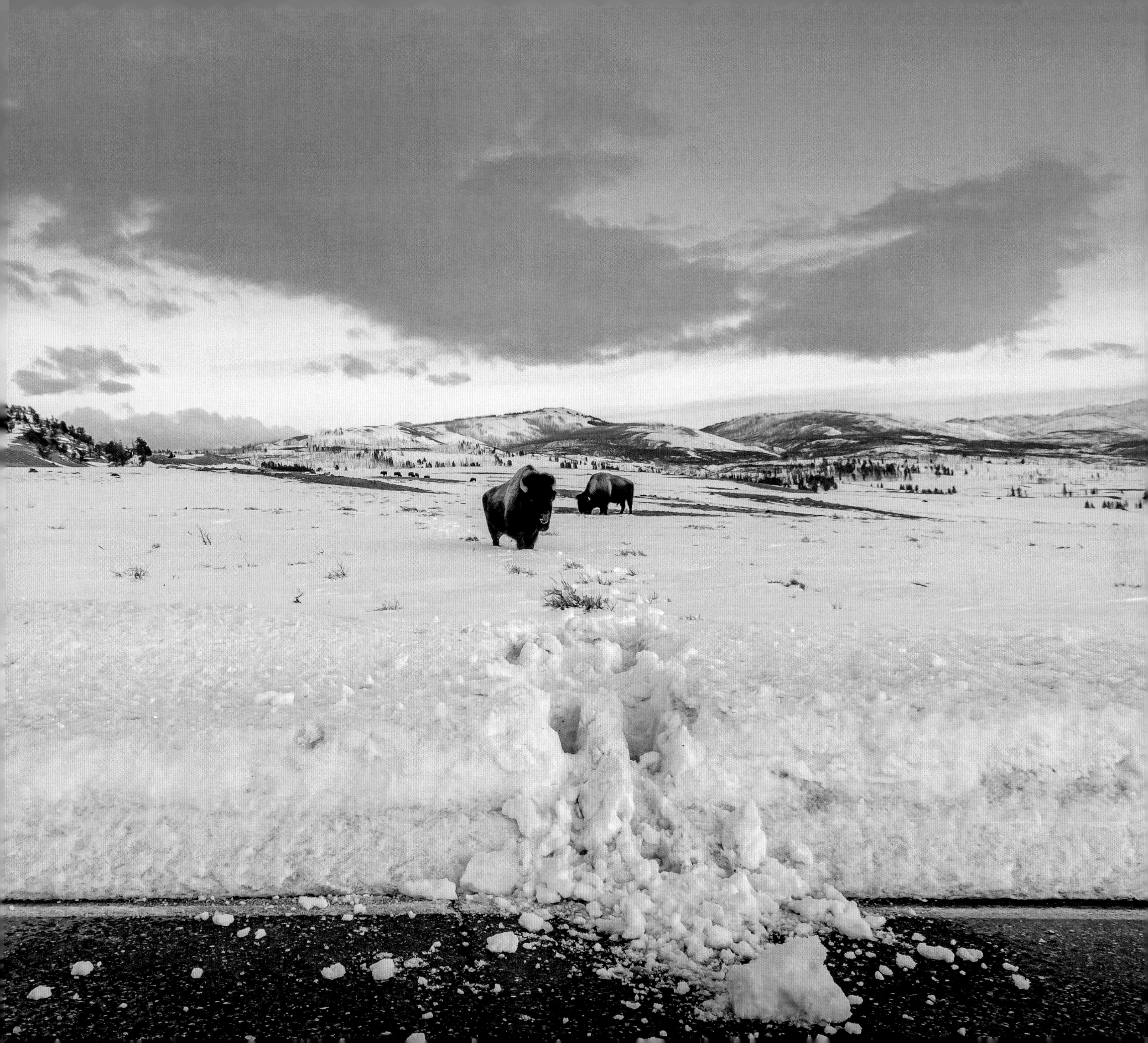

I'm grateful to a new friend, Joanne Dornan, who was instrumental in the conversation and communication with Yvon and Malinda Chouinard. We appreciate the spirit that the Chouinards have brought to conservation, ethics, and their passion for bison. Thank you, Joanne, Malinda, and Yvon.

Chase Reynolds Ewald brings a level of commitment and professionalism that is appreciated by everyone involved. She's detail-oriented, creative, and thoughtful. This is our sixth book together, and I couldn't imagine doing one without her. We've become close friends over the years, and doing books together always ensures an adventure or two, and keeps us on our toes.

Michelle Branson, our editor at Gibbs Smith, answered my calls and emails at all hours of the day. She's the person who keeps everything on track, moving forward, and on time. When I first pitched the idea of this book, she understood the vision and signed on, even though this kind of a book is a departure for the publisher. Thank you, Michelle, for taking the leap with us!

Jeffrey Cross, a travel editor at *Sunset* magazine, sent me to many fascinating places on assignment, including Thermopolis, Wyoming, and Hot Springs State Park which borders the quirky Western town. I'm grateful for the images from the *Sunset*

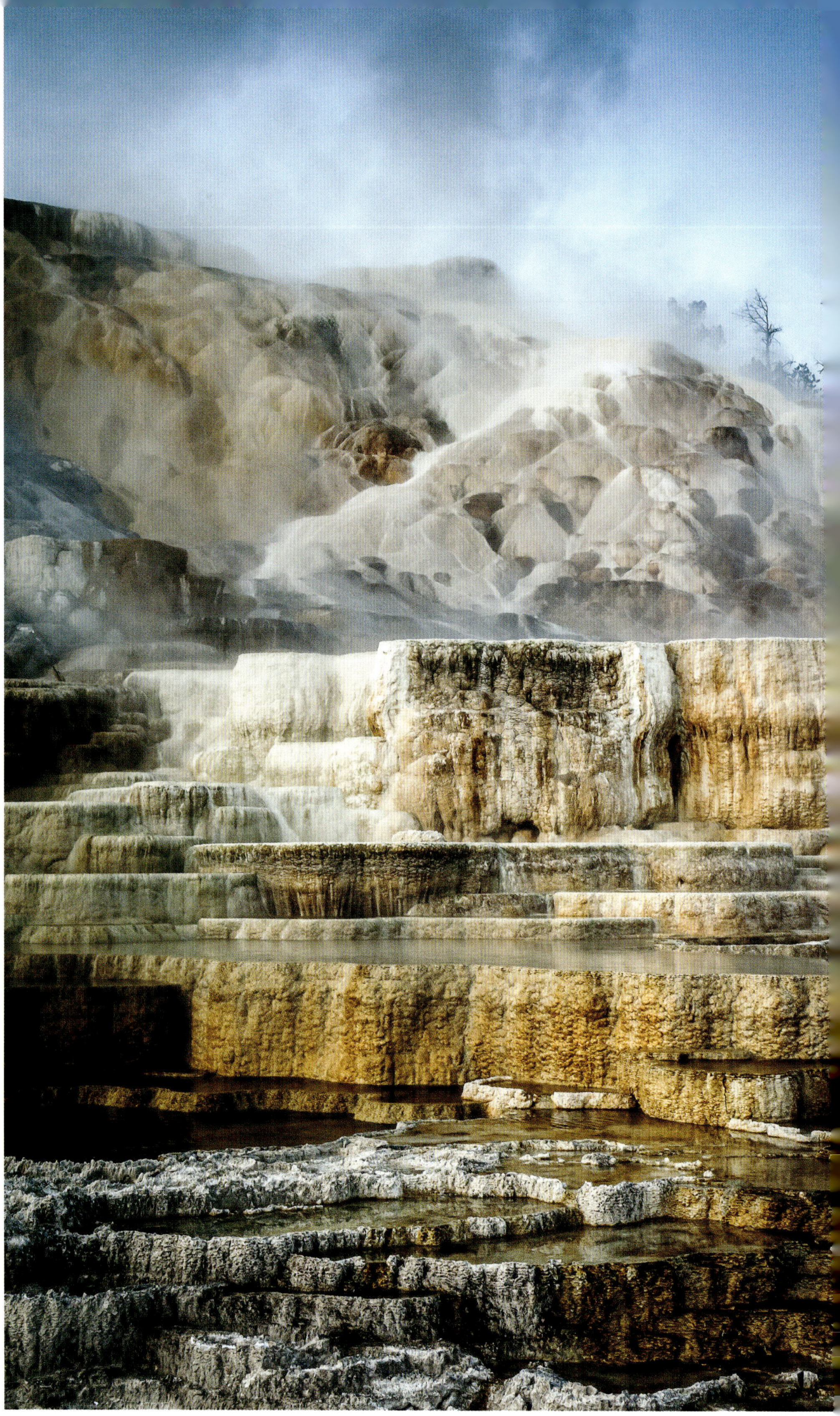

feature that are a part of the book, and to Jeffrey for all the incredible photo "scavenger hunts" sprinkled across the Rockies.

Michael Green, the photo editor of *Preservation* magazine, assigned me to travel and explore the area of the Black Hills in South Dakota. I was fortunate to photograph bison in the Bear Butte State Park and include them in this book as well.

Tom Rodgers of the Blackfeet Nation encouraged and supported my first trips to Browning and the Eastern front of Glacier National Park. Tom has been such a great friend and patron over the years in his understanding of the vision of my work.

I met Jill O'Brien at the Diamond 4D out of the sheer luck of timing. Her passion for bison and photography radiates as brightly as the colorful setting sun. Getting to spend time with her in a field full of bison during magic hour was an extra special treat.

Paul and Jillian Bertelli have been friends for years. They both have supported my work and vision deeply and were kind enough to help facilitate the communication with Mike Phillips and the Flying D Ranch. Thank you, Paul, Jillian, and Mike. I'm also grateful to Becky Traucht. She keeps the lines of communication flowing between all the collaborative projects with the Bertellis.

My journey as an artist has been inspired by my mother, Odete "Coco" Andren. Her continued support, encouragement, and advice has helped direct my decision making and confidence to pursue a life filled with creativity.

My husband, Todd Harris, never bats an eye when I announce that I'm embarking on a creative endeavor which takes me away from home for stretches of time. His support and willingness to be a second set of eyes on the images I curate and the words I write are not just appreciated, but offer an essential perspective I am grateful for.

Thank you to our family in Kansas City, the Harrises, and the Passetts. I appreciate their love, encouragement, and road trips to Montana to visit and connect. I'm grateful for this extended family, always kind and supportive.

I finally thank the animals, the bison themselves, who have been with me since childhood. I've witnessed life's full cycle of birth and death, the seasons cold to warm, watching as soft fur shed onto the prairie or iced over from the freezing steam of the thermal features in Yellowstone. I've learned so much about my home through them. Montana, and its surrounds, are indeed full of incredible scenery, but these animals and still-wild places are what give the region its depth, soul, and identity.

—AUDREY HALL

Working on *Bison: Portrait of an Icon* afforded me an extraordinary opportunity to take a deep dive into a subject that is seemingly straightforward, but is actually wide, broad, and surprisingly nuanced. I couldn't have done it without the bison ranchers, cultural interpreters, and public-policy shapers who gave so willingly of their time and expertise, including:

- George Horse Capture, Fort Belknap Indian Community Council representative and cultural tour guide
- Ervin Carlson, InterTribal Buffalo Council member and bison rancher
- Dr. Ian Thompson, bison rancher and Choctaw tribal authority on traditional foods
- Shawn Henderson, Herd Manager for the Quapaw Tribe in Oklahoma
- Terrance Guardipee, Blackfeet Ledger Artist and Painter
- Ron Brownotter, bison rancher on the Standing Rock Indian Reservation
- Cathy Smith, Emmy Award–winning costume designer and authority on Plains Indian clothing and gear
- Jason Baldes, National Wildlife Federation's Tribal Buffalo Coordinator and ecologist and adviser for the Eastern Shoshone tribe's buffalo herd on Wyoming's Wind River Reservation
- Dennis Jorgensen, World Wildlife Fund's Bison Initiative Coordinator and Program Officer for the Northern Great Plains
- Jim Matheson, Assistant Director, National Bison Association
- Colton Jones, Sourcing Manager, Wild Idea Buffalo Company
- Chris Bechtold, Manager, Diamond 4D Ranch
- Evan Moran, Habitat Manager and Range Specialist, Diamond 4D Ranch
- Matt Skoglund, Founder/Owner, North Bridger Bison Company
- Daniel Kinka, PhD, Wildlife Restoration Manager, American Prairie Reserve
- Damien Austin, Reserve Superintendent, American Prairie Reserve

I am grateful to authors Dan O'Brien, Michael Punke, Kurt Repanshek, and Steven Rinella. Their excellent books *Buffalo for the Broken Heart: Restoring Life to a Black Hills Ranch; Last Stand: George Bird Grinell, The Battle to Save the Buffalo, and the Birth of the New West; Re-Bisoning the West: Restoring an American Icon to the Landscape;* and *American Buffalo: In Search of a Lost Icon* were instrumental in helping me understand the bison story.

Audrey and I owe a huge debt to our editor, Michelle Branson, who brought boundless enthusiasm to this project, and to Rita Sowins and Virginia Snow who pulled it together so beau-

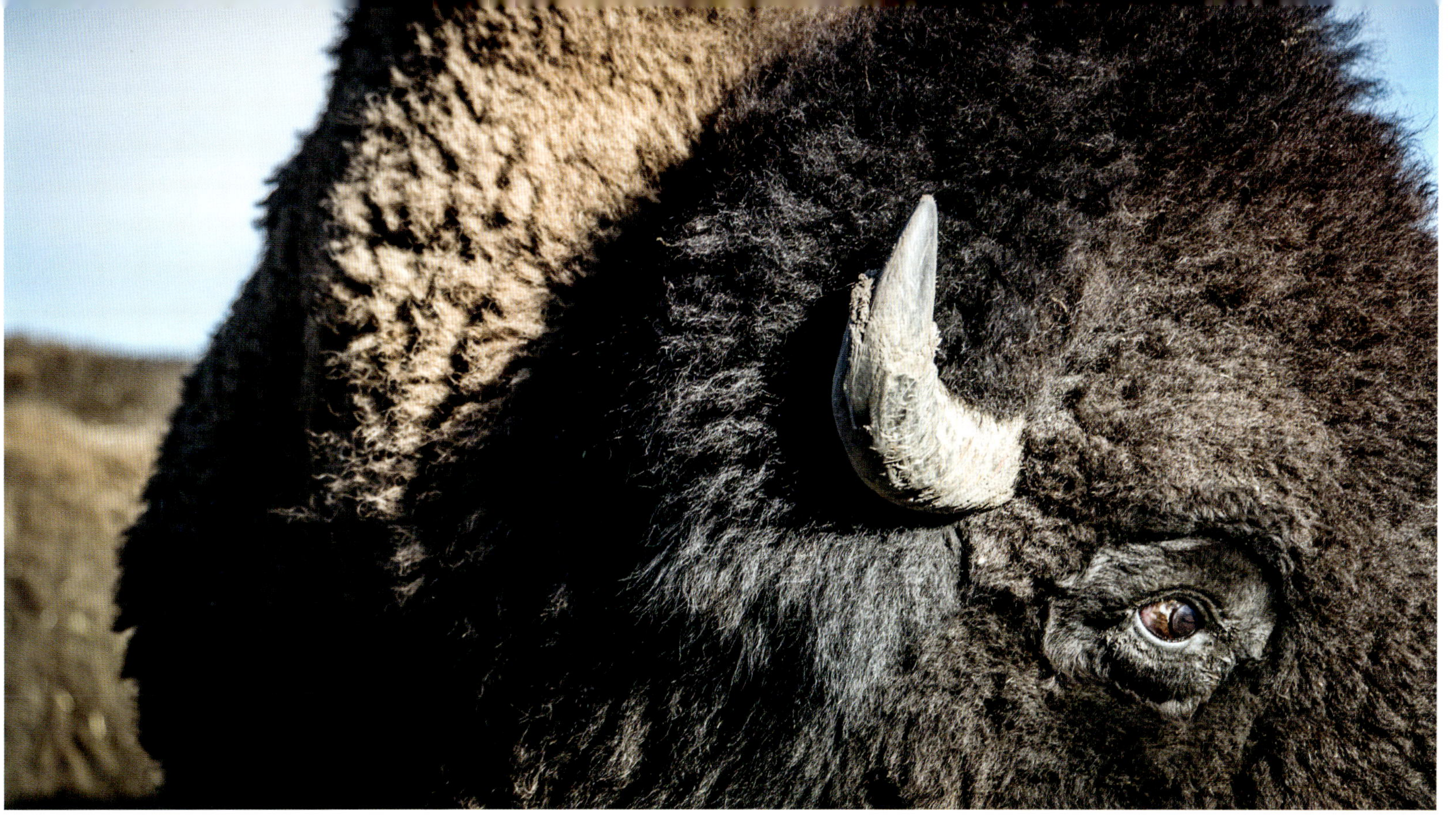

tifully. This is our sixth collaboration with Gibbs Smith and we are so grateful to have a publishing home that shapes our words and photographs into timeless, beautiful and enduring books.

I'd like to send a huge western shout out to Deb Stegelman, Lisa Diekmann, Sue Simpson Gallagher, Carrie Ballantyne, Hannah Ballantyne, John Mortensen, Hilary Heminway, John Heminway, and Kathryn Wilmerding Heminway for providing 'homes-away-from-home' in the Greater Yellowstone area; to Missy Janes for organizing a life-changing canoe trip through the Upper Missouri River Breaks National Monument and visit to the American Prairie Reserve; and to Charles, Addie, Jessie, Ross, and Katherine for patiently enduring countless Yellowstone "bison jams" during our road trips across the American West.

It has been an enormous privilege to work side by side with the multi-talented Audrey Hall. Singer, songwriter, photographer, fine artist, dog whisperer, and all-round wonderful person, she approaches each assignment with enthusiasm, passion, and meticulous attention to detail. It has been so fun — and challenging — to shift gears from our many design projects to a book like *Bison*, which resides at the place where natural history meets art. I was honored to be asked to collaborate on this project with an artist reaching the peak of her powers. To contemplate Audrey's images of buffalo in their natural habitat is to understand both their ancient presence and their immediate relevance as they continue to occupy a crucial niche in the Western landscape.

—CHASE REYNOLDS EWALD

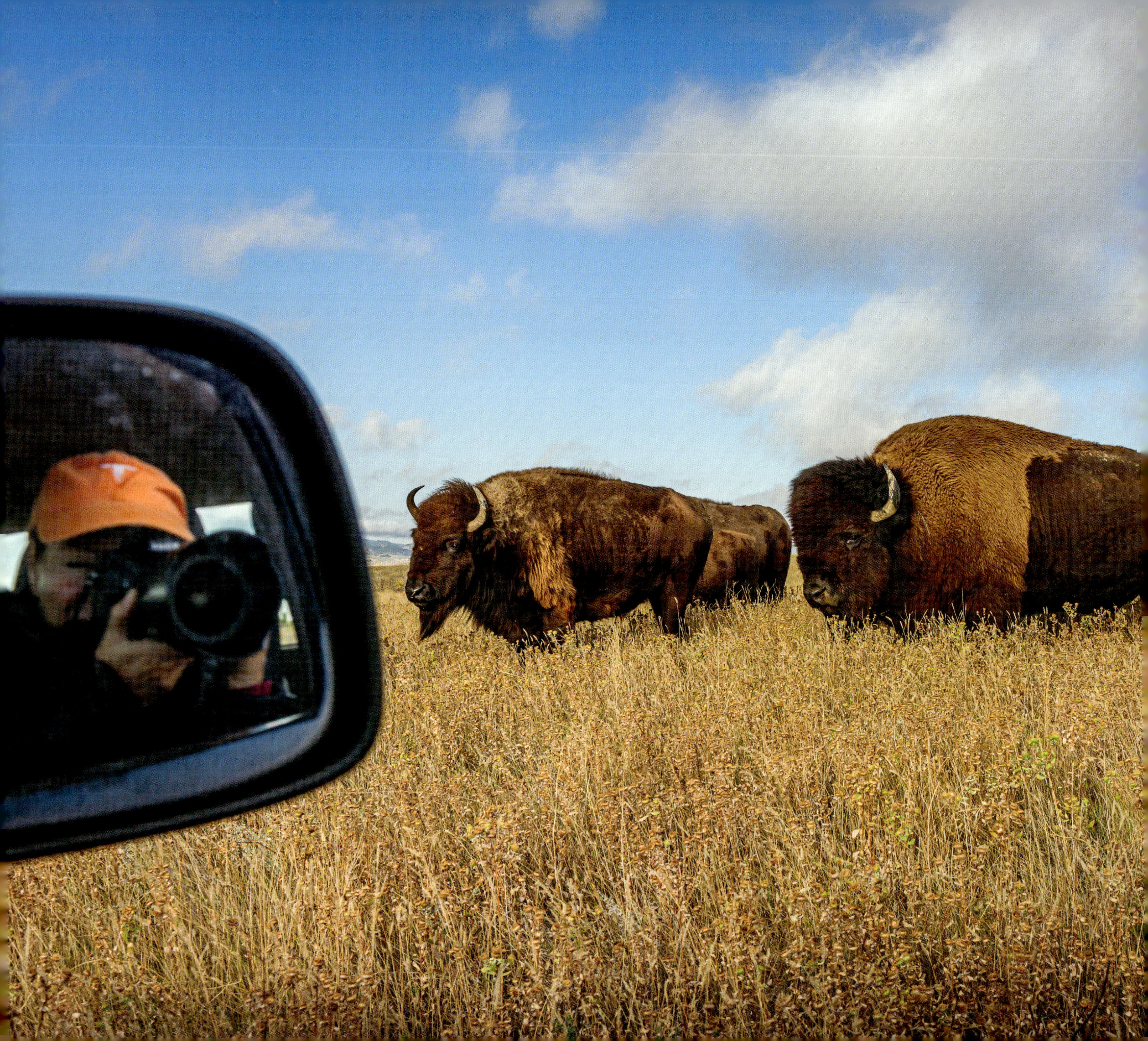

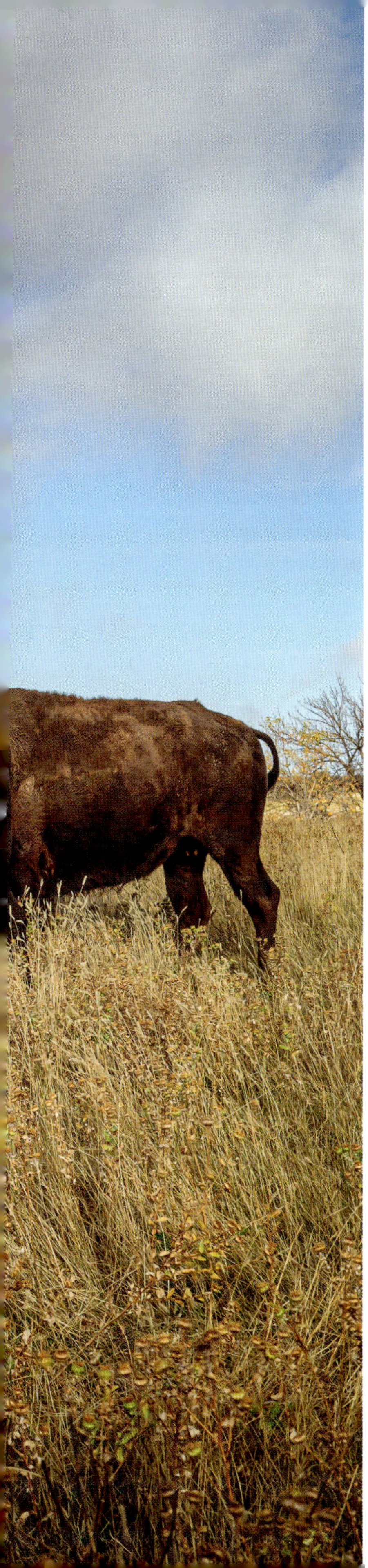

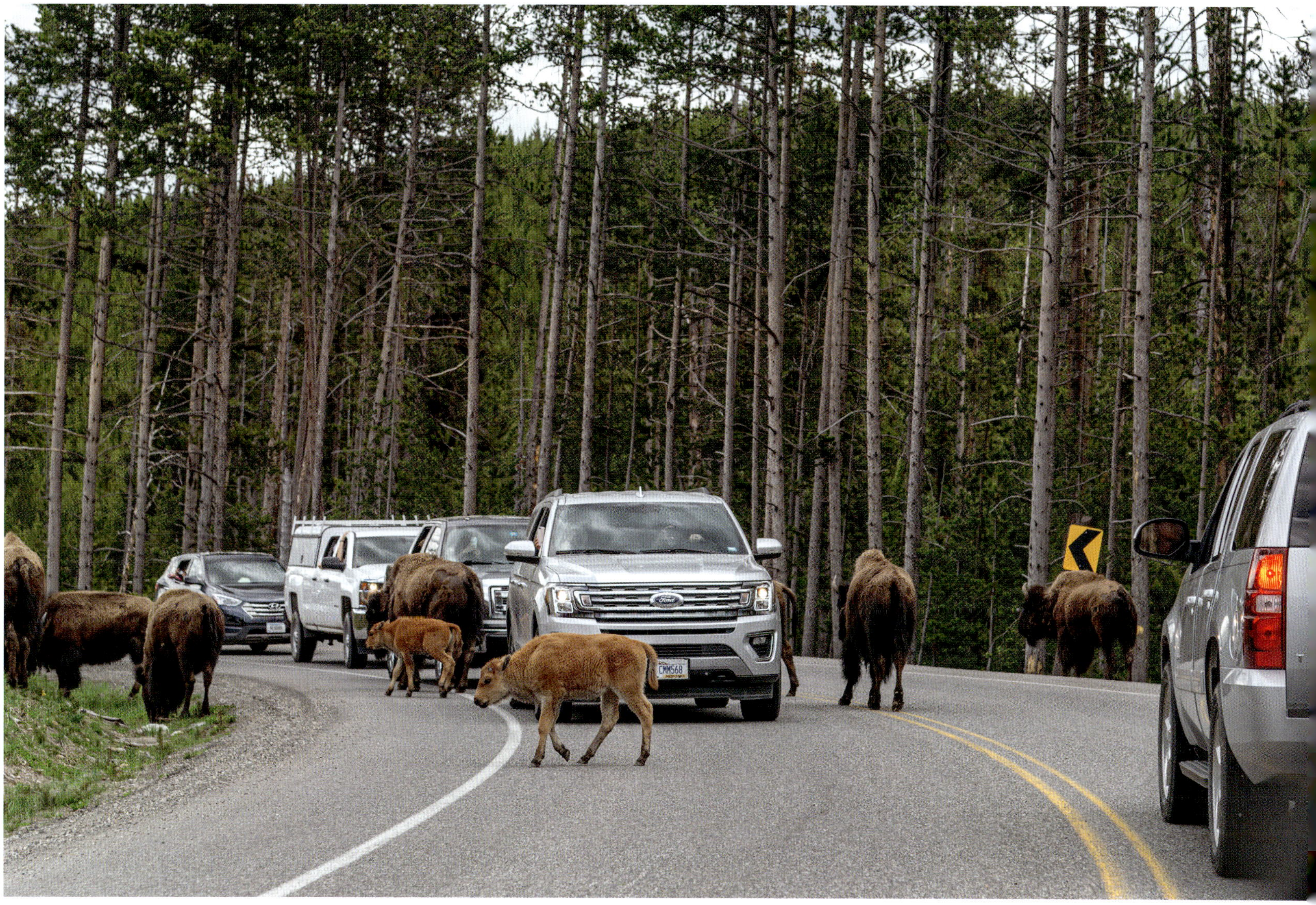

RESOURCES

INFORMATION

American Prairie Reserve
americanprairie.org

American Bison Society
ambisonsociety.org

Diamond 4D Ranch
diamond4dranch.com

Intertribal Bison Cooperative
itbcbuffalonation.org

Nan Awaya Heritage Farmstead
nanawaya.com

National Bison Association
bisoncentral.com

The Nature Conservancy
nature.org

North Bridger Bison
northbridgerbison.com

Wild Idea Buffalo Company
wildideabuffalo.com

Wolokota Buffalo Range
rosebudbuffalo.org

World Wildlife Fund Bison Initiative
worldwildlife.org/species/plains-bison

Yellowstone Forever
yellowstone.org

Yellowstone National Park
nps.gov/yell

VIEWING

American Prairie Reserve, Montana
americanprairie.org/bison-viewing

Antelope Island State Park, Utah
stateparks.utah.gov/parks/antelope-island

Bear Butte State Park, South Dakota
gfp.sd.gov/parks/detail/bear-butte-state-park

Custer State Park, South Dakota
gfp.sd.gov/parks/detail/custer-state-park

Catalina Island, California
catalinaconservancy.org

Fort Belknap Buffalo Reserve
Aaniiih Nakoda Tours, Montana
visitmt.com

Golden Gate Park, California
goldengatepark.com/buffalo-paddock

Hot Springs State Park, Wyoming
thermopolis.com/attractions/hot-springs-state-park

National Bison Range, Montana
fws.gov/refuge/national_bison_range

Neal Smith National Wildlife Refuge, Iowa
fws.gov/refuge/Neal_Smith

Theodore Roosevelt National Park, North Dakota
nps.gov/thro

Wichita Mountains Wildlife Refuge, Oklahoma
fws.gov/refuge/wichita_mountains

Wind Cave National Park, South Dakota
nps.gov/wica

Yellowstone National Park, Montana and Wyoming
nps.gov/yell

READING

American Buffalo: In Search of a Lost Icon, Steve Rinella

Buffalo for the Broken Heart: Restoring Life to a Black Hills Ranch, Dan O'Brien

A Field Guide to the Plains Bison, Wes Olson

The Journals of Lewis and Clark, Meriwether Lewis and William Clark

Re-Bisoning the West: Restoring an American Icon to the Landscape, Kurt Repanshek

Last Stand: George Bird Grinnell, the Battle to Save the Buffalo, and the Birth of the New West, Michael Punke

Last Stand: Ted Turner's Quest to Save a Troubled Planet, Todd Wilkinson

Portraits of the Bison: An Illustrated Guide to Bison Society, Wes Olson

Theodore Roosevelt and Bison Restoration on the Great Plains, Keith Aune and Glenn Plumb

We Pointed Them North, Teddy Blue Abbott with Helen Huntington Smith

The West, Geoffrey C. Ward

ORGANIZATIONS

American Bison Society
Originally established by the Director of the New York Zoological Society; reestablished by the Wildlife Conservation Society and others in 2006 "to continue the passion of these pioneers to save the wild bison from extinction." ambisonsociety.org

InterTribal Buffalo Council
A sixty-tribe collective whose mission is to return bison to reservations. itbcbuffalonation.org

National Bison Association
Represents 1,100 ranchers, processors, marketers, and bison enthusiasts in fifty states and ten foreign countries. bisoncentral.com

National Wildlife Federation
A conservation education and advocacy group with more than six million members and supporters. nwf.org

Wildlife Conservation Society
Founded as the New York Zoological Society in 1895, the organization's mission is to save wildlife and wild places. wcs.org

World Wildlife Fund's Bison Initiative
Identifying opportunities where bison can thrive in large herds of 1,000 animals on vast landscapes in the Northern Great Plains; partnering with Oglala Lakota to create the nation's first tribal national park, working to conserve genetically important bison; and conducting research and outreach. worldwildlife.org/species/plains-bisonResources

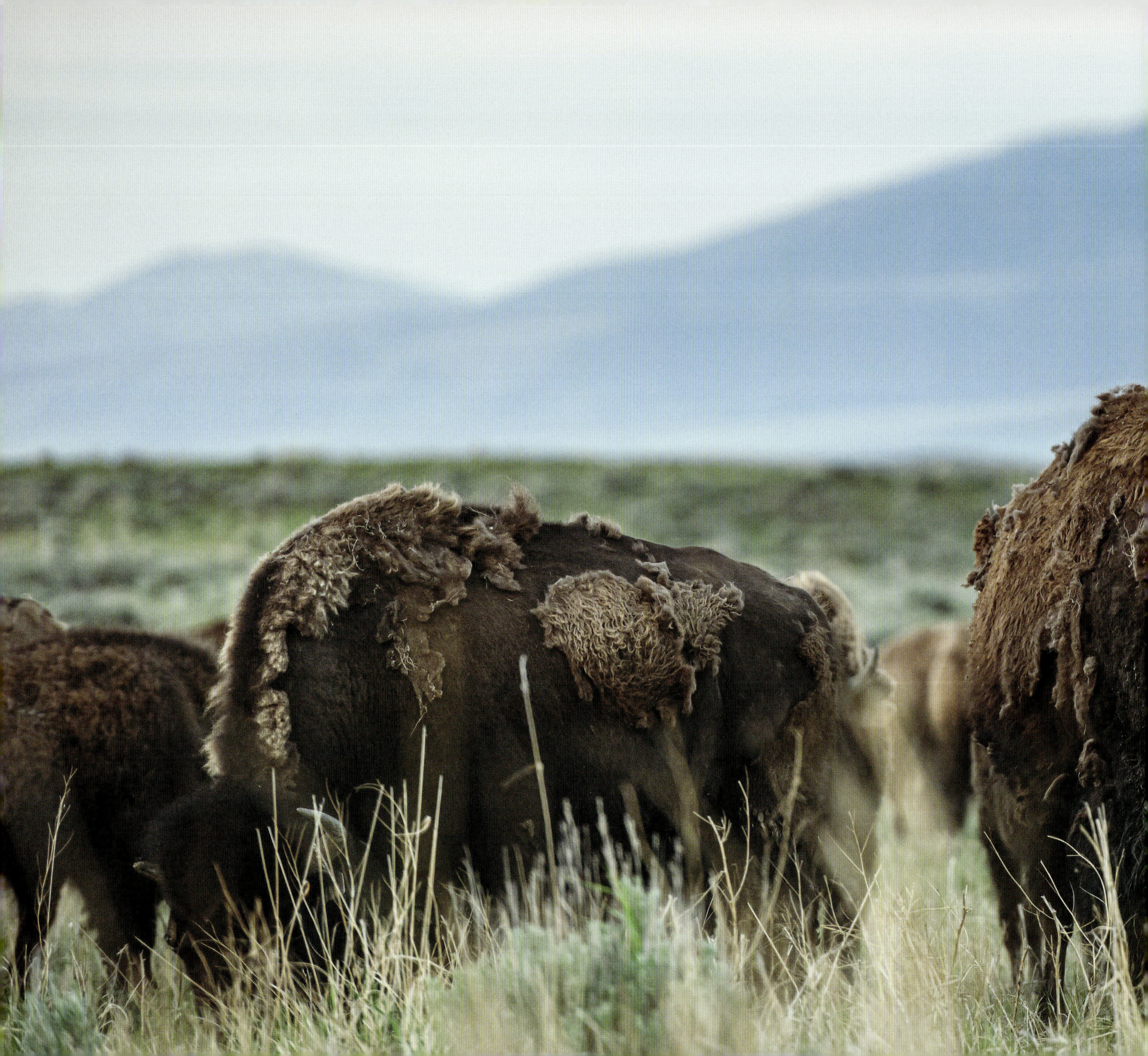

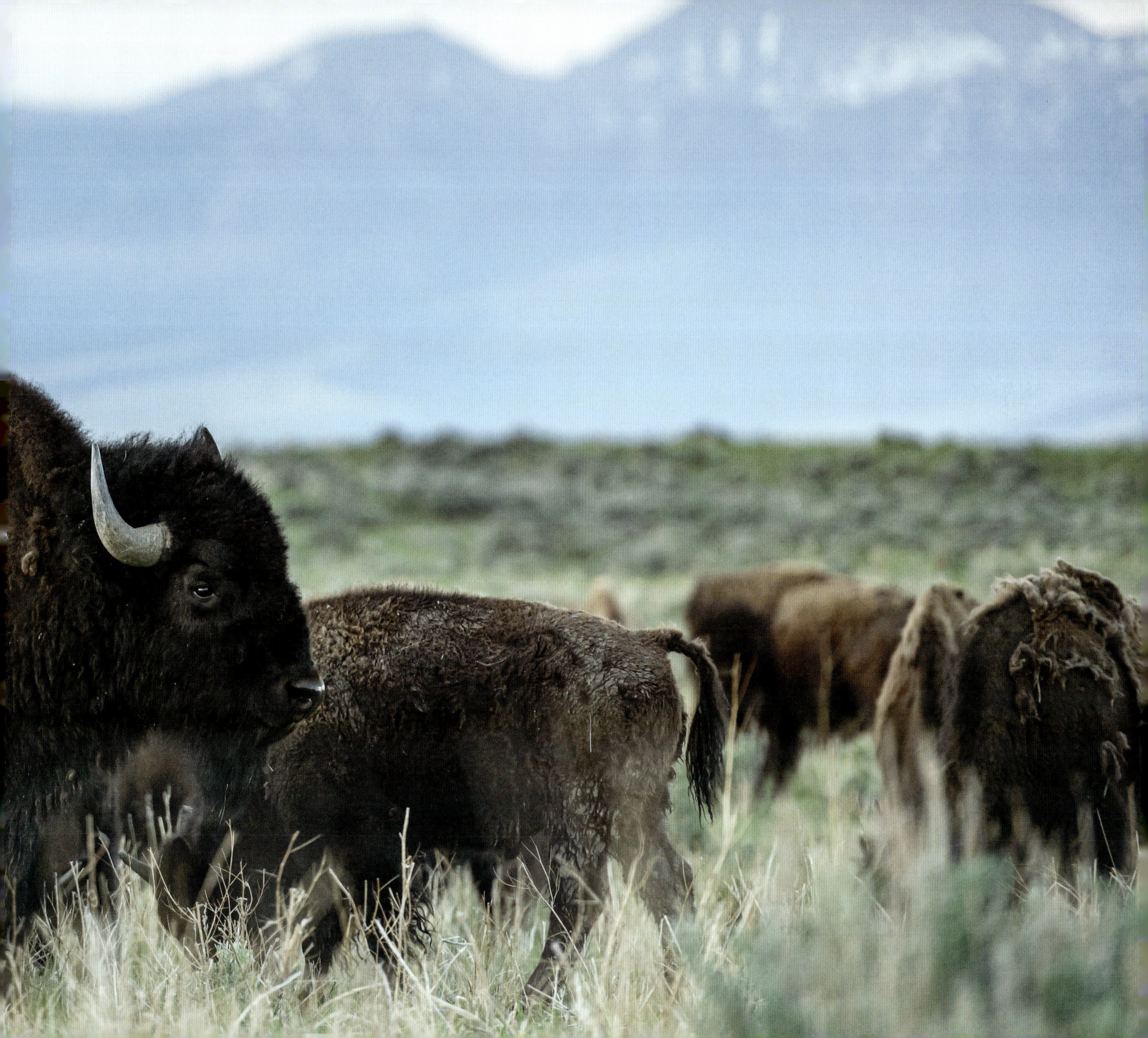

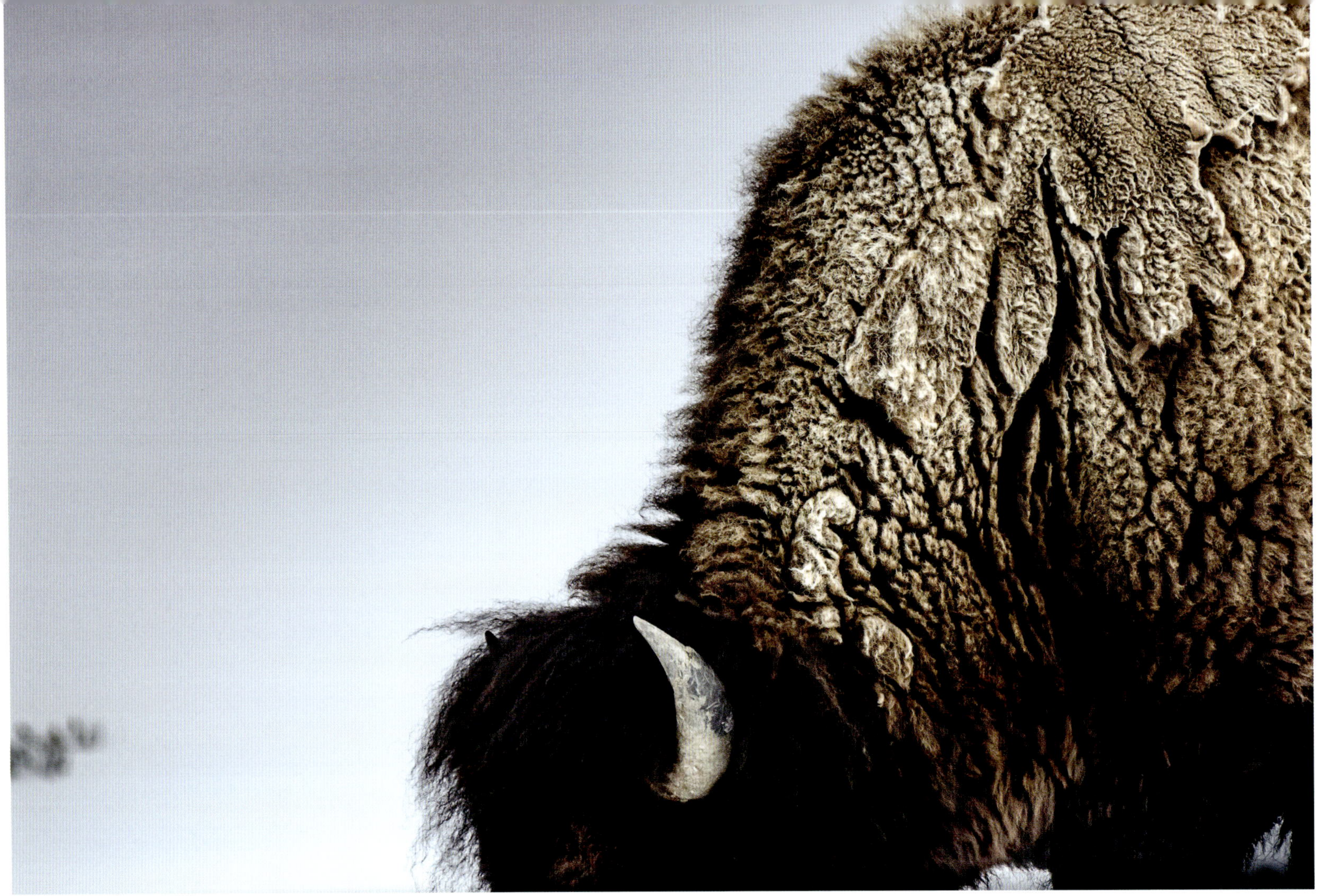

Henry Real Bird was born on the Crow Indian Reservation in Montana and raised in the traditions of the tribe in which he is enrolled, the *Upsauloeka*, or Crow (he still speaks Crow as his primary language). He holds a MS in general education from Montana State University-Billings, and before retiring, taught at reservation schools from the pre-K to college level, and served as president of Little Big Horn College from 2000 to 2002. His body of work consists of twelve children's books that he also illustrated, and six poetry collections, and his poetry has appeared in several anthologies and been recorded on spoken word albums. His honors include winning the Western Heritage Award for Poetry Book from the National Cowboy Hall of Fame (for poems anthologized by editor Anne Heath Widmark in *Between Earth and Sky: Poets of the Cowboy West*, serving as Montana Poet Laureate from 2009 to 2011, being featured on NPR's *All Things Considered*, and being recognized with both the High Plains Book Award for Poetry and the Academy of Western Artists' Will Rogers Medallion Award for Cowboy Poetry for his book *Horse Tracks*. His latest poetry collection is *Wolf Teeth*, and he is currently writing a book of Crow oral history under the working title *Many Ghosts of My Mind*. He lives with his wife, Alma, near Lodge Grass, Montana, raising bucking horses, continuing to mentor local youth, and caring for the grandchildren from two sons and one daughter.

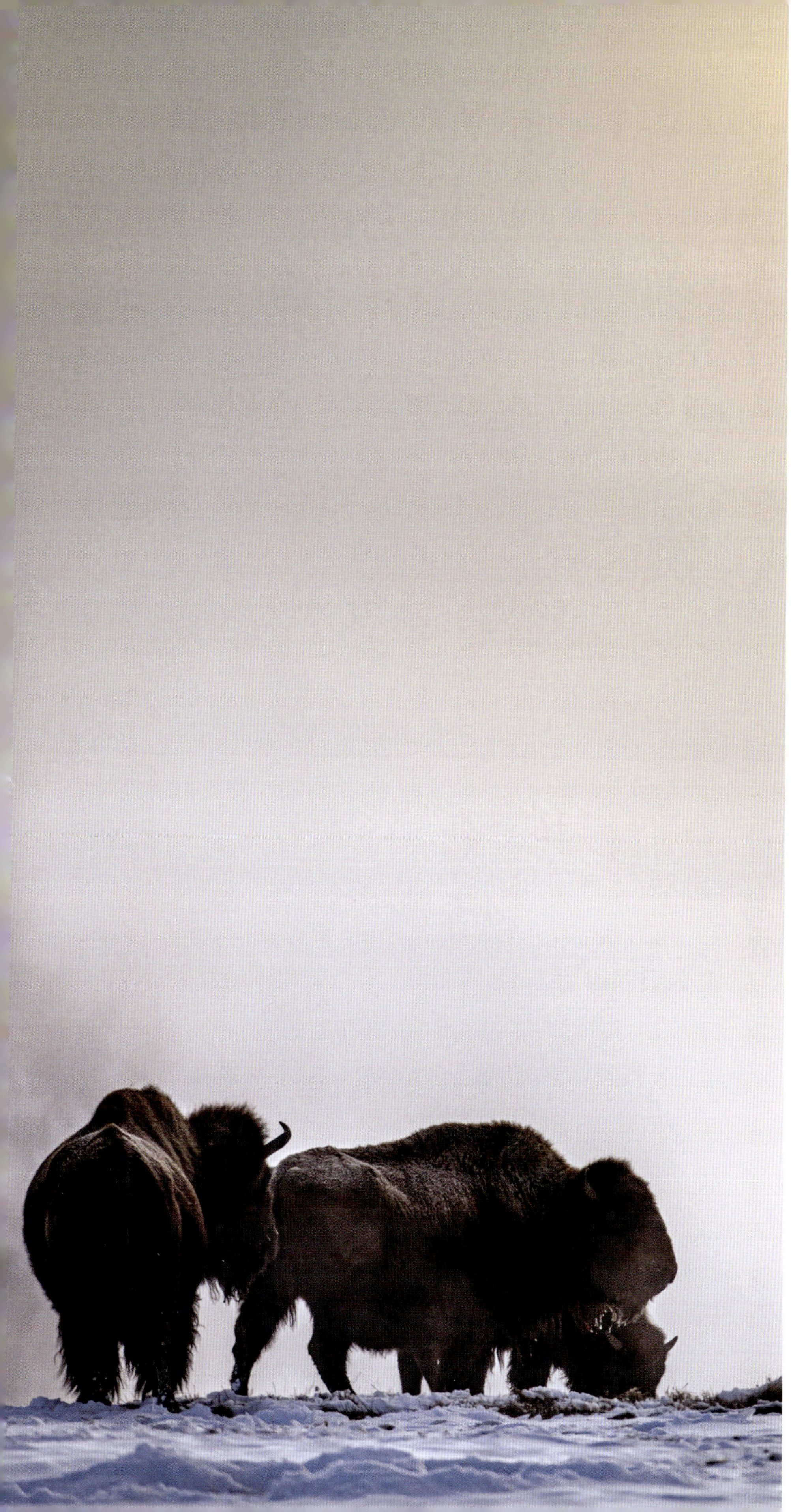

Harrison Stevens is an editorial assistant who collaborated with Henry Real Bird on his "Thought Ways: Buffalo Jump Preparation" excerpt, "Buffalo Talk," "The Seven Buffalo Bulls Become the Big Dipper," and "The Disappearance and Return of the Buffalo." He graduated from the University of Montana with a concentration in creative writing, holds an MA in English from Montana State University, and is a former writing tutor at the University of Oregon's Center for Teaching Writing. He lives just outside the Crow Indian Reservation town of Lodge Grass, Montana, where he has worked for several years as a K-12th grade English teacher and librarian.

John Heminway is a writer and filmmaker whose body of work has focused on nature, science, history, biography, Africa, and the American West. Heminway's filmmaking career at PBS, the Discovery Channel, National Geographic, and other production companies spans five decades and several hundred films and has garnered many awards, including two Emmys, two Peabodys and a DuPont Columbia Journalism award. Heminway's sixth book, *In Full Flight: A Story of Africa and Atonement*, was published in 2018.

First Edition
29 28 27 26 9 8 7 6 5

Published by
Gibbs Smith
570 N. Sportsplex Dr.
Kaysville, Utah 84037

www.gibbs-smith.com

The authorized representative in the EEA is Simon and Schuster Netherlands BV, Herculesplein 96 3584 AA Utrecht, Netherlands, info@simonandschuster.nl

Designed by Rita Sowins/Sowins Design
Design production by Virginia Snow

Gibbs Smith books are printed on paper produced from sustainable PEFC-certified forest/controlled wood source. Learn more at www.pefc.org.
Printed and bound in China

Library of Congress Cataloging-in-Publication Data
Names: Ewald, Chase Reynolds, 1963- author. | Hall, Audrey, photographer (expression) Title: Bison : portrait of an icon / Chase Reynolds Ewald ; photographs by Audrey Hall. Description: First edition. | Layton, Utah : Gibbs Smith, [2021] | Identifiers: LCCN 2020033017 | ISBN 9781423653752 (hardcover) | ISBN 9781423653776 (epub)
Subjects: LCSH: Bison—West (U.S.)—Pictorial works. | LCGFT: Essays. Classification: LCC QL737.U53 E93 2021 | DDC 599.64/3--dc23
LC record available at https://lccn.loc.gov/2020033017

This product is made of FSC®-certified and other controlled material.

FSC www.fsc.org MIX Paper | Supporting responsible forestry FSC® C208677